Contents

Public Policies for Environmental Protection

Second Edition

Editors
Paul R. Portney
Robert N. Stavins

Contributors
A. Myrick Freeman III
Molly K. Macauley
Jason F. Shogren
Hilary Sigman
Michael A. Toman
Margaret A. Walls

Resources for the Future
Washington, DC

Printed in the United States of America.

An RFF Press Book
Published by Resources for the Future
1616 P Street, NW, Washington, DC 20036–1400
www.rff.org

Library of Congress Cataloging-in-Publication Data

Public policies for environmental protection / Paul R. Portney and Robert N. Stavins, editors.—2nd ed.
p. cm.
Includes bibliographical references (p.) and index.
ISBN 1–891853–03–1
1. Environmental policy—United States. 2. Environmental law—United States. 3. Pollution—United States. 4. Pollution—Law and legislation—United States. I. Portney, Paul R. II. Stavins, R. N. (Robert N.), 1948–
GE180 .P83 2000
363.7′056′0973—dc21 00–032334

f e d

The paper in this book meets the guidelines for permanence and durability of the Committee on Production Guidelines for Book Longevity of the Council on Library Resources.

This book was copyedited by Pamela Angulo. The text was designed and typeset in Minion and Barmeno by Betsy Kulamer. The cover was designed by Debra Naylor Design.

About Resources for the Future and RFF Press

Founded in 1952, Resources for the Future (RFF) contributes to environmental and natural resource policymaking worldwide by performing independent social science research.

RFF pioneered the application of economics as a tool to develop more effective policy about the use and conservation of natural resources. Its scholars continue to employ social science methods to analyze critical issues concerning pollution control, energy policy, land and water use, hazardous waste, climate change, biodiversity, and the environmental challenges of developing countries.

RFF Press supports the mission of RFF by publishing book-length works that present a broad range of approaches to the study of natural resources and the environment. Its authors and editors include RFF staff, researchers from the larger academic and policy communities, and journalists. Audiences for RFF publications include all of the participants in the policymaking process—scholars, the media, advocacy groups, NGOs, professionals in business and government, and the general public.

About the Authors

A. Myrick Freeman III is the William D. Shipman Research Professor of Economics at Bowdoin College, Brunswick, Maine. He is also author of *The Measurement of Environmental and Resource Values: Theory and Methods.*

Molly K. Macauley is a senior fellow at Resources for the Future. Her recent research on waste management issues focuses on modeling the interstate shipment of municipal solid waste and evaluating proposed policies to manage these shipments. She has coauthored research on using economic incentives to regulate toxic substances and on the use of new technologies to monitor and enforce environmental regulations.

Paul R. Portney is president and a senior fellow of Resources for the Future. He is the author, editor, or coeditor of several books, including *Discounting and Intergenerational Equity* and *Footing the Bill for Superfund Cleanups: Who Pays and How?*

Jason F. Shogren is Stroock Distinguished Professor of Natural Resource Conservation and Management and also a professor of economics at the University of Wyoming. His research focuses on the behavioral underpinnings of private choice and public policy, especially for environmental and natural resources.

Hilary Sigman is an assistant professor of economics at Rutgers University in New Brunswick and a faculty research fellow of the National Bureau of

Economic Research. She currently serves on the Environmental Economics Advisory Committee of the U.S. Environmental Protection Agency's Science Advisory Board and the Editorial Council of *The Journal of Environmental Economics and Management.*

Robert N. Stavins is Albert Pratt Professor of Business and Government and faculty chair of the Environment and Natural Resources Program at the John F. Kennedy School of Government at Harvard University. He is also chair of the Environmental Economics Advisory Committee of EPA's Science Advisory Board. His recent publications include *Economics and the Environment: Selected Readings.*

Michael A. Toman is a senior fellow and the director of the Energy and Natural Resources division at Resources for the Future. He is the coauthor, coeditor, and editor of several books, including *Pollution Abatement Strategies in Central and Eastern Europe, Assessing Surprises and Nonlinearities in Greenhouse Warming,* and *Technology Options for Electricity Generation.*

Margaret A. Walls is an associate professor in the School of Economics and Finance at Victoria University of Wellington and a university fellow with Resources for the Future. She has published extensively in the area of environmental and resource economics, including several articles on the economics of solid waste and recycling.

one

Introduction

Paul R. Portney and Robert N. Stavins*

In the decade that has passed since the appearance of the first edition of *Public Policies for Environmental Protection*, significant changes have occurred in U.S. environmental policies, related policy debates, and the context in which they have occurred. In this chapter, we briefly review these policy developments and describe how the book has evolved to reflect them. We also describe the scope and level of the book, and highlight ways in which it can be used as a complement to texts and other readings in the science, economics, and politics of the environment. Finally, we provide brief previews of the book's chapters.

Environmental Policy Developments and Trends since 1989

Six trends, of varying importance, stand out. First, interest in market-based instruments for environmental protection has greatly increased, as evidenced by the creation of the sulfur dioxide allowance trading program in the 1990 amendments to the Clean Air Act (CAA). Second, there has been a proliferation of information provision programs, such as the expansion of the Toxics Release Inventory. Third, there has been a moderate expansion in the use of benefit–cost analysis under several environmental statutes and executive

*The authors would like to thank Sheila Cavanagh for valuable research assistance.

orders. Fourth, distributional issues on both the benefit and cost sides of the regulatory equation have gained heightened attention, often under the rubric of "environmental justice." Fifth, concerns about global climate change have emerged as an important focal point of many policy debates. Sixth and finally, there has been an upsurge of recycling activity and a related new focus of federal waste management policy.

Market-Based Instruments

The most striking change that has taken place since the first edition of this book is with regard to the employment of economic-incentive or market-based environmental policy instruments, approaches that encourage behavior through market signals rather than through explicit directives regarding pollution control levels or methods. These policy instruments, such as tradable permits, pollution charges, and deposit–refund systems, are often characterized as "harnessing market forces" because, if they are well designed and implemented, they encourage firms (and/or individuals) to undertake pollution control efforts that are in their own interests and that collectively meet policy goals. In political terms, market-based instruments have by now moved center stage, and policy debates look very different from when these ideas were characterized as "licenses to pollute" or dismissed as impractical. Indeed, market-based instruments often seem to have become the new conventional wisdom among policymakers in the environmental realm, at least in the United States.

In 1989, the federal government set up a tradable permit system and levied an excise tax on specific chlorofluorocarbons to meet international obligations established under the Montreal Protocol that limit the release of chemicals that deplete stratospheric ozone. One year later, the U.S. Environmental Protection Agency (EPA) began to allow averaging, banking, and trading of credits for nitrogen oxide and particulate emissions reductions among eleven heavy-duty truck and bus engine manufacturers. Also enacted in 1990 was the most important application ever made of a market-based instrument for environmental protection: the tradable permit system intended to reduce sulfur dioxide emissions by 10 million tons below 1980 levels. A robust market of bilateral sulfur dioxide permit trading gradually emerged, resulting in cost savings on the order of $1 billion annually (Carlson and others 2000). Subsequently, twelve northeastern states and the District of Columbia, under EPA guidance, implemented a regional nitrogen oxide cap-and-trade system. Potential compliance cost savings of 40–47% have been estimated for the period 1999–2003 (Farrell and others 1999).

In addition, considerable action with market-based instruments has taken place at the state and local level. The South Coast Air Quality Management District, which is responsible for controlling emissions in a four-county area of southern California, launched a tradable permit program in January 1994 to reduce nitrogen oxide and sulfur dioxide emissions in the Los Angeles area. One prospective analysis predicted 42% cost savings, amounting to $58 million annually (Anderson 1997). Also since 1989, California, Colorado, Georgia, Illinois, Louisiana, Michigan, and New York have established emissions credit programs for nitrogen oxides and volatile organic compounds, authorized under EPA's emissions trading program framework (Bryner 1999).

Information Programs

Over the past decade, information-based environmental policies have proliferated. Most prominently, the U.S. Toxics Release Inventory (TRI), mandated in 1986 under the Emergency Planning and Community Right-to-Know Act, requires firms to report to local emergency planning agencies information on use, storage, and release of hazardous chemicals. Such information reporting serves compliance and enforcement purposes, but also may increase public awareness of firms' actions; in turn, this could encourage firms to alter their behavior, although the evidence is mixed (Konar and Cohen 1997; Hamilton and Viscusi 1999).

The U.S. Energy Policy and Conservation Act (EPCA) of 1975 requires that some household appliances carry labels with information on energy efficiency and estimated annual energy costs and that new cars carry labels indicating fuel efficiency. The Energy Policy Act of 1992 added fluorescent and incandescent lamps to the list of products requiring labels, and it expanded the EPCA labeling requirements to include water flow information for showerheads, faucets, and toilets. Since 1996, EPA also requires uniform labeling of certain types of rechargeable batteries (U.S. EPA 2000).

Notification requirements extend to the public sector, as well. The 1996 Amendments to the Safe Drinking Water Act require all community drinking water systems to mail to each customer an annual report containing information about source water quality and the levels of various contaminants.

Expanded Use of Benefit–Cost Analysis and Distributional Issues

Although the use of benefit–cost analysis in environmental regulation has not increased dramatically since 1989, it has been expanded by Presidential executive orders and legislation. (The Flood Control Act of 1936 may include the first legislative mandate to use benefit–cost analysis. Since then, several statutes have been interpreted as restricting the ability of regulators to consider benefits and costs, while others clearly require regulators to consider them; see Arrow and others 1996.) Presidents Carter, Reagan, Bush, and Clinton all introduced formal processes for reviewing economic implications of major environmental, health, and safety regulations. In 1993, President Clinton replaced Executive Orders 12291 and 12498 (issued by President Reagan) with Executive Orders 12866 and 12875, whereby regulation is considered appropriate only upon "reasoned justification that benefits justify costs," and benefit–cost analysis is required for all "significant regulatory actions."

Congress has supported requirements for benefit–cost analysis only in selected contexts. Section 812 of Title VII of the CAA amendments of 1990 requires EPA to conduct a comprehensive analysis of the retrospective benefits and costs of CAA from 1970 to 1990, in addition to biennial analyses of the benefits and costs of the 1990 amendments, which must include future projections. EPA issued its final retrospective report in October 1997, following six years of controversial development and review. The agency's first prospective report, covering the period 1990–2010, was released in November 1999 (U.S. EPA 1999).

The 1996 amendments to the Safe Drinking Water Act allow EPA to consider overall risk reduction when setting standards and direct EPA to conduct benefit–cost analyses for new regulations. Further, these amendments allow EPA to adjust maximum contaminant levels in light of the results of benefit–cost analysis. More broadly, in 1995, Congress enacted the Unfunded Mandates Reform Act, which requires quantitative comparison of benefits and costs for all proposed and final rules, including environmental regulations, with an expected annual cost greater than or equal to $100 million. In addition, this act mandates that agencies choose the least-cost regulatory alternative or explain why the least-cost alternative was not chosen; evidence indicates, however, that these benefit–cost policies have had only limited effects on agency rulemaking (U.S. GAO 1998; Hahn and others 2000).

Distributional concerns have long been the focus of political debates and, in recent years, have become an explicit element in required economic analyses. Clinton's Executive Orders require examination of "distributive impacts" and "equity." In 1994, Executive Order 12898 formalized the President's position by instructing federal agencies to identify and address "disproportionately high and adverse human health or environmental effects of its programs, policies, and activities on minority populations and low-income populations." In addition, the Small Business Regulatory Enforcement Fairness Act of 1996 requires EPA (and other affected agencies) to prepare regulatory flexibility analysis of all rules with "significant economic impact" on a "substantial number" of small entities (businesses, nonprofits, and small government organizations).

examines the question of whether it has contributed to the observed decline in toxic chemical releases.

Finally, in Chapter 8, Molly Macauley and Margaret Walls review federal solid waste policies. They describe the composition of municipal solid waste in the United States, its generation, and its regulation. They discuss the rationale for such regulation, identify specific externalities, and suggest corrective mechanisms, with cost-effectiveness as one important criterion. Macauley and Walls appraise other potential goals of solid waste policies, such as resource conservation, meeting demand for secondary materials, reducing greenhouse gas emissions, and addressing life-cycle externalities. They find, however, that solid waste policy should not attempt to address these concerns directly. They conclude by suggesting that pricing solid waste collection and disposal directly can be a first-best solution for market failures in solid waste management, if illegal disposal is not a serious problem, and that among alternative (second-best) policies, deposit–refund systems may reduce waste disposal at least cost.

An Invitation to Readers

As we emphasized at the outset, environmental policy is very much a moving target. Furthermore, from different vantage points there will be differing views of what would be the most useful level and the most appropriate scope for a book such as this. Hence we invite all readers of this edition of *Public Policies for Environmental Protection*, whether practitioners, teachers, or students, to send us or RFF Press any thoughts or suggestions for future editions. We intend to keep this book as up to date as is permitted by that scarcest of contemporary scarce resources, time.

References

Anderson, R. 1997. *The U.S. Experience with Economic Incentives in Environmental Pollution Control Policy.* Washington, DC: Environmental Law Institute.

Arrow, K.J., M.L. Cropper, G.C. Eads, R.W. Hahn, L.B. Lave, R.G. Noll, P.R. Portney, M. Russell, R. Schmalensee, V.K. Smith, and R.N. Stavins. 1996. Is There a Role for Benefit–cost Analysis in Environmental, Health, and Safety Regulation? *Science* 272(12 April):221–222.

Bryner, G.C. 1999. *New Tools for Improving Government Regulation: An Assessment of Emissions Trading and Other Market-Based Regulatory Tools.* Arlington, VA: PricewaterhouseCoopers Endowment for the Business of Government.

Carlson, C., D. Burtraw, M. Cropper, and K. Palmer. 2000. Sulfur Dioxide Control by Electric Utilities: What Are the Gains from Trade? Discussion paper 98-44-REV. Washington, DC: Resources for the Future.

Farrell, A., R. Carter, and R. Raufer. 1999. The NO_x Budget: Market-Based Control of Tropospheric Ozone in the Northeastern United States. *Resource and Energy Economics* 21:103–124.

Hahn, R.W., J.K. Burnett, Y.I. Chan, E.A. Mader, and P.R. Moyle. 2000. Assessing the Quality of Regulatory Impact Analyses. Working Paper 00-1. Washington, DC: AEI–Brookings Joint Center for Regulatory Studies.

Hamilton, J., and K. Viscusi. 1999. *Calculating Risks? The Spatial and Political Dimensions of Hazardous Waste Policy.* Cambridge, MA: MIT Press.

Konar, S., and M.A. Cohen. 1997. Information as Regulation: The Effect of Community Right-to-Know Laws on Toxic Emissions. *Journal of Environmental Economics and Management* 32:109–24.

Miranda, M.L., S. LaPalme, and D.Z. Bynum. 1998. Unit-Based Pricing in the United States: A Tally of Communities. Report to the U.S. Environmental Protection Agency. July.

U.S. EPA (Environmental Protection Agency). 1999. *The Benefits and Costs of the Clean Air Act Amendments of 1990.* Final Report to Congress. EPA 410-R-99-001. Washington, DC: U.S. EPA.

———. 2000. *Environmental Labeling: A Comprehensive Review of Issues, Policies and Practices Worldwide.* Washington, DC: U.S. EPA.

U.S. GAO (General Accounting Office). 1998. *Unfunded Mandates: Reform Act Has Little Effect on Agencies' Rulemaking Actions.* GAO/GGD-98-30. Washington, DC: U.S. Government Printing Office.

two

EPA and the Evolution of Federal Regulation

Paul R. Portney

Compared with many other federal regulatory agencies, the Environmental Protection Agency (EPA) and its siblings of the 1970s are relative newcomers. Indeed, Congress created the Interstate Commerce Commission (ICC) to regulate surface transportation industries more than a century ago (in 1887), and the Federal Reserve Board and Federal Trade Commission were created to regulate commercial banks and deceptive trade practices, respectively, nearly ninety years later. The first great burst of federal regulatory activity took place in the 1930s, when Congress created the Federal Power Commission, the Food and Drug Administration (FDA), the Federal Home Loan Bank Board, the Federal Deposit Insurance Corporation (FDIC), the Securities and Exchange Commission (SEC), the Federal Communications Commission (FCC), the Federal Maritime Commission, and the Civil Aeronautics Board (CAB).

Between 1938 and 1970, little took place in the way of new federal regulatory activity.[1] However, following this lull came the second major burst of federal regulatory activity. In quick order, EPA, the National Highway Traffic Safety Administration, the Consumer Product Safety Commission, the Occupational Safety and Health Administration (OSHA), the Mine Safety and Health Administration, the Nuclear Regulatory Commission (NRC), the Commodity Futures Trading Commission, and the Office of Surface Mining Reclamation and Enforcement were created. With the exception of the Commodity Futures Trading Commission, these agencies and the FDA have come to be known as the "social" regulatory agencies. Generally speaking, these agencies regulate environmental protection and the safety and health of consumers and workers.

Besides age, at least three important features distinguish the "old-line" agencies from the relatively newer social regulators like EPA (Eads and Fix 1984). The first has to do with their reasons for being. In principle at least, all regulatory agencies have been created to remedy a perceived failure of the free market to allocate resources efficiently. Except for the FDA, the older agencies were meant to either control "natural monopolies" or protect individuals from fraudulent advertising or unsound financial practices on the part of financial intermediaries or depository institutions.[2] (The latter justification was clearly a reaction to the calamitous Great Depression, during which many of these agencies were created.)

Federal intervention in the areas of environmental protection and the safety and health of workers and consumers has a very different rationale. Here, it is argued, the government must intervene because of externalities or imperfect information. Externalities arise when the production of a good or service results in some costs (such as pollution damage) that, in the absence of regulation, are unlikely to be borne by the producer. In such cases, the prices of products will not reflect what society must give up to have them, so that Adam Smith's "invisible hand" will steer us awry. Workers or consumers who have imperfect information are only dimly aware of the health hazards associated with their occupations or with the food and nonfood products they purchase. Because they are uninformed, they are unable to trade off higher risks for higher wages or lower prices; as a result, the market solution does not result in the right amount or the correct distribution of risk.

The newer regulatory agencies differ from the older ones in another important way, which has to do with the specificity of their focus. The older agencies can be thought of as dealing with a single industry—the ICC with surface transportation, the FCC with communications, the CAB with airlines, and so on. In contrast, the newer social regulatory agencies cover a focused aspect across a range of industries. For instance, OSHA regulates workplace conditions in various industries, from chemicals to agriculture. Similarly, EPA regulates emissions that contribute to air and water pollution from the electric utility, steel, food processing, petroleum refining, and many other industries. The broad mandate of the social regulatory agencies may be difficult to satisfy, because it requires each agency to become knowledgeable about and sensitive to the special problems and production technologies in many different industries.

A third distinction between the old-line agencies and their newer counterparts concerns recent developments in the scope of their activities.

Although it is always difficult to generalize over members of such a diverse group, the thrust of 1980s legislation and administrative actions at the older regulatory agencies was toward curtailing intervention in the markets they regulated. This strategy was most pronounced in the case of the CAB, which was legislated out of existence in 1985 after it was recognized that the airline industry was ripe for competition. The ICC suffered a similar fate in 1997 for the same reason. The FCC proposed significant reductions in the scope of its own activity, as has the Federal Maritime Commission. The Federal Home Loan Bank board was abolished and replaced by the Office of Thrift Supervision. The other financial regulatory agencies and the Federal Energy Regulatory Commission (FERC, the successor to the Federal Power Commission) also have seen their powers eroded somewhat over time.

In general, such has not been the case at the newer, social regulatory agencies. Even though many critics have questioned the way these agencies have pursued their goals, relatively few have suggested that the agencies have become obsolete. It would be difficult to argue that unfettered competition among firms would lead to the right amount of pollution, product safety, and workplace risk as long as the problems (external effects or imperfect information) characterize the conditions of production or consumption. Thus, although the agitation for regulatory reform continues—about which more is said below—there are very few calls for the abolition of EPA, OSHA, FDA, or any other social regulatory agency. Indeed, in the case of EPA at least, new responsibilities and regulatory programs have been added to the old almost continually since the agency's creation. Trying to initiate new programs while struggling to master the existing ones is one of the problems EPA has had to contend with constantly.

The Creation and Growth of EPA

The vanguard of the new social regulatory agencies, EPA got its start on July 9, 1970, when President Richard Nixon submitted Reorganization Plan No. 3 of 1970 for congressional approval. That reorganization plan proposed to consolidate under one roof—EPA's—various functions being performed at that time by the Departments of the Interior, Health, Education and Welfare, and Agriculture as well as the Atomic Energy Commission, the Federal Radiation Council, and the Council on Environmental Quality. By December 1970, the plan had been approved by Congress, and EPA was in action.

Because it was created out of existing programs, EPA was never a very small agency. In 1971, its first full year of existence, it had about 5,700 employees and a budget of $4.2 billion. By 1980, the agency had grown to more than 13,000 employees. Its budget by that time was more than $7 billion. By 1999, EPA had more than 18,000 employees and a budget of $7.6 billion. (All financial data are expressed in 1999 dollars and were supplied by EPA.) The EPA budget, like that of most federal agencies, shrank during the Reagan years but resumed its growth under Presidents George Bush and Bill Clinton.

One must be careful not to ascribe too much importance to the budget of EPA (or of any other regulatory agency). Although the budget provides some guidance as to the agency's capabilities, its spending authority is less important than the costs incurred by those subject to the agency's various regulations. Those costs, called compliance costs by economists, never show up in the agency's budget but can dwarf its operating costs. For example, in 1998, the budget for EPA was $7.4 billion. That same year, those individuals, corporations, and other parts of government subject to EPA's regulations under the Clean Air and Clean Water Acts, the Safe Drinking Water Act, and four other major environmental statutes were forced to spend $127 billion to comply with the regulations (U.S. EPA 1990). These expenditures—for pollution control equipment, cleaner fuels, sludge removal, additional manpower—give a much more accurate picture of the economic importance of EPA than does its budget. (As each of the subsequent chapters point out, environmental regulations may reap substantial benefits as well. One must look at both costs and benefits before passing judgment on the overall worth of a particular regulatory program.)

The rapid growth of off-budget environmental compliance costs (estimated to be only $33 billion in 1972) led to periodic concern about the possible effects of pollution control spending on the overall performance of the economy. Although a detailed review of the relevant literature is beyond the scope of this chapter, it may be useful to summarize them briefly (see Portney 1981). With a fair degree of consistency, studies have found that pollution control spending has had a relatively minor impact on macroeconomic performance. It has exacerbated inflation somewhat and slowed the growth rate of productivity and of the gross domestic product (GDP). However, studies also have found that pollution control spending appears to have very modestly stimulated employment, at least during the time when spending for sewage treatment plant construction was high. None of this should be too surprising. Although annual expenditures of hundreds of billions of dollars are hardly trivial, they are relatively small in comparison to an $8 trillion GDP like that of the United States in 2000. Thus, it is unlikely that environmental or other regulatory programs on the present scale ever will exert a significant impact on the measured performance of the economy.[3]

Fundamental Choices in Environmental Regulation

It may prove helpful in understanding the remainder of this book to review some of the basic choices that must be made in environmental regulation. The first issue is one often taken for granted: whether to regulate at all. Although today we tend to take the need as a given, there are several possible alternatives to environmental regulation. They include the private use of our legal system as well as private negotiation or mediation.

A Legal Approach

Both alternatives depend on a clear prior definition of property rights. Imagine that it was clearly understood that any citizen had an absolute right to be compensated fully for the damages from any kind of pollution. If the smoke from a neighbor's woodstove or a factory were ruining your laundry business, you could take the offending party to court. If your damages could be accurately assessed and if the polluter were held liable for them, the legal approach would create the right incentives for polluters. They would undertake certain pollution-control measures if the costs of these measures were less than the damages they would have to pay you. And they would continue to pollute—and pay you for the damage it does—if it were more expensive to control than to reimburse you. Economists like such solutions because they minimize the total costs (control expenditures plus residual damages) associated with pollution control. Until the start of the twentieth century, all air pollution problems were handled under the nuisance and trespass provisions of common law.

Unfortunately, the real world is much more complicated than this example would suggest, and the added complexity makes a purely legal approach to environmental protection much less practical. First, it is not perfectly clear where property rights in clean air are, or ought to be, vested in our society. This concept may seem puzzling, because most people accept the right of the citizenry to be free from pollution. Yet even this apparently sensible proposition seems strained, say, in the case of a laundry owner who deliberately moves from a clean location to one directly adjacent to a factory and then demands compensation for smoke damage. In other words, it seems to make some difference who was there first. Furthermore, if property values and rents are lower in polluted areas than in clean ones (as they generally are), the laundry owner seems to be on shakier ground still, because he or she would have already reaped some savings in operating costs (lower rent) by virtue of the new location. In one sense, the laundry owner would be getting double benefits if the factory were forced to curtail its operations. Not surprisingly, arguments like these are often advanced in defense of firms resisting pollution control. They are not wholly without merit.[4]

Even if property rights were clearly defined, environmental protection via the legal system or private negotiation would not be without difficulty. For instance, pollution rarely occurs on a one-to-one basis as in the simple example above. There are often many polluters, thus making it difficult or impossible to know which factory, car, or woodstove is responsible for which damages. Also, there are generally many "pollutees," no one of whom may be suffering sufficient damages to merit taking legal action or initiating mediation or negotiation alone. Legal transaction costs may be so high that they inhibit the filing of class action suits, even though aggregate damages across all pollutees may be significant.

Additionally, some of the damaging effects of pollution may be both more subtle yet more serious than mere dirty laundry. For example, pollution may be one of the causes of cancer and other serious illnesses. Yet the long latency period between exposure and manifestation, coupled with the possibility of other causes, makes it virtually impossible to assess liability satisfactorily in a courtroom or arbitration chamber. Add the difficulty of valuing the pain and suffering from such illnesses, and one can quickly understand the possible shortcomings of alternative approaches to government intervention. Finally, at the start of the twenty-first century, many people believe that too much litigation takes place in the United States. Expanding the role of litigation, even though a rationale may appear to exist for it, would no doubt strike some as foolish.

Having said this much, I should make one more quite important point. Proponents of government action, regulatory or otherwise, are quick to point to the imperfections inherent in free markets as justification for intervening. Public goods, externalities, natural monopolies, and imperfect information are all problems economists recognize as standing in the way of efficient resource allocation. Yet regulation seldom goes exactly as planned when it is substituted for the forces of the market (Wolf 1988). It often is poorly conceived, time-consuming, arbitrary, and manipulated for political purposes completely unrelated to its original intent. Thus, the real comparison one must make in contemplating a regulatory intervention is that between an admittedly imperfect market and what will inevitably be imperfect regulation. Until we recognize this dilemma before us, we will be dissatisfied with either approach.[5]

Designing Intervention

If government intervention is deemed desirable, one must then ask, at what level should it take place? Although environmental protection is very important, so too are public school quality, police and fire protection, income assistance, and criminal justice. Yet the latter are functions that are entrusted largely to local or state governments in our federal system. Thus, it is not obvious that all or even most environmental regulation should take place at the federal level. For air and water pollution problems that tend to cross state borders, the rebuttable presumption should be for federal authority. However, it is less clear that the federal government should be setting uniform national standards for landfills or even drinking water contaminants. Few if any interstate "spillovers" exist for such situations, so the states may be better equipped to regulate them.

Once the decision has been made to intervene at some level of government, we must decide how much protection the government should provide. Several frameworks can be used to help answer this question, and Congress has directed different social regulatory agencies to use different frameworks in establishing levels of protection (see Lave 1981). In fact, even within EPA, the approach differs depending on the regulatory program in question.

Zero-Risk Approach. One way to select the degree of protection might be called the zero-risk or safe-levels approach. The administrator of the EPA would be directed to set a particular environmental standard at a level that would ensure against any adverse health (or other kind of) effect. As later chapters illustrate, this approach is not uncommon. And at face value, it certainly seems reasonable. After all, would we want a standard to be set at a level that poses some recognizable threat to health? Perhaps surprisingly, the answer is maybe.

Science and economics contribute to this unexpected response. Accumulated research in physiology, toxicology, and other health sciences suggests that for many environmental pollutants, particularly carcinogens, there may be no threshold concentrations below which exposures are safe. These data imply that standards for these pollutants must be set at zero concentrations if the populace really is to be protected against all risks. Here economics intrudes in a jarring way. Simply put, it is impossible to eliminate all traces of environmental pollution without simultaneously shutting down all economic activity, an outcome which neither the Congress nor the public would abide. Yet this is where the zero-risk framework often appears to lead if interpreted literally.

Having raised this disquieting possibility, let me push it further to consider another interesting case. Suppose a particular pollutant was harmful at ambient (or outdoor) levels to one very large group of people—the one-third of American adults who choose to smoke cigarettes—but only this group. If the costs of reducing ambient concentrations of the contaminant were very large, might society not decide to forgo this health protection? It might well, in view of the role that the sensitive population has played in predisposing itself to environmental illness. Here too, then, the zero-risk approach would cause problems.

Perhaps more realistically, decisions to live with some risk might be reached even if the group at risk had done nothing to create its sensitive status. At some point, the costs of additional protection might be judged by society to be too great if the added health benefits are relatively small. Painful though such decisions may be, they are the rule rather than the exception in environmental and other policy areas. The problem with the zero-risk approach is that it prevents such trade-offs from being made. EPA grappled with this uncomfortable problem when it announced new air quality standards in 1997 (see Chapter 4, Air Pollution Policy).

Technology-Based Approach. Another way to decide how much protection to provide is a variant of the zero-risk approach. It is often referred to as the technology-based approach. Under this framework, the only pollution permitted is that remaining after sources have installed "best available" or other state-of-the-art control technology. The underlying idea is simply that all technologically feasible pollution control measures will be required, and only after they have been implemented will residual risks be accepted. This approach is somewhat weaker than a strict zero-risk approach because it admits the possibility of some risks. But in its strictest form, it is uncompromising with respect to trading off cost savings for less strict pollution control. For this reason, it appeals to many.

However, in its application, the technology-based approach faces several drawbacks. First, there is no unambiguously "best" technology; emissions always can be reduced further, at additional cost. In the limit, of course, sources could be closed down entirely—an ultimate, perhaps draconian, form of best technology. Moreover, implicit in the technology-based approach is the assumption that the resulting control must be worth the cost. It might well depend on the details of the situation. For instance, very strict control may be deemed essential for polluters in densely populated areas but much less important for those in remote, unpopulated regions. Yet the uniform technological approach has the liability of precluding the trade-offs necessary to decide how to handle such variables. In addition, the technology-based approach suffers in a dynamic setting because it locks sources into specific means of control. It is unlikely that a firm required to meet this year's best technology will be told to scrap that equipment if next year's is even better. This approach may deprive us of the opportunity to reduce the costs and increase the efficiency of pollution control over time. Nevertheless, Congress chose it for the regulation of certain hazardous air pollutants in the 1990 amendments to the Clean Air Act.

Balancing Approach. The final framework for standard-setting discussed here, the balancing approach, formalizes the weighing of competing effects and incorporates it into environmental law. The relevant statutory language might direct the administrator of the EPA to set standards to protect health and other values while taking account of the costs and other adverse consequences of the regulations. The advantage of this approach is that it makes possible—indeed, mandatory—the kinds of trade-offs we have suggested might be desirable. On the other hand, it also forces the administrator to make very difficult decisions. Moreover, if all favorable and unfavorable effects are supposed to be expressed in dollars, so that precise benefit–cost ratios are required, this approach would impose a burden that economic analysis is not prepared to bear. Despite progress in valuing environmental benefits and costs (see Freeman 1993), the science is still far short of being able to make such comparisons in a precise way. For this reason, the balancing approach is best left in a qualitative or judgmental form.

At this point, some might reasonably chafe at the balancing framework. Why compromise citizens' health or welfare so that corporate or other polluters might remain economically healthy? This very natural question deserves a straightforward answer, one which proceeds along the following lines. Although it would be nice if there were, there are no disembodied corporate entities into whose deep pockets we can reach for pollution control spending without imposing losses on ourselves or our fellow citizens at the same time. The reason is that corporations are merely legal creations, the financial returns to which all accrue to individuals in one capacity or another. Thus, if corporations spend more for pollution control, these costs may be passed on to consumers in the form of higher product prices.[6] Alternatively, if costs cannot be shifted to consumers, then stockholders, laborers, or the management of the corporations will suffer reduced earnings.

Thus far, the trade-off may still seem appealing, because it is expressed in terms of dollars versus health. However, the reduced incomes of consumers, stockholders, or employees eventually will mean less spending on goods and services they value. At this point, then, the trade-off becomes more stark. Pollution control spending sometimes can protect health, but at an eventual cost to society of forgone health, education, shelter, or other valued things (see Graham and Wiener 1995 and Keeney 1990). The real trick in environmental policy—and any other area of government intervention—is to ensure that the value of the resulting output is greater than that which must be sacrificed. And this sacrifice will take place regardless of the framework for setting standards that is used. In the economist's view, the balancing approach is desirable not only because it is a natural way to make decisions but also because it brings the terms of trade out in the open, so to speak. If we dislike the compromises being made by our regulatory officials, we can demand their removal.

Encouraging Attainment

After selecting environmental standards (or ambient standards, as they sometimes are called), we must choose the means of attainment. In other words, how do we control the sources of pollution so that the environmental goals are met? Although other possibilities exist, the two most common approaches are via direct centralized regulation and through an incentive-based decentralized system.[7] Under the first, individual polluters are assigned specific emissions reductions; under the latter, they are given more latitude.

Centralized Approaches. With a centralized approach, the regulating authority has considerable discretion in apportioning the emission reductions required to meet the ambient standard. Only when there is but one source of pollution is it clear where emissions must be reduced.[8] More typically, there are multiple sources; in such cases, the authority has to decide how much each source must curtail its offending activities. There are several ways this decision can be made.

Equal Percentage. If aggregate emissions must be reduced by 25% to meet the environmental standard, for example, each source could be required to cut back its own emissions by 25%. This equiproportional rule has the very attractive feature of *appearing* fair, but on reflection, we discover that it really is not. Why only the appearance of fairness? Because of the very great diversity of sources for many environmental contaminants, which range from neighborhood dry cleaners and car repair shops to steel mills and large chemical plants. The different characteristics and technological circumstances of these sources mean that one source may be able to reduce its emissions by 25% quite inexpensively, perhaps by switching quite easily to a less polluting fuel or altering its manufacturing technique slightly. Yet another source might find that it can meet its 25% reduction only by installing expensive control technology, or by abandoning a profitable product line. Thus, a requirement for equal-percentage reductions may mean very unequal financial burdens.

Affordability. Under another approach, emission reductions might be apportioned on the basis of affordability—that is, the largest cutbacks might be required of those in the best financial shape. This approach has some obvious appeal. Indeed, under our present individual income tax system, we ask those in higher income brackets to pay a higher percentage of their income in taxes, and this approach seems to extend that principle to pollution control.

On closer inspection, however, assigning emission reductions on the basis of ability to pay has several potentially serious drawbacks. First, it would penalize successful, well-managed firms while rewarding laggards that may well be largely responsible for their own poor financial state. In this sense, the approach sends exactly the wrong set of signals to firms and slows the replacement of failing enterprises with newer, more efficient ones. Second, there may be no relationship whatsoever between the magnitude of a firm's emissions and the firm's financial condition. Thus, a very profitable firm may have very low emissions (particularly if it has continually modernized) but, under the affordability criterion, still would be forced to spend heavily on additional emission reductions. Meanwhile, a smoke-belching firm in perilous financial condition would be let off lightly because it could not afford to implement the appropriate pollution-control technology. For these reasons, an affordability criterion is less attractive than it may at first appear.

Cost Minimization. Finally, the regulatory authority could try to apportion emission reductions among sources in such a way that the required aggregate reduction was accomplished at the least additional cost to society. (This would no doubt be a regulator who had taken a good undergraduate course in economics!) In other words, the regulator would look at all sources and ask where the first ton of emissions might be reduced most inexpensively, then require it to be removed there. The second ton of emissions reductions then would be assigned, again to the source that could accomplish it most cheaply. This procedure would be followed until the aggregate emissions goal had been met.

This approach has the advantage of ensuring that society (through the affected sources) gives up as little as possible to get the desired emission reductions. But the regulator would require an awful lot of information, namely, the marginal costs of pollution reduction for all sources of the pollutant in question. This is a great deal to ask, it goes without saying. This approach also raises the possibility of another sort of inequity. Suppose that one source, among a large number of polluters in a particular area, was always the lowest cost abater? This is unlikely to be the case in reality, but it might hold true in certain circumstances. It hardly would seem fair to place the entire burden of emissions control on that source merely because it could reduce pollution more inexpensively than the other sources. Thus, although this cost-minimization approach has some very obvious appeal, it is not ideal.

Decentralized Approaches. Decentralized approaches, on the other hand, do address some of these problems, although they present difficulties of their own.

Effluent Charge. Perhaps the best-known of the decentralized approaches is the effluent charge or pollution tax.[9] Under this scheme, the regulatory authority imposes a tax or fee on each unit of the environmental contaminant discharged. In its purest form, the charge would be set to reflect the damage done to human health or to ecosystems by each unit of emissions. Rather than tell each firm how much to reduce emissions, the authority would leave it to the firm to respond to the charge however the firm best sees fit. Some sources would reduce their emissions immediately—the ones that can do so at unit cleanup costs less than the amount of the charge. By doing so, they save the difference between their per-unit cost of control and the per-unit charge. Other sources will find it economical to continue discharging—the ones that find it cheaper to pay the tax than to incur the required control costs.

Such an approach has several advantages. First, it ensures that the sources that elect to take control measures are those with the lowest control costs. In other words, it mimics the least-cost centralized approach but without requiring the central authority to specify emission reductions for each and every source. (This is one example of the "invisible hand" of which Adam Smith wrote, and it is truly miraculous.) Second, and perhaps more important, it provides a continuing incentive for firms to reduce their costs of pollution control. Because they must continue to pay the per-unit charge, it continues to be economical for them to find ways to reduce emissions for less than that charge. Third, this system requires something from all sources—they must either reduce pollution to escape the charge, or continue to pay it. No one gets off scot-free.

Modified Effluent Charge. As might be expected, the effluent-charge route has shortcomings, which at least some economists have been slow to recognize or acknowledge.[10] For one thing, it is no picnic to determine the damage done by each unit of pollution. How does one put a value on a slight increase in the risk of, say, contracting a chronic respiratory disease or causing damage to a saltwater marsh? In practice, the dollar amount could only be approximated at best. Some have suggested that the difficulty of "monetizing" damages is such a liability of the charge approach that a modified version of the approach should be used (Baumol and Oates 1971).

Under this variant, the central authority would first select the desired level of environmental quality and then set the charge at a level sufficient to induce the emissions control that would achieve it. Yet even such a variant would require some trial and error, and the uncertainty it might create could make firms reluctant to come to their initial emissions control decisions.

The effluent charge routes also present one serious political problem. The emissions that sources are free to discharge under the current permit system would be subject to the charge. Thus, many sources that presently complain about overregulation would have a new complaint: a major effluent-tax liability.

Permits. A second variant of the incentive-based approach involves marketable pollution "rights" or permits. This approach could work in one of two basic ways. Under one version, the central authority would first decide how much total pollution was consistent with the predetermined environmental goal. It then would print up individual discharge permits, the total quantity of which added up to the maximum amount permitted. The permits could be allocated among sources in one of several ways. First, a sale might be held at which all of the permits were auctioned off to the highest bidders. Alternatively, the permits could be distributed free of charge on some predetermined (or even random) basis—perhaps on the basis of historical levels of pollution. Either way, the permits would be marketable anytime after the initial distribution, and no one without a permit would be allowed to discharge the regulated pollutants.

The incentive effect under a system of marketable permits is not unlike that of the effluent charge discussed above. Those sources that currently pollute but could reduce pollution for less than the cost of a permit would take control measures. Those sources that found it very expensive to reduce pollution would buy discharge permits instead. Thus, as if guided by the same invisible hand, the emission reductions would take place at the low-cost sources, thereby minimizing the costs associated with a given reduction in emissions. Similarly, those firms that bought permits would have a continuing incentive to reduce their costs of pollution control: As soon as they could do so, they could stop buying permits and save themselves money in the process.

The permit approach has one major advantage over the effluent charge approach: it looks more like the current U.S. regulatory system, which involves permits issued by EPA or state environmental authorities, than does the latter. It may sound strange, but radical change is almost always more difficult to accommodate than gradual change. Because the marketable permit system is capable of accomplishing most of the same things as the effluent charge, why not advance it if it will be easier to put in place? This logic appears to have prevailed, and most of the inroads made by incentive-based approaches in environmental policy over the past ten years have featured permits.

Marketable permits are not without shortcomings, of course. One concern has to do with the possibility that certain sources might buy up all the permits as an anticompetitive tactic. For instance, if a certain product could not be produced without the emission of a particular pollutant and one company obtained the lion's share of the discharge permits available for this pollutant, then the company could effectively control entry to the market for that product. Although this tactic ought to be rectifiable through governmental antitrust actions, in practice, such actions might take time. Another issue concerns the initial distribution of permits before the development of secondary markets. If all the permits are auctioned off, then this approach would fall prey to the same political problems that arise under an effluent approach: Some sources would have to pay for emissions that they are granted free under the existing system. Thus, political problems could become formidable. If the initial permits are to be distributed free of charge, how should they be allocated? On the basis of previous emissions? To all citizens equally? To environmental and industry groups? This is a potentially thorny problem, although not an insurmountable one.

Monitoring for Compliance

A fifth and final question that arises in environmental policy is often overlooked: How do we monitor for compliance with the standards we set and take enforcement actions against those in violation? Suffice it to say now that the choice of both environmental and individual source discharge standards ought to be (but often is not) influenced by the realities of monitoring and enforcement. Key issues involve the extent of reliance on financial penalties for noncompliance, the comparative strengths and weaknesses of civil as opposed to criminal penalties, and the choice of a monitoring strategy in a world of limited resources.

During the late 1980s and 1990s, political administrations began using numbers of civil suits and criminal prosecutions as measures of their commitment to the environment. This development is not wholly satisfactory. Although serious violations of environmental laws *always* should be pursued, and violators appropriately punished, one gets the sense that not all suits brought are serious cases. Rather, it appears that at least on some occasions, relatively minor infractions related to paperwork or reporting requirements brought forth litigation from an administration anxious to show that it was "tougher" than its predecessors. We all would be better off if this competition were based on measured improvement in environmental quality rather than the number of lawsuits initiated and won.

A Hybrid Approach

The fundamental issues that I present in this chapter have been approached in an eclectic, hybrid way as environmental policy has evolved in the United States. With respect to the topic of intervention, the federal, state, and local governments have all decided to intervene. Environmental statutes exist at all three levels—in fact, special districts have been formed in many areas around environmental problems. Thus, the decision was made long ago not to entrust environmental problems and disputes solely to markets, the courts, or mediation services.

This observation also suggests the level at which intervention has taken place—every level. Even federal environmental laws reserve important functions for state and local governments. For instance, under the Clean Air and Clean Water Acts, the federal government (as embodied in EPA) sets important ambient environmental and source discharge standards, yet the monitoring and enforcement of these standards is left largely to the states and localities. In fact, some ambient and source discharge standards are reserved for lower levels of government in certain important cases. In other words, even federal laws are "federalist" in nature. As a new century begins, the balance of power between the states and the federal government is shaping up as a potentially quite controversial issue.

In choosing pollution standard goals, U.S. environmental laws embody a range of approaches. A number of the most important environmental laws, or parts thereof, reflect the zero-risk (or threshold) philosophy. For instance, the Clean Air Act directs that ambient standards for common air pollutants be set at levels that provide an "adequate margin of safety" against adverse health effects, whereas standards for the so-called hazardous air pollutants are to provide an "ample margin of safety." Under the Clean Water Act, ambient water quality standards—which are left to the states rather than the federal government to establish—also are to include a margin of safety for the protection of aquatic life.

Other environmental standards are based on the technological approach to goal setting. This is true of the Clean Air and Clean Water Acts, the Resource Conservation and Recovery Act, and the Safe Drinking Water Act, as later chapters will discuss in some detail. The notion of best or maximum available technology—along with its cousins "best conventional" technology, "reasonably available" technology, "lowest-achievable emissions," and others—play a big role in U.S. environmental policy, even in those statutes which in other places embrace the zero-risk goal.

Even the balancing framework favored by economists is alive and well in environmental policy. Although balancing appears to be prohibited under certain sections of the Clean Air and Clean Water Acts, it is *mandated* under the most important parts of the Toxic Substances Control Act and the basic pesticide law, the Federal Insecticide, Fungicide, and Rodenticide Act (FIFRA). Moreover, in 1997, the Safe Drinking Water Act was amended to give the EPA administrator some latitude to balance the health risk reductions with the incremental cost of attaining more stringent standards. You may be tempted to throw up your hands and say, "I'll never make sense of it all!" One of the purposes of this book is to help you do exactly that.

In one important respect, the environmental laws have been rather uniform. When it comes to the means of pursuing environmental goals, the centralized or command-and-control approach has been given precedence over incentive-based approaches until very recently. Congress has rather consistently written regulations that direct EPA to establish emissions standards for sources (with the help of the states and localities) and to issue and enforce permits that specify those standards. Various factors have influenced the emission reductions that EPA has required. Although the agency often claims to have pursued a least-cost strategy, the uniform rollback and ability-to-pay criteria clearly have dominated in apportioning emission cutbacks.

The effluent charge approach has never really gotten off the ground in U.S. environmental policy, but a tax on sulfur emissions into the air was proposed by the Nixon administration in EPA's first year and again in 1988. Then, in the 1990 amendments to the Clean Air Act, Congress imposed a tax on the production of chlorofluorocarbons (CFCs), chemicals that deplete ozone in the Earth's stratosphere. However, the standards-and-permits approach to air pollution control has evolved in the direction of marketable permits, and there is talk of applying this approach more widely in air pollution control as well as in other regulatory programs. In fact, those same 1990 amendments also mandated the creation of a system of marketable "allowances" to discharge sulfur dioxide (see Chapter 3, Market-Based Environmental Policy, and Chapter 4, Air Pollution Policy). This program, which was intended to control pollutants implicated in acid rain, has been an unqualified success. Because of this and a growing number of other successes, marketable permits apparently will play a larger role in future U.S. environmental policy. Needless to say, this is an important development, and each of the chapters in this book touches on the possibilities for an expanded role for economic incentives.

Problems Facing the Agency

Before concluding this chapter, I should mention several generic problems that have arisen in the hybrid U.S. environmental policy since 1970. The problems that face EPA are of four sorts (Crandall and Portney 1984, 47–55). The first has to do with the tremendous complexity of our environmental laws and their penchant for promising a very great deal in a very short time. For instance, the clean air and clean water laws promise "safe" air and water quality, call for the establishment of literally tens of thousands of discharge standards, mandate the creation of comprehensive monitoring networks, and impose numerous other important tasks on the EPA administrator. Yet, the laws allotted just 180 days for completion of many of these responsibilities. Today, nearly thirty years after passage of the laws, many of those assignments have yet to be carried out.

Similarly, the Toxic Substances Control Act calls for the promulgation of separate testing rules for each new chemical. Although such chemicals come on the market at the rate of 1,000 per year, EPA has issued testing rules for only a few substances. Of the more than 50,000 existing chemicals in commerce, only a small fraction have been tested for carcinogenicity or other harmful effects. Each of the environmental laws provides examples like this one, where Congress either misunderstood the time required to issue careful regulations or disregarded it in the rush to get legislation on the books. EPA has tried to run faster and faster since its creation but has fallen farther and farther behind because of its impossible burden and an occasional lack of will.

The second kind of problem has to do with the spotty compliance with those standards that have been issued and our poor ability to know which standards are being violated and which sources are responsible. Two situations appear to be responsible for these problems. First, monitoring both ambient environmental quality and the emissions from individual sources is much more complicated and expensive than one would imagine. Monitoring is not a straightforward matter, as might be supposed from reading the laws. Second, monitoring and enforcement have always been poor stepsisters in the eyes of Congress. Apparently, it is more fashionable to write new laws and call attention to problems with existing laws than it is to engage in the dirty work of fashioning an enforceable and scientifically meaningful set of standards. Thus, enforcement programs have always suffered financially at the expense of new and emerging regulatory programs.

The third sort of generic problem concerns the frequent emphasis on absolutist goals in environmental statutes. Waters are to be "fishable and swimmable" as one step toward a world of "zero discharges" into rivers, lakes, and the oceans. Conventional and hazardous air pollutants are to be at "safe" levels, as are drinking water contaminants. Such an approach has obvious political appeal; voters are comforted to hear that they will be safe from all environmental threats. But for many—perhaps most—pollutants, that simply will not be the case unless standards are to be set at zero, which is often impossible. Thus, although it is surely done, the balancing of environmental versus other important goals, economic and otherwise, is done implicitly. This approach has resulted in some standards being set at levels that appear to be hard to justify on any rational basis.

Finally, and perhaps inevitably, the fourth sort of problem is that environmental statutes have become contaminated by redistributive goals that often work against the environmental programs in which they are nested. For example, newly built electric power plants are prohibited from switching from high-sulfur to low-sulfur coal in their efforts to reduce sulfur dioxide emissions. This prohibition, which protects the jobs of a small number of high-sulfur coal miners, exists despite the tremendous cost savings that might be reaped if the plants were permitted to change fuels. Similarly, federal subsidies for the construction of sewage treatment plants have been continued, even though the plants seem to have had a questionable impact on water quality in many areas and even though the federal subsidy has crowded out state and local spending for these same plants. Apparently, the pork-barrel aspects of the program have proved too attractive to eliminate. Water pollution from farms, animal feedlots, and other nonpoint sources has been overlooked almost completely until recently, because the political power of the farmers and other parties would be affected by tighter controls. Some evidence also suggests that environmental regulations have been structured to protect declining regions of the country from the effects of further economic growth in faster growing Sun Belt states.

Although all these contortions of environmental policy are understandable, they also stand in the way of an effective and less costly approach to environmental protection. As such, they deserve to be starkly highlighted.

Notes

1. Exceptions include the creation of the Federal Aviation Administration, the Federal Highway Administration, and the Federal Railroad Administration to oversee safety in those respective industries.

2. A natural monopoly is said to exist when the per-unit cost of producing a good or service continues to fall with increases in output. It is argued that in such a case, consumers will enjoy the lowest prices if one firm serves the whole market instead of sharing it with two or more competitors, as long as the single provider is regulated so as not to abuse its monopoly position. Natural monopolies are thought to arise most often when the fixed costs of doing business are a large proportion of total costs. The traditional examples include local telephone service, electricity distribution, and natural gas pipelines.

3. If regulation is the barrier to entry and discouragement to new growth that some maintain it is, and if these effects could be quantified, then this conclusion could change.

4. In the real world, in fact, property rights appear to be shared. Even under existing regulation, firms are generally permitted to emit at least some pollution without being held responsible for the damage it may cause. In addition, tax breaks are provided for some investments in pollution control equipment. In effect, this system shifts some of the costs of pollution control to the taxpayer, as would be the case if the property right were initially vested in the polluter.

5. Although it probably is inappropriate in the case of environmental problems, a third alternative to regulation exists in cases involving occupational hazards or dangerous consumer products: The government could limit its role to the provision of information about the risks inherent in different jobs and/or products. Workers and consumers then could hold out for higher wages for risky jobs (as they do now) or pay low prices for risky products. Employers or producers would have to decide whether it was in their interest to continue to pay higher wages or receive lower prices rather than reduce the hazards. In this way, too, market forces would work toward optimal risk. It should be noted that this model would be successful only if workers and consumers were fully informed about such risks and only if they had a range of jobs and products from which to choose.

6. This is the point of such regulations, because the higher prices discourage consumers from purchasing products whose production generates pollution.

7. Other possibilities are moral suasion and direct government purchase of pollution control equipment (see Baumol and Oates 1979).

8. Even in this apparently simple case, the matter is not so straightforward, because nature accounts for a share of many major pollutants. For instance, particulate matter can be blown from fields or roadways just as it can be generated by steel mills or cement plants. In such cases, the human contribution must be assessed before required cutbacks can be determined.

9. For a comprehensive description and discussion, see Bohm and Russell 1985 and Anderson and others 1977.

10. For an interesting analysis of attitudes toward effluent charges, see Kelman 1981.

References

Anderson, Frederick R., Allen V. Kneese, Phillip D. Reed, Serge Taylor, and Russell B. Stevenson. *Environmental Improvement through Economic Incentives.* Washington, DC: Resources for the Future, 1977.

Baumol, William J., and Wallace E. Oates. 1971. The Use of Standards and Prices for Protection of the Environment. *Swedish Journal of Economics* 73(March):42–54.

Baumol, William J., and Wallace E. Oates. 1979. *Economics, Environmental Policy, and the Quality of Life.* Englewood Cliffs, NJ: Prentice-Hall, 217–24.

Bohm, Peter, and Clifford S. Russell. 1985. Comparative Analysis of Alternative Policy Instruments. In *Handbook of Natural Resource and Energy Economics*, edited by Allen V. Kneese and James L. Sweeney. Vol. 1. New York: North-Holland, 395–460.

Crandall, Robert W., and Paul R. Portney. 1984. Environmental Policy. In *Natural Resources and the Environment: The Reagan Approach*, edited by Paul R. Portney. Washington, DC: Urban Institute.

Eads, George C., and Michael Fix. 1984. *Relief or Reform? Reagan's Regulatory Dilemma.* Washington, DC: Urban Institute Press, 12–15.

Freeman, A. Myrick III. 1993. *The Measurement of Environmental and Resource Values: Theory and Methods.* Washington, DC: Resources for the Future.

Graham, John, and Jonathan Wiener, eds. 1995. *Risk vs. Risk.* Cambridge, MA: Harvard University Press.

Keeney, Ralph L. 1990. Mortality Risks Induced by Economic Expenditures. *Risk Analysis* 10(1):147–59.

Kelman, Steven. 1981. *What Price Incentives?* Boston, MA: Auburn House.

Lave, Lester B. 1981. *The Strategy of Social Regulation.* Washington, DC: Brookings Institution.

Portney, Paul R. 1981. The Macroeconomic Impacts of Federal Environmental Regulation. *Natural Resources Journal* 21(July):459–88.

Wolf, Charles Jr. 1988. *Markets or Government.* Cambridge, MA: MIT Press.

three

Market-Based Environmental Policies

Robert N. Stavins*

Nearly all environmental policies consist of two components, either explicitly or implicitly: the identification of an overall goal (either general or specific, such as a degree of air quality or an upper limit on emission rates) and some means to achieve that goal. In practice, these two components often are linked within the political process, because both the choice of a goal and the mechanism for achieving that goal have important political ramifications.[1] The focus of this chapter is exclusively on the second component, that is, the means—the "instruments"—of environmental policy. In particular, I consider economic-incentive, or market-based, policy instruments.

What Are Market-Based Policy Instruments?

Market-based instruments are regulations that encourage behavior through market signals rather than through explicit directives regarding pollution control levels or methods.[2] These policy instruments, which include tradable permits and pollution charges, often are described as "harnessing market forces" (see Stavins 1988; 1991; OECD 1989; 1991; U.S. EPA 1991)[3] because

*Quindi Franco provided excellent research assistance. Dallas Burtraw, Robert Hahn, Paul Portney, and Tom Tietenberg gave very helpful comments on a previous version of this chapter. The author alone is responsible for any errors.

if they are well designed and implemented, they encourage firms (and/or individuals) to undertake pollution-control efforts that both are in those firms' (or individuals') interests and collectively meet policy goals.

In contrast, conventional approaches to regulating the environment are often referred to as "command and control" regulations, because they allow relatively little flexibility in the means of achieving goals. Early environmental policies, such as the Clean Air Act of 1970 and the Clean Water Act of 1972, relied almost exclusively on these approaches. (For descriptions of the use of command-and-control instruments for various environmental problems, see Chapters 2, 4, and 6–8 in this volume.)

In general, command-and-control regulations tend to force firms to shoulder similar shares of the pollution-control burden, regardless of the relative costs to them. However, various command-and-control standards do this in different ways (see Helfand 1991). Command-and-control regulations typically set uniform standards for firms, the most prevalent of which are technology-based and performance-based standards. Technology-based standards specify the method, and sometimes the actual equipment, that firms must use to comply with a particular regulation. For example, all electric utilities might be required to use a specific kind of scrubber to remove particulates. A performance standard sets a uniform control target for firms while allowing some latitude in how this target is met. For example, a regulation might limit the number of allowable units of a pollutant released in a given time period but not dictate the means by which this result is achieved.

Holding all firms to the same target can be expensive and, in some circumstances, counterproductive. Although standards may effectively limit emissions of pollutants, they typically exact relatively high costs in the process by forcing some firms to resort to unduly expensive means of controlling pollution. Because the costs of controlling emissions may vary greatly among firms, and even among sources within the same firm, the appropriate technology in one situation may be inappropriate in another. Thus, control costs can vary enormously because of a firm's production design, its physical configuration, the age of its assets, or other factors. One survey of eight empirical studies of air pollution control found that the ratio of actual aggregate costs of the conventional command-and-control approach to the aggregate costs of least-cost benchmarks ranged from 1.07 for sulfate emissions in the Los Angeles area to 22.0 for hydrocarbon emissions at all domestic DuPont plants (Tietenberg 1985).

Furthermore, command-and-control regulations tend to freeze the development of technologies that might otherwise result in greater levels of control. Little or no financial incentive exists for businesses to exceed their control targets, and both technology-based and performance-based standards discourage the adoption of new technologies. A business that adopts a new technology may be "rewarded" by being held to a higher standard of performance, but it is not given the opportunity to benefit financially from its investment—except to the extent that its competitors have even more difficulty reaching the new standard.

Characteristics

The two most notable advantages that market-based instruments offer over traditional command-and-control approaches are cost-effectiveness and dynamic incentives for technology innovation and diffusion.

In theory, properly designed and implemented market-based instruments allow pollution to be reduced to any desired level at the lowest possible cost to society. Incentives to reduce the greatest amounts of pollution are provided to the firms that can achieve those reductions most cheaply.[4] Rather than equalizing pollution levels among firms (as with uniform emission standards), market-based instruments equalize the incremental amount that firms spend to reduce pollution (their marginal cost).[5]

It is important to recognize that command-and-control approaches could—theoretically—achieve this cost-effective solution, but different standards would have to be set for each pollution source and, consequently, policymakers would have to obtain detailed information about the compliance costs each firm faces. Such information simply is not available to government. Market-based instruments provide for a cost-effective allocation of the pollution control burden among sources without requiring the government to have this information.

In contrast to command-and-control regulations, market-based instruments have the potential to provide powerful incentives for companies to adopt cheaper and better pollution-control technologies. Market-based instruments benefit the firms that identify and adopt low-cost pollution-control methods (technologies or processes).[6]

Categories

Market-based instruments can be grouped in four major categories: pollution charges, tradable permits, market barrier reductions, and government subsidy reductions. (For general information, see OECD 1994a; 1994b; 1994c.)

Pollution charge systems assess a fee or tax on the amount of pollution that a firm or source generates.[7] Consequently, it is worthwhile for the firm to reduce emissions to the point where its marginal abatement cost is equal to the tax rate. Thus, firms will control pollution to differing degrees, with high-cost controllers controlling less and low-cost controllers controlling more. One challenge with charge systems is identifying the appropriate tax rate. Ideally, it should be set equal to the incremental benefits of cleanup at the efficient level. However, policymakers are more likely to think in terms of a desired level of cleanup and do not know beforehand how firms will respond to a given level of taxation.

A special case of pollution charges is a *deposit–refund system*, whereby consumers pay a surcharge when they purchase potentially polluting products and receive a refund when they return the product (or its packaging) to an approved center for recycling or disposal. Many states have implemented this approach through "bottle bills" to control litter from beverage containers and to reduce the flow of solid waste to landfills; the concept also has been applied to lead-acid batteries (Bohm 1981; Menell 1990).

Tradable permits, like pollution charges, can allocate the control burden at minimum cost; however, they also set a maximum level of overall emissions.[8] Under a tradable permit system, an allowable overall level of pollution is established and allocated among firms in the form of permits.[9] Firms that keep their emission levels below their allotted level may sell their surplus permits to other firms or use them to offset excess emissions in other parts of their facilities.

Market barrier reductions also can serve as market-based policy instruments. In such cases, substantial gains can be made in environmental protection by simply removing existing explicit or implicit barriers to market activity. Three kinds of market barrier reductions stand out: market creation, liability rules, and information programs.

Government subsidy reductions are the fourth and final category of market-based instruments. Subsidies are the mirror image of taxes and, in theory, can provide incentives to address environmental problems. However, in practice, many subsidies promote economically inefficient and environmentally unsound practices. This market distortion received much attention in the 104th U.S. Congress under the rubric of "corporate welfare," an example of which is the below-cost sale of timber by the U.S. Forest Service (Stavins 1988).

In the simplest models, pollution taxes and tradable permits are symmetrical, but that symmetry begins to break down in implementation (Stavins and Whitehead 1992). First, permits fix the level of pollution control, whereas charges fix the costs of pollution control. Second, in the presence of technological change and without additional government intervention, permits freeze the level of pollution control, whereas charges increase it. Third, with permit systems as typically adopted, resources are transferred only within the private sector, whereas with ordinary pollution charges, resources are transferred from the private sector to the public sector. Fourth, whereas both systems increase costs on industry and consumers, charge systems tend to make those costs more obvious to both groups. Fifth, permits adjust automatically for inflation, whereas some kinds of charges do not. Sixth, permit systems may be more susceptible to strategic behavior than charges (Hahn 1984; Malueg 1989; Misolek and Elder 1989). Seventh, significant transaction costs can drive up the total costs of compliance, having a negative effect under either system, but particularly with tradable permits (Stavins 1995). Eighth and finally, in the presence of uncertainty, either permits or charges can be more efficient, depending on the relative slopes of the marginal benefit and marginal cost functions (Weitzman 1974; Adar and Griffin 1976; Tisato 1994) and any correlation between them (Stavins 1996).

The degree of abatement achieved by a pollution tax and the tax's effect on the economy will depend in part on what is done with the tax revenue. Agreement that revenue recycling (that is, using pollution tax revenues to lower other taxes) can significantly lower the costs of a pollution tax is widespread (Jorgenson and Wilcoxen 1994; Goulder 1995). Furthermore, some researchers have suggested that all of the abatement costs associated with a pollution tax can be eliminated through revenue recycling by cutting taxes on labor (Repetto and others 1992). But pollution taxes can exacerbate distortions associated with remaining taxes on investment or labor. Environmental taxes impose their own distortions that are at least as great as those from labor taxes (Bovenberg and de Mooij 1994; Goulder 1995; Parry 1995; Bovenberg and Goulder 1996). Using revenues from an environmental tax—or from the auction of pollution permits (Fullerton and Metcalf 1997; Goulder and others 1997)—to reduce labor taxes can reduce the efficiency costs of the environmental tax. However, in most cases, the potentially beneficial environmental consequences of such a tax will be tempered by its costs. Thus, the primary justification for environmental taxes should be their environmental benefits, not reform of the tax system per se.

U.S. Experience with Tradable Permit Programs

The most frequently used market-based environmental instruments in the United States have been tradable permit systems (see U.S. EPA 1992; Tietenberg 1997). They include the Emissions Trading Program of the U.S. Environmental Protection Agency (EPA), the leaded gasoline phasedown, water-quality permit trading, chlorofluorocarbon (CFC) trading, the sulfur dioxide (SO_2) allowance system for acid rain control, the Regional Clean Air Incentives Market (RECLAIM) program in the Los Angeles metropolitan region, and tradable development rights for land use (Table 3-1).[10]

EPA's Emissions Trading

Beginning in 1974, EPA experimented with emissions trading as part of the Clean Air Act's program for improving local air quality. Firms that reduced emissions below the level required by law received "credits" usable against higher emissions elsewhere. Companies could use "netting" and "bubbles" to trade emissions reductions among sources within the firm, as long as total combined emissions did not exceed an aggregate limit.[11] Emissions from all the components of an industrial plant were combined and considered a single source for purposes of regulation.[12]

The offset program, which began in 1976, goes further in allowing firms to trade emission credits. Firms that wish to establish new sources in areas that are not in compliance with ambient standards must offset new emissions with emissions reductions in existing facilities. It can be accomplished through internal sources or through agreements with other firms.

Finally, under the banking program, firms may store earned emission credits for future use. Banking allows for either future internal expansion or the sale of credits to other firms.

EPA codified these programs in its Emissions Trading Program in 1986 (U.S. EPA 1986), but the programs have not been widely used. States are not required to use the program, and uncertainties about its future course seem to have made firms reluctant to participate (Liroff 1986). Nevertheless, companies such as Armco, DuPont, USX, and 3M have traded emissions credits, and a market for transfers has long since developed (Main 1988). Even this limited degree of participation in EPA's trading programs may have saved between $5 billion and $12 billion over the life of the program (Hahn and Hester 1989b).

Lead Trading

The purpose of the lead trading program, developed in the 1980s, was to allow gasoline refiners greater flexibility in meeting emissions standards at a time when the lead content of gasoline was reduced to 10% of its previous level. In 1982, the EPA authorized inter-refinery trading of lead credits (U.S. EPA 1982). If refiners produced gasoline with a lower lead content than was required, they earned lead credits. In 1985, EPA initiated a program allowing refineries to bank lead credits, and firms made extensive use of this program.[13] EPA terminated the program at the end of 1987, when the lead phasedown was completed.

TABLE 3-1. Major Federal Tradable Permit Systems.

Program	*Traded commodity*	*Years of operation*	*Effects: Environmental*	*Effects: Economic*
Emissions trading program	Criteria air pollutants under the Clean Air Act	1974–present	Performance unaffected	Total savings of \$5–12 billion
Lead phasedown	Rights for lead in gasoline among refineries	1982–1987	More rapid phaseout of leaded gasoline	\$250 million annual savings
Water-quality trading	Point and nonpoint sources of nitrogen and phosphorus	1984–1986	NA[a]	NA[a]
CFC trading for ozone protection	Production rights for some CFCs, based on depletion potential	1987–present	Targets achieved ahead of schedule	Effect of tradable permit system unclear
RECLAIM program	Local SO_2 and NO_x emissions trading among stationary sources	1994–present	NA[b]	NA[b]
Acid rain reduction	SO_2 emission reduction credits; mainly among electric utilities	1995–present	Targets achieved ahead of schedule	Annual savings of up to \$1 billion

Notes: The Regional Clean Air Incentives Market (RECLAIM) program in the Los Angeles metropolitan area is a regional initiative intended to achieve federal and state targets. CFCs = chlorofluorocarbons; NA = not applicable.

[a] No trading occurred because ambient standards not binding.

[b] Unknown.

Sources: Hahn 1989; Hahn and Hester 1989b; Schmalensee and others 1998.

The lead program was clearly successful in meeting its environmental targets. And although the benefits of the trading scheme are more difficult to assess, the level of trading activity (Kerr and Maré 1997) and the rate at which refiners reduced their production of leaded gasoline (Nichols 1997) suggest that the program was relatively cost-effective. The high level of trading between firms far surpassed levels observed in earlier environmental markets.[14] EPA estimated savings from the lead trading program of approximately 20% over alternative programs that did not provide for lead banking (U.S. EPA 1985), a cost savings of about \$250 million per year.

Water Quality Trading

The United States has had very limited experience with tradable permit programs for controlling water pollution; however, nonpoint sources—particularly agricultural and urban runoff—may constitute the major remaining American water pollution problem (Peskin 1986). An experimental program to protect the Dillon Reservoir in Colorado demonstrates how tradable permits could be used, in theory, to reduce nonpoint-source water pollution.

Dillon Reservoir is the major source of water for the city of Denver. Nitrogen and phosphorus loading threatened to turn the reservoir eutrophic, despite the fact that point sources from surrounding communities were controlled to best-available technology standards (U.S. EPA 1984). Rapid population growth in Denver and the resulting increase in urban surface water runoff further aggravated the problem. In response, state policymakers developed a point-/nonpoint-source control program to reduce phosphorus flows, mainly from nonpoint urban and agricultural sources. The program, implemented in 1984 (Kashmanian 1986), allowed publicly owned sewage treatment works to finance the control of nonpoint sources in lieu of upgrading their own treated effluents to drinking water standards (Hahn 1989, 103). EPA estimated that the plan could save more than $1 million per year (Hahn and Gordon 1989a, 395) because of differences in the marginal costs of control between nonpoint sources and the sewage treatment facilities. However, very limited trading took place under the program, for various reasons, which included the implementation of other regulations that reduced the nonpoint-source runoff, lower-than-expected costs for installing additional treatment facilities, and high regional precipitation that diluted pollutant concentrations in the reservoir.

CFC Trading

A market in tradable permits was used in the United States to help comply with the Montreal Protocol, an international agreement aimed at slowing the rate of stratospheric ozone depletion.[15] The protocol called for reductions in the use of CFCs and halons, the primary chemical groups thought to lead to ozone depletion. The market limits both the production and consumption of CFCs by issuing allowances that limit these activities. The Montreal Protocol recognizes the fact that different kinds of CFCs are likely to have different effects on ozone depletion, and so each CFC is assigned a different weight on the basis of its depletion potential. If a firm wishes to produce a given amount of CFC, it must have an allowance to do so (Hahn and McGartland 1989), calculated on this basis.

From 1986 to mid-1991, thirty-four firms participated in the market, and eighty trades took place (Feldman 1991).[16] However, the overall efficiency of the market is difficult to determine, because no studies were conducted to estimate cost savings. The timetable for the phaseout of CFCs was subsequently accelerated, and a tax on CFCs was introduced.[17] Indeed, the tax may have become the binding (effective) instrument.[18] Nevertheless, relatively low transaction costs associated with trading in the CFC market suggest that the system was relatively cost-effective.

SO_2 Allowance Trading

One centerpiece of the Clean Air Act Amendments of 1990 is a tradable permit system that regulates SO_2 emissions, the primary precursor of acid rain. Title IV of the act reduces sulfur dioxide and nitrous oxide emissions by 10 million tons and 2 million tons, respectively, from 1980 levels.[19] The first phase of sulfur dioxide emissions reductions was achieved by 1995, and a second phase of reduction was accomplished by 2000.

In Phase I, individual emissions limits were assigned to the 263 units that generated the highest levels of SO_2 emissions per unit of energy generated at 110 plants, operated by 61 electric utilities, and located largely at coal-fired power plants east of the Mississippi River. After January 1, 1995, these utilities could emit SO_2 only if they had adequate allowances to cover their emissions. (Under specified conditions, utilities that had installed coal scrubbers to reduce emissions could receive two-year extensions of the Phase I deadline plus additional allowances.) During Phase I, the EPA allocated each affected unit, on an annual basis, a specified number of allowances related to its share of heat input during the baseline period (1985–87) plus bonus allowances available under a variety of special provisions.[20] Cost-effectiveness is promoted by permitting allowance holders to transfer their permits amongst themselves or bank them for later use.

Under Phase II of the program, beginning January 1, 2000, almost all electric power-generation units are brought within the system. Certain units are exempted to compensate for potential restrictions on growth and to reward units that already are unusually clean. If trading permits represent the carrot of the system, then the stick is a penalty of $2,000 per ton of emissions that exceed any year's allowances (and a requirement that such excesses be offset the following year).

A robust market of bilateral SO_2 permit trading has emerged, resulting in cost savings (defined as the difference between the costs experienced with the allowance trading program and what the costs would otherwise have been; hence, any estimate of cost savings is sensitive to the choice of counterfactual for comparison purposes) on the order of $1 billion annually compared with the costs under command-and-control regulatory alternatives. Although the program had low levels of trading in its early years (Burtraw 1996), trading levels increased significantly over time.[21] Concerns have been expressed that state regulatory authorities would hamper trading in order to protect their domestic coal industries, and some research indicates that state public utility commission cost-recovery rules have provided poor guidance for compliance activities (Bohi 1994; Rose 1997). Other analysis suggests that state regulatory authorities have not been a major problem (Bailey 1996). Similarly, in contrast to early assertions that the structure of EPA's small permit auction market would cause problems (Cason 1995), evidence now indicates that it has had little or no effect on the vastly more important bilateral trading market (Joskow and others 1998).

RECLAIM

The South Coast Air Quality Management District (SCAQMD), which is responsible for controlling emissions in a four-county area of southern California, launched a tradable permit program in January 1994 to reduce nitrogen oxide and sulfur dioxide emissions in the Los Angeles area.[22] One prospective analysis predicted a 42% cost savings, which would amount to $58 million annually (Anderson 1997). As of June 1996, 353 participants in the RECLAIM program had traded more than 100,000 tons of nitrogen oxide (NO_x) and SO_2 emissions, at a value of more than $10 million (Brotzman 1996).[23] The RECLAIM program, which operates through the issuance of permits that authorize specified decreasing levels of pollution over time, governs stationary sources that have emitted more than 4 tons of NO_x and SO_2 annually since 1990; some sources, such as equipment rental facilities and "essential public services" (including landfills and wastewater treatment facilities), are excluded. The SCAQMD has considered expanding the program to allow trading between stationary and mobile sources (Fulton 1996).

Transferable Development Rights

Local governments in the United States have a considerable history of using transferable development rights to balance some of the attributes and amenities ordinarily addressed by zoning provisions with the demands of economic growth and change (for example, Field and Conrad 1975; Bellandi and Hennigan 1977; Mills 1980). A relatively recent application of the same general instrument with an environmental focus has been for the protection of wetlands.

Certain development activities in wetlands are regulated in the United States by Section 404 of the Clean Water Act, which establishes conditions and procedures for such activities. Firms or individuals must apply for permits for activities that will have negative impacts on wetlands. In some cases, compensating mitigation is required of potential developers, and applicants are allowed to purchase mitigation credits from land banks to meet these obligations (Tripp and Dudek 1989; Scodari and others 1995; Voigt and Danielson 1996). These mitigation banks have been established in several states, including California, Florida, Minnesota, New Jersey, and North Carolina.

U.S. Experience with Charge Systems

The conventional wisdom is that U.S. environmental policy has made increasing use of tradable permit systems while essentially ignoring the option of taxes or charges. This view is not strictly correct, and if one defines charge systems broadly, a significant number of applications can be identified. These applications can be categorized as effluent charges, deposit–refund systems, user charges, insurance premiums, sales taxes, administrative charges, and tax differentiation.

Most applications of charge systems in the United States have probably not had the incentive effects associated with a Pigovian tax, either because of the structure of the systems or because of the low levels at which charges have been set.[24] Nevertheless, it appears that some of these systems may have affected behavior.

Effluent Charges

In the United States, the closest that any charge system comes to operating as a Pigovian tax may be the unit-charge approach to financing municipal solid waste collection, where households (and businesses) are charged the incremental costs of collection and disposal (see Chapter 8, Solid Waste Policy). So called "pay-as-you-throw" policies, where users pay in proportion to the volume of their waste, now are used in well over 100 jurisdictions (U.S. OTA 1995). This experience provides evidence that unit charges have been somewhat successful in reducing the volume of household waste generated (McFarland 1972; Wertz 1976; Stevens 1978; Efaw and Lanen 1979; Skumatz 1990; Lave and Gruenspecht 1991; Repetto and others 1992; Miranda and others 1994; Fullerton and Kinnaman 1996).

Deposit–Refund Systems

As the costs of legal disposal increase, incentives for improper (illegal) disposal also increase. Hence, waste-end fees designed to cover the costs of disposal, such as unit curbside charges, can lead to increased incidence of illegal dumping (Fullerton and Kinnaman 1995). For waste that poses significant health or ecological impacts, cleanup of illegally disposed waste is an especially unattractive option, and the prevention of improper disposal is particularly important. One alternative might seem to be a front-end tax on waste precursors, because such a tax would give manufacturers incentives to find safer substitutes and to recover and recycle taxed materials. But substitutes may not be available at reasonable costs, and once wastes are generated, incentives that affect choices of disposal methods are still problematic.

This dilemma can be resolved with a special front-end charge (deposit) combined with a refund payable when quantities of the substance in question are turned in for recycling or disposal. This refund can provide an incentive to follow rules for proper disposal (and to prevent losses in the process in which the substance is used). The mechanics of the system vary by product, but the general framework is that producers or initial users of regulated materials pay a deposit when those materials enter the production process. In principle, the size of the deposit is based on the social cost of the product being disposed of illegally. As the product changes hands in the production and consumption process (starting with wholesalers and distributors), the responsibility for proper disposal also is passed on. The process continues until the consumer of the good turns in the product (or its packaging) to a certified collection center responsible for recycling or proper disposal.

Deposit–refund systems (Bohm 1981) are most likely to be appropriate when the incidence and the consequences of improper disposal are great (Macauley and others 1992), but these systems have frequently been portrayed as mechanisms to foster greater levels of recycling. In general, properly scaled deposit–refund systems can be attractive for three reasons. First, government's monitoring problem is converted from the nearly impossible one of preventing illegal dumping of small quantities of waste at diverse sites in the environment to what may be the more manageable problem of ensuring that products being returned for refund are what they are purported to be. Second, the system can provide firms with incentives to prevent losses of the material in the industrial process in which it is used. Third, because of inevitable net losses in the production and consumption processes, incentives exist for firms to look for substances that cause less damage to the environment—substances to which the deposit–refund system does not apply.[25] For some products, a nationwide approach may be appropriate if firms face national markets, products are easily transportable, toxicity problems associated with improper disposal do not vary greatly by geographic area, and the national approach is likely to be less costly for manufacturers and recyclers than various state or local programs.

The major application of this approach in the United States has been in the form of state-level "bottle bills" for beverage containers (Table 3-2). A brief examination of these systems provides some insights into the merits and the limitations of the approach. Deposit–refund systems on beverage containers have been implemented in ten states to reduce littering and reduce the flow of solid waste to landfills, but since the initial enthusiasm in the late 1970s and early 1980s, no other states have taken action.

In most programs, consumers pay a deposit at the time of purchase which can be recovered by returning the empty container to a redemption center. Typically, the deposit is the same regardless of the kind of container. In some respects, these bills seem to have accomplished their objectives; in Michigan, for example, the return rate of containers one year after the program was implemented was 95% (Porter 1983); in Oregon, littering was reduced, and long-run savings in waste management costs were achieved (U.S. GAO 1990).

By charging the same amount for each kind of container material, these programs do not encourage consumers to choose containers with the lowest product life-cycle costs (including those of disposal). In particular, if bottle bills do not include deposits and refunds for metal, plastic, and glass containers, they may encourage a shift of consumer purchases from easily recyclable to less-recyclable containers. Furthermore, by requiring consumers to separate containers and deliver them to redemption centers, deposit–refund systems can foster net social losses, rather than gains. Additionally, by removing some of the most profitable elements from the waste stream, bottle bills may undermine the viability of more comprehensive alternatives, such as curbside programs.

Analysis of the effectiveness—let alone the cost-effectiveness or efficiency—of deposit–refund systems for beverage containers has been limited. A major cost of bottle bills is associated with labor and capital required for implementation, which includes, for example, the area set aside and labor used at grocery stores for collection purposes. Also of economic significance are the personal inconvenience costs of returning containers to retail outlets. These inconvenience costs may be quite significant, and the few rigorous studies that have been carried out of the benefits and costs of bottle bills have found that the social desirability of deposit law depends critically on the value of the time it takes consumers to return empty containers and the willingness to pay for reduced litter (see Porter 1978).

TABLE 3-2. State Deposit–Refund Systems for Two Regulated Products.

State	*Year of initiation*	*Amount of deposit ($)*
Specified beverage containers		
Oregon	1972	0.05[a]
Vermont	1973	0.05
Maine	1978	0.05
Michigan	1978	0.10
Iowa	1979	0.05
Connecticut	1980	0.05
Delaware	1983	0.05
Massachusetts	1983	0.05
New York	1983	0.05
California	1987	0.025–0.06[b]
Auto batteries		
Minnesota	1988	5.00
Maine	1989	10.00
Rhode Island	1989	5.00
Washington	1989	5.00
Arizona	1990	5.00
Connecticut	1990	5.00
Michigan	1990	6.00
Idaho	1991	5.00
New York	1991	5.00
Wisconsin	1991	5.00
Arkansas	1991	10.00

Notes: Florida had an advance disposal fee (an upfront processing fee charged to plastic and glass containers) that is no longer in effect.

[a] $0.02 for refillable containers.

[b] Deposits are $0.025 and $0.05, respectively, for aluminum and bimetal beverage containers smaller than 24 ounces and $0.03 and $0.06, respectively, for containers 24 ounces and larger.

Sources: U.S. GAO 1990; U.S. EPA 1992.

Deposit–refund systems are most likely to be appropriate where

- the objective is to reduce illegal disposal rather than reduce the waste stream or increase recycling in general and
- a significant asymmetry exists between the costs of legal disposal and the costs of cleanup (of illegal disposal or litter).

For these reasons, deposit–refund systems may be among the best policy options to address disposal problems associated with hazardous waste that can be transported in containers (for example, lead) (Sigman 1995).

As a means of reducing the quantity of lead entering unsecured landfills and other potentially sensitive sites, several states have enacted deposit–refund programs for lead acid motor vehicle batteries (Table 3-2).[26] Under these systems, a deposit is collected when manufacturers sell batteries to distributors, retailers, or original equipment manufacturers; likewise, retailers collect deposits from consumers at the time of battery purchase. Consumers can collect their deposits by returning used batteries to redemption centers; these redemption centers, in turn, redeem deposits from battery manufacturers.

The programs are largely self-enforcing, because participants have incentives to collect deposits on new batteries and obtain refunds on used ones, but a potential problem inherent in the approach is an increase in incentives for battery theft. The higher the deposit, the greater the incentive for theft, particularly if one only needs to show up at a redemption center with a battery to claim a refund. An alternative is to require a sales receipt on redemption or to permit refunds only for those exchanging an old battery for a new one. However, either of these alternatives will reduce the comprehensiveness of the program.[27] In any event, a deposit of $5 to $10 per battery exceeds the typical market value of used batteries. Thus, it may be small enough to avoid much of the theft problem, but large enough to encourage a substantial level of return.

User Charges

User charges raise funds for the management and maintenance of resources. Charges of the magnitude necessary to fully cover costs have not been implemented at the federal level, with the possible exception of an experimental fee program for the national parks, initiated in 1996. However, various federal recreation and transportation taxes can be considered user charges because their revenues are dedicated to support usage (Table 3-3).

Recreation and entrance fees in the National Park System and other federally managed recreational areas have been legally mandated since 1951,[28] but the revenues from these fees historically have gone to the U.S. Treasury, to be reappropriated to the park system as a whole. In 1996, Congress approved a three-year experimental program, the Recreation Fee Demonstration Program, which permits fifty specified parks to raise entrance fees and keep up to 80% of incremental revenues. Some of the fee increases have been quite substantial; fees in Yosemite, Grand Canyon, and Yellowstone National Parks, for example, doubled from $10 to $20. In addition, two states—New Hampshire and Vermont—have created nearly self-financing park systems (see Reiling and Kotchen 1996).

TABLE 3-3. Federal User Charges.

Item taxed	Year: First enacted	Year: Modified	Rate	Use of revenues
Sport fishing equipment	1917	1984	10%[a]	Sport Fishing Restoration Account of Aquatic Resources Trust Fund
Trucks and trailers (excise tax)	1917	1984	12%	Highway Trust Fund/ Mass Transit Account
Firearms and ammunition	1918	1969	10%	Federal Aid to Wildlife Program
Noncommercial motorboat fuels	1932	1992	$0.183/gallon	Aquatic Resources Trust Fund
Motor fuels	1932	1993	$0.183/gallon	Highway Trust Fund/ Mass Transit Account
Nonhighway recreational fuels and small-engine motor fuels	1932	1993	$0.183/gallon gasoline, $0.243/ gallon diesel	National Recreational Trails Trust Fund and Wetlands Account of Aquatic Resources Trust Fund
Annual use of heavy vehicles	1951	1993	$100–500/vehicle	Highway Trust Fund/ Mass Transit Account
Bows and arrows	1972	1984	11%	Federal Aid to Wildlife Program
Inland waterways fuels	1978	1993	$0.233/gallon	Inland Waterways Trust Fund

[a] Except 3% for outboard motors.

Source: Barthold 1994.

Insurance Premium Taxes

Several federal taxes are levied on industries or groups to fund insurance pools against potential environmental risks associated with the production or use of taxed products (Table 3-4). Such taxes can encourage firms to internalize environmental risks in their decisionmaking, but in practice, these taxes frequently have not been targeted at the respective risk-creating activities. For example, to support the Oil Spill Liability Trust Fund, all petroleum

TABLE 3-4. Federal Insurance Premium Taxes.

	Year			
Item or action taxed	*First enacted*	*Modified*	*Rate*	*Use of revenues*
Coal production	1977	1987	$1.10/ton underground; $0.55/ton surface	Black Lung Disability Trust Fund
Chemical production	1980	1986	$0.22–4.88/ton	Superfund (CERCLA)
Petroleum production	1980	1986	$0.097/barrel crude	
Corporate income	1986		0.12%[a]	
Petroleum-based fuels, except propane	1986	1990[b]	$0.001/gallon	Leaking Underground Storage Trust Fund
Petroleum and petroleum products	1989	1990	$0.05/barrel	Oil Spill Liability Trust Fund

Note: CERCLA = The Comprehensive Environmental Response, Compensation, and Liability Act of 1980.

[a] 0.12% of "alternative minimum taxable income" that exceeds $2 million.

[b] Expired 1995.

Source: Barthold 1994.

products are taxed, regardless of how they are transported. This across-the-board tax may create small incentives to use less petroleum but not to use safer ships or other means of transport.

An excise tax on specified hazardous chemicals is used to fund (partially) the cleanup of hazardous waste sites through the Superfund program. The tax functions as an insurance tax to the extent that funds are used for future cleanups (Barthold 1994). The Leaking Underground Storage Trust Fund, established in 1987, is replenished through taxes on all petroleum fuels,[29] and the Oil Spill Liability Trust Fund, established subsequent to the Exxon Valdez oil spill, receives revenue from a tax on petroleum and petroleum products. The fund can be used to meet unrecovered claims from oil spills.[30] Finally, the Black Lung Disability Trust Fund was established in 1954 to pay miners who became sick and unable to work because of prolonged exposure to coal dust in mines. Since 1977, the fund has been financed by excise taxes on coal from underground and surface mines.[31]

Sales Taxes

It has been argued that only two federal sales taxes have affected behavior in the manner of a Pigovian tax: the "gas guzzler" tax on new cars and the excise tax on ozone-depleting chemicals (Barthold 1994). However, it is far from clear that the CFC tax actually affected business decisions (Table 3-5).

The Energy Tax Act of 1978 established a tax on the sale of new vehicles that fail to meet statutory fuel efficiency levels (gas guzzlers), set at 22.5 miles per gallon. The tax ranges from $1,000 to $7,700 per vehicle, based on fuel efficiency; however, the tax does not depend on actual performance or on mileage driven. The tax is intended to discourage the production and purchase of fuel-inefficient vehicles,[32] but it applies to a relatively small set of luxury cars—light trucks, which include sport utility vehicles, are fully exempt from the tax (see Bradsher 1997)—and so has had limited effects.

To meet international obligations established under the Montreal Protocol to limit the release of chemicals that deplete stratospheric ozone, the federal government set up a tradable permit system and levied an excise tax on specific CFCs in 1989. Producers are required to have adequate allowances, and users pay a fee (set proportional to a chemical-specific ozone-depleting factor). Considerable debate regards which mechanism should be credited with the successful reduction in the use of these substances (Hahn and McGartland 1989; U.S. OTA 1995).[33]

Additionally, several states impose taxes on fertilizers and pesticides, but at levels below that required to affect behavior significantly. The taxes generate revenues that are used to finance environmental programs. For example, the Iowa Groundwater Protection Act of 1987 imposes taxes on fertilizers and pesticides (0.1% on pesticide sales at the retail level, 0.2% of manufacturer sales, and $0.75 per ton of nitrogen fertilizer). Revenues fund statewide programs for sustainable agriculture and for testing and research on public

TABLE 3-5. Federal Sales Taxes.

	Year			
Item or action taxed	*First enacted*	*Modified*	*Rate ($)*	*Use of revenues*
New tires	1918	1984	0.15–0.50/pound	U.S. Treasury
New automobiles not meeting fuel efficiency standards	1978	1990	1,000–7,700/auto	U.S. Treasury
Ozone-depleting substances	1989	1992	4.35/pound	U.S. Treasury

Source: Barthold 1994.

water supplies (Morandi 1992; International Institute for Sustainable Development 1995).

Administrative Charges

These charges raise revenues to help cover the administrative costs of environmental programs (Table 3-6). Although the charges are not intended to change behavior, this method of raising public funds is broadly consistent with the so-called "polluter pays principle." For example, under the National Pollution Discharge Elimination System of the Clean Water Act, charges for discharge permits are based on the quantity and kind of pollutant discharged. Likewise, the Clean Air Act Amendments of 1990 allow states to tax regulated air pollutants to recover administrative costs of state programs and allow areas in extreme noncompliance to charge higher rates. Under this structure, the SCAQMD in Los Angeles has the highest permit fees in the country (U.S. OTA 1995).

Tax Differentiation

I use the term "tax differentiation" to refer to credits, tax cuts, and subsidies that serve as implicit taxes on environmentally undesirable behavior (Table 3-7). Several federal and state taxes have been implemented in attempts to encourage the use of renewable energy sources, implicitly taking into account externalities associated with fossil fuel energy generation and use. In the Energy Policy Act of 1992, for example, electricity produced from wind and biomass fuels receives a credit of $0.015 per kilowatt-hour (kWh), and solar and geothermal investments can receive up to a 10% tax credit. Although economists' natural response to energy-related externalities is to advise that fuels or energy use be taxed, econometric evidence indicates that energy-

TABLE 3-6. Administrative Charges.

Item or action taxed	*Year first enacted or modified*	*Rate*	*Use of revenues*
Water pollutant discharges	1972	Varies by substance	State administrative cost of National Pollution Discharge Elimination System, Clean Water Act
Criteria air pollutants	1990	Varies by implementing state	State administrative cost of state clean air programs under Clean Air Act

Source: U.S. OTA 1995.

TABLE 3-7. Federal Tax Differentiation.

Item or action taxed	*Provision*	*Year*		*Rate*
		First enacted	*Modified*	
Motor fuels excise tax exemptions[a]	Natural gas	1978	1990	$0.07/gallon
	Methanol	1978	1990	$0.06/gallon
	Ethanol	1978	1990	$0.054/gallon
Income tax credits	Alcohol fuels	1980	1990	$0.60/gallon methanol; $0.54/ gallon ethanol
	Business energy	1980	1990	10% solar
			1990	10% geothermal
	Nonconventional fuels	1980	1990	$3.00/Btu-barrel equivalent of oil
	Biomass production	1992		$0.01.5/kWh
	Electric automobiles	1992		10% credit
	Wind production	1992		$0.01.5/kWh
Other income tax provisions	Van pools	1978		Tax-free employer-provided benefits
	Mass transit passes	1984	1992	
	Utility rebates	1992		Exclusion of subsidies from utilities for energy conservation measures
Tax-exempt private activity bonds	Mass transit	1968	1986	Interest exempt from federal taxation
	Sewage treatment	1968	1986	
	Solid waste disposal	1968	1986	
	Waster treatment	1968	1986	
	High-speed rail	1988	1993	

[a] From the motor fuels excise tax of $0.183/gallon (see Table 3-3).

Source: Barthold 1994.

efficiency technology adoption subsidies may be more effective—in some circumstances—than proportional energy taxes (Jaffe and Stavins 1995).

Since 1979, employers can provide employees with commuting subsidies such as free van pools and tax-free mass transit passes. Likewise, subsidies from utilities for energy conservation investments have been excludable from individual income taxes. On the state and local level, many jurisdictions offer subsidies and various kinds of tax relief to encourage investments in technologies that use recycled products (see Chapter 8, Solid Waste Policy).

U.S. Experience with Reducing Market Barriers

In some situations, environmental protection can be fostered by reducing explicit or implicit barriers to market activity. Three kinds of market barrier reductions are

- market creation, whereby governments facilitate the evolution of new markets;
- liability rules, which encourage firms to consider the potential environmental damages of their decisions; and
- information programs, which improve the functioning of markets by requiring or otherwise encouraging the producers of goods and services to provide certain kinds of information to the consumers of those goods and services.

Market Creation

Two examples of using market creation as an instrument of environmental policy stand out: measures that facilitate the voluntary exchange of water rights and thus promote more efficient allocation and use of scarce water supplies, and policies that facilitate the restructuring of electricity generation and transmission. Both are considered in this section.

First, the western United States has long been plagued by inefficient use and allocation of its scarce water supplies, largely because users do not have incentives to take actions consistent with economic and environmental values. Voluntary market-oriented transfers of water rights have begun to address this problem by encouraging rational conservation measures, better allocation of supplies among competing users, and water quality improvements.

For more than a decade, economists have noted that federal and state water policies have been aggravating, not abating, these problems (Anderson 1983; El-Ashry and Gibbons 1986; Frederick 1986; Wahl 1989). For example, as recently as 1990, in the Central Valley of California, some farmers were paying as little as $10 for enough water to irrigate an acre of cotton while just a few hundred miles away in Los Angeles, local authorities were paying up to $600 for the same quantity of water. This dramatic disparity provided evidence that increasing urban demands for water could be met at relatively low cost to agriculture or the environment (that is, without constructing new, environmentally disruptive dams and reservoirs). Subsequent reforms allowed markets in water to develop, so that voluntary exchanges could take place that made both parties better off. For example, an agreement was reached to transfer 100,000 acre-feet of water per year from the farmers of the Imperial Irrigation District (IID) in southern California to the Metropolitan Water District (MWD) in the Los Angeles area.[34] Subsequently, policy reforms spread throughout the West, and transactions emerged elsewhere in California as well as in Colorado, New Mexico, Arizona, Nevada, and Utah (MacDonnell 1990).

A second example of market creation is the current revolution in electricity restructuring that is motivated by economic concerns but may have significant environmental impacts.[35] For many years, utilities—closely overseen by state public utility commissions (PUCs)—have provided electricity within exclusive service areas. The utilities were granted these monopoly markets and guaranteed a rate of return on their investments, conditional on their setting reasonable rates and meeting various social objectives, such as universal access. The Energy Policy Act of 1992 took a major step toward opening the industry up to competition by allowing independent electricity-generation companies to sell power directly to utilities, and in 1996, the Federal Energy Regulatory Commission (FERC) moved things further by issuing regulations that require utilities with transmission lines to transmit power for other parties at reasonable rates (FERC 1996).

The purpose of these regulatory changes was to encourage competition at the wholesale (electricity-generation) level, but several states—including California, Illinois, Massachusetts, and New Hampshire—have taken this strategy further by facilitating competition at the retail level, so that consumers can contract directly for their electricity supplies. Legislation has been introduced in the U.S. Congress to establish guidelines for retail competition throughout the nation.[36]

These changes have environmental implications. First, as electricity prices fall in the new competitive environment, electricity consumption is expected to increase. This result might be expected to increase pollutant emissions, but to whatever degree electricity substitutes for other, more polluting forms of energy, the overall effect may be environmentally beneficial. Second, deregulation will unquestionably make it easier for new firms and sources to enter markets. Because new power plants tend to be both more efficient and less polluting (relying more on natural gas), environmental impacts may decrease.[37] Third, more flexible and robust markets for electricity can be expected to increase the effectiveness of various market-based incentives for pollution control, such as the SO_2 allowance trading system.[38]

Liability Rules

Liability rules can provide strong incentives to firms to consider the potential environmental damages of their decisions and thereby can change those decisions.[39] In theory, a liability rule can be cost-effective as a policy instrument because technologies or practices are not specified. For example, taxing hazardous materials or their disposal creates incentives for firms to reduce their use of those materials but does not provide overall incentives for firms to reduce societal risks from those materials. An appropriately designed liability rule can do just this (Revesz 1997). On the other hand, transaction costs associated with litigation may make liability rules most appropriate only for acute hazards. It is in such situations, in fact, that this approach has been used at the federal level: liability for toxic waste sites and hazardous material spills (see Chapter 7, Hazardous Waste Policy).

The Comprehensive Environmental Response, Compensation, and Liability Act (CERCLA) of 1980 established retroactive liability for companies that are found responsible for the contamination of a site requiring cleanup. Governments can collect cleanup costs and damages from waste producers, transporters, and handlers as well as current and past owners and operators of a site. Similarly, the Oil Pollution Act makes firms liable for cleanup costs, natural resource damages, and third-party damages caused by oil spills onto surface waters, and the Clean Water Act makes responsible parties liable for cleanup costs for spills of hazardous substances.

Information Programs

Because well-functioning markets depend in part on the existence of well-informed producers and consumers, information programs can, in theory, help foster market-oriented solutions to environmental problems.[40] One approach to government improving the set of information available to consumers is a product-labeling requirement (Table 3-8). The Energy Policy and Conservation Act of 1975 specifies that certain appliances and equipment (including air conditioners, washing machines, and water heaters) carry labels with information regarding products' energy efficiency and estimated annual energy costs (U.S. OTA 1992; CFR 1995). More recently, EPA and the U.S. Department of Energy (DOE) developed the Energy Star Program, under which energy-efficient products can display an Energy Star label. The label does not provide specific information about the product but signals to consumers that the product is, in general, energy-efficient. This program is much broader in its coverage than the appliance-labeling program; by 1997, more than 13,000 product models carried the Energy Star label (U.S. Department of State 1997). There has been little rigorous economic analysis of the efficacy of such programs, but limited econometric evidence suggests that product labeling (specifically, appliance efficiency labels) can have significant effects on efficiency improvements, essentially by making consumers (and therefore producers) more sensitive to energy price changes (Newell and others 1997).

TABLE 3-8. Federal Information Programs.

Information program	*Year of implementation*	*Enabling legislation*
Energy Efficiency Product Labeling	1975	Title V of the Energy Policy and Conservation Act
NJ Hazardous Chemical Emissions	1984	New Jersey Community Right-to-Know Act
Toxic Release Inventory	1986	Emergency Planning and Community Right-to-Know Act
CA Hazardous Chemical Emissions	1987	California Air Toxics Hot Spots, Information Assessment Act
CA Proposition 65	1988	California Safe Drinking Water Act, Toxic Enforcement Act
Energy Star	1993	Joint program of the U.S. Environmental Protection Agency and the U.S. Department of Energy

A second kind of government information program is a reporting requirement. The country's first such program was New Jersey's Community Right-to-Know Act, passed in 1984. Two years later, a similar program was established at the national level. The Toxics Release Inventory (TRI) was initiated under the Emergency Planning and Community Right-to-Know Act (EPCRA) (see Chapter 7, Hazardous Waste Policy). The TRI requires firms to report (to local emergency planning agencies) information regarding use, storage, and release of hazardous chemicals. Such information reporting serves compliance and enforcement purposes but may also increase public awareness of firms' actions, which may be linked with environmental risks. This public scrutiny can encourage firms to alter their behavior.

The Safe Drinking Water Act and Toxic Enforcement Act were adopted in California as a ballot initiative (Proposition 65) in 1986. The law, which covers consumer products and facility discharges, requires firms to provide a "clear and reasonable warning" if they expose populations to certain chemicals. In 1987, California enacted its Air Toxics Hot Spots Information and Assessment Act, which sets up an emissions reporting system to track the emissions of more than 700 toxic substances. The law requires the identification and assessment of localized risks of air contaminants and provides information to the public about the possible impact of those emissions on public health.

U.S. Experience with Reducing Government Subsidies

A final category of market-based instruments is government subsidy reduction. Because subsidies are the mirror image of taxes, they can, in theory, provide incentives to address environmental problems. But in practice, various subsidies are believed to promote economically inefficient and environmentally unsound practices. Here, I consider two examples: the below-cost sale of timber by the U.S. Forest Service, and explicit and implicit subsidies that are conveyed to suppliers of energy.

Below-Cost Timber Sales

The public lands of the United States, which encompass more than 25% of the nation's entire land base, contain valuable natural resources—timber, minerals, coal, oil, and natural gas—all of which are valued (and priced) in the market place. These lands also provide various public goods, which tend not to be fully valued and priced in the market—wilderness, fish and wildlife habitats, watersheds, and recreational opportunities. Because it is difficult for individual landowners to provide these public goods profitably, the burden for providing such environmental amenities tends to fall on the public lands.

Subsidies that benefit selected extractive industries may impede the provision of such amenities on public lands. Below-cost timber sales (whereby the U.S. Forest Service does not recover the full cost of making timber available) constitute an important case in point. It has been estimated that removal of these subsidies would foster environmental protection and could save taxpayers up to $1.2 billion over five years (U.S. Congressional Budget Office 1990).

Congress has mandated that the Forest Service pursue a policy of multiple-use management for timber, recreation, wildlife habitat, and watershed purposes (Bowes and Krutilla 1989). But the Forest Service is not under legal or regulatory requirements to sell its timber at a price that will recover the government's costs of growing and marketing that timber, and a substantial amount of publicly owned timber is sold below cost. That is, the commercial activity of moving timber from public lands into the marketplace frequently costs the federal government more than it gets in return. This implicit subsidy has most frequently been in the form of credits to private lumber companies for road building.[41]

Claims have been made that the Forest Service's disregard of timber production costs has led to excessive logging in unproductive national forests (Hyde 1981; Repetto and Gillis 1988). In response to such concerns, several administrations and congresses have considered various initiatives to deal with the problem, each of which would essentially direct more attention to economic considerations when managing and selling federal timber. But through 1997, no significant action had been taken.[42]

Fossil Fuel Energy Subsidies

Because of concerns about global climate change, increased attention has been given to federal subsidies and other programs that promote the use of fossil fuels. One EPA study indicates that eliminating these subsidies would have a significant effect on reducing carbon dioxide (CO_2) emissions (Shelby and others 1997). The federal government is involved in the energy sector through the tax system and through a range of individual agency programs. One other study indicates that these activities together cost the government $17 billion annually (Alliance to Save Energy 1993).

A substantial share of these subsidies and programs were enacted during the "oil crises" to encourage the development of domestic energy sources and reduce reliance on imported petroleum. They favor energy supply over energy efficiency.[43] Although there is an economic argument for government policies that encourage new technologies that have particularly high risk or long-term payoffs, mature and conventional technologies currently receive nearly 90% of the subsidies. Furthermore, within the fossil fuels, natural gas—the most environmentally benign fuel—receives only about 20% of the subsidies.

It also should be recognized that federal user charges (Table 3-3) and insurance premium taxes (Table 3-4) include significant levies on fossil fuels, and that federal tax differentiation has tended to favor renewable energy sources and unconventional fossil fuels (Table 3-7). In any event, the Clinton administration's 1997 proposal to address global climate change includes a $5 billion program (over five years) of government-funded research and development and private-industry tax credits for renewable energy sources and energy efficiency (Easterbrook 1997).

Why Have There Been Relatively Few Applications of Market-Based Instruments?

Despite the great interest given to market-based instruments by politicians in recent years and the great progress that has been made, market-based instruments have yet to fundamentally transform the landscape of U.S. environmental policy. For the most part, these instruments still exist only at the fringes of regulation.

A Stock Flow Problem

Market-based instruments represent only a trivial portion of existing regulation, for many reasons. Perhaps the most obvious is that there has not been a great deal of new environmental regulation. Since 1990, the Clean Air Act and Safe Drinking Water Act are the only major environmental regulations to be reauthorized. Given that Title 40 of the Code of Federal Regulations, entitled Protection of the Environment, contains more than 14,310 pages of environmental regulations, it could take a very long time indeed for market instruments to become the core of environmental policy, unless Congress is willing to use them for "old" problems as well as new ones.

Resistance from Interest Groups

Within the government environmental bureaucracy exists a desire to see effective environmental regulation adopted, but traditional regulatory programs require regulators with a technical or legal skill set, whereas market-based instruments require market-trained thinkers, including people with MBAs, economists, and others. Members of the government bureaucracy may rationally be resisting the dissipation of their human capital (Hahn and Stavins 1991).

Although some environmental groups have increasingly welcomed the selective use of market-based instruments,[44] others are concerned that increased flexibility in environmental regulation will result in the reduction of the overall degree of environmental protection. And in parts of the environmental community, the sentiment remains that environmental quality is an inalienable right and that market-based programs inappropriately condone the "right to pollute." Finally, some environmental professionals—like their government counterparts—may be resisting the dissipation of their human capital.

The ambivalence of the regulated community also has served to retard the use of market-based instruments. Many industries and companies have applauded market-based instruments in the abstract because they promise flexibility and cost-effectiveness. But few businesses actually have supported the adoption of new applications. One factor is reluctance to promote any regulation, no matter how flexible or cost-effective. Businesses may believe that political forces beyond their control might unfavorably distort the design and implementation of these instruments. First, cost savings might be taken away from them by an increase in the stringency of standards. Second, the design of instruments may limit their flexibility. Third, the rules may change over time. For businesses to optimize environmental investments, regulations have to be not only flexible but also predictable. Fourth, some firms remain concerned that "buying the right to pollute" could lead to negative publicity. Fifth and finally, private industry representatives may resist these reforms to prevent the dissipation of their human capital.

Public Resistance

The slow penetration of market-based instruments into environmental policies also may be due to these instruments not being well-understood by the general public. The benefits to consumers of market instruments typically are not visible, and the perceived costs can be transparent. Under traditional command-and-control policies, consumers may see prices go up, but they clearly find it difficult to associate those price increases with environmental regulations. For example, it is not readily apparent to consumers that gasoline and electricity prices are lower than they otherwise would have been because of the use of market-based programs to phase out lead or reduce SO_2 emissions. At the same time, market-based instruments—especially charges—may suffer from making environmental costs more transparent. Although encouraging individuals to consciously link environmental costs and benefits may be a good thing, it certainly can undermine the enthusiasm with which market-based instruments are embraced.

Why Has the Performance Record Been Mixed?

When market-based environmental policy instruments have been used, they have not always performed as predicted. Why?

Inaccurate Predictions

One reason market-based instruments have sometimes fallen short in delivering predicted cost savings is that the predictions themselves often have been unrealistic, based on perfect performance under ideal conditions. That is, these predictions have assumed that the cost-minimizing allocation of the pollution control burden among sources would be achieved and that marginal abatement costs would be equated across all sources. In a frequently cited table, Tietenberg calculated the ratio of the cost of an actual command-and-control program to a least-cost benchmark (Tietenberg 1985), but others have mistakenly used this ratio as an indicator of the likely gains of adopting specific market-based instruments. The more appropriate comparison would be between actual command-and-control programs and either actual or reasonably constrained theoretical market-based programs (Hahn and Stavins 1992).

In addition, predictions made during policy debates typically have ignored factors that can adversely affect performance: transaction costs involved in implementing market-based programs, uncertainty as to the property rights bestowed under programs, uncompetitive market conditions, a preexisting regulatory environment that does not give firms incentives to participate, and the inability of firms' internal decisionmaking capabilities to fully take advantage of program opportunities.

Design Problems

Many of the factors cited for the unpredictable success of market-based systems suggest the need for changes in the design of future market-based instruments. Whereas some program design elements have reflected miscalculations of market reactions, others were known to be problematic at the time the programs were enacted but nevertheless were incorporated into programs to ensure adoption by the political process. One striking example is the "20% rule" under EPA's Emissions Trading Program (Hahn 1990). This rule, adopted at the insistence of the environmental community, stipulates that each time a permit is traded, the amount of pollution authorized thereunder must be reduced by 20%. Because permits that are not traded retain their full quantity value, this regulation discourages permit trading and thereby increases regulatory costs.

Limitations in Firms' Structure

A third explanation for the mixed performance of implemented market-based instruments is that firms are simply not well-equipped internally to make the decisions necessary to fully implement these instruments. Because market-based instruments have been used on a limited basis only and because firms are not certain that these instruments will be a lasting feature on the regulatory landscape, most companies have chosen not to reorganize their internal structure to fully exploit the cost savings these instruments offer. Rather, most firms stick with what they know—minimizing the costs of complying with command-and-control regulations, not making the strategic decisions allowed by market-based instruments.[45]

The focus of environmental, health, and safety departments in private firms has been primarily on problem avoidance and risk management, rather than on the creation of opportunities made possible by market-based instruments. Because of the strict rules companies have faced under command-

and-control regulation, they have built skills and developed processes that comply with regulations but do not help them benefit competitively from environmental decisions. Absent significant changes in structure and personnel, the full potential of market-based instruments will not be realized.

The Changing Politics of Market-Based Instruments

Given the historical lack of receptiveness by the political process to market-based approaches to environmental protection, why has there been a recent rise in the use of market-based approaches?[46] It would be gratifying to believe that increased understanding of market-based instruments had played a large part in fostering their increased political acceptance, but how important has this really been? In 1981, political scientist Steven Kelman surveyed Congressional staff members and found that support and opposition to market-based environmental policy instruments was based largely on ideological grounds: Republicans who supported the concept of economic-incentive approaches offered as a reason the assertion that "the free market works," or "less government intervention" is desirable, without any real awareness or understanding of the economic arguments for market-based programs. Likewise, the Democrats' opposition was largely based on analogously ideological factors, with little or no apparent understanding of the real advantages or disadvantages of the various instruments (Kelman 1981). What would happen if we were to replicate Kelman's survey today? My refutable hypothesis is that we would find increased support from Republicans, greatly increased support from Democrats, but insufficient improvements in understanding to explain these changes.[47] So, what else has mattered?

First, one factor has surely been increased pollution control costs, which have led to greater demand for cost-effective instruments. By the late 1980s, even political liberals and environmentalists were beginning to question whether command-and-control regulations could produce further gains in environmental quality. During the previous twenty years, pollution abatement costs had continually increased, as stricter standards moved the private sector up the marginal cost-of-control function. By 1990, U.S. pollution control costs had reached $125 billion annually, nearly a 300% increase in real terms from 1972 levels (U.S. EPA 1990; Jaffe and others 1995).

Second, one factor that became important in the late 1980s was strong and vocal support from some segments of the environmental community.[48] By supporting tradable permits for acid rain control, the Environmental Defense Fund (EDF) seized a market niche in the environmental movement and successfully distinguished itself from other groups.[49] Related to this, a third factor was that the SO_2 allowance trading program, the leaded gasoline phasedown, and the CFC phaseout were all designed to *reduce* emissions, not simply to *reallocate* them cost-effectively among sources. Market-based instruments are most likely to be politically acceptable when proposed to achieve environmental improvements that would not otherwise be feasible (politically or economically).

Fourth, deliberations regarding the SO_2 allowance system, the lead system, and CFC trading differed from previous attempts by economists to influence environmental policy in an important way: the separation of ends from means, that is, the separation of consideration of goals and targets from the policy instruments used to achieve those targets. By accepting—implicitly or otherwise—the politically identified and potentially inefficient goal (for example, the ten-million ton reduction of SO_2 emissions), economists were able to focus successfully on the importance of adopting a cost-effective means of achieving that goal. The risk, of course, was "designing a fast train to the wrong station."

Fifth, acid rain was an unregulated problem until the SO_2 allowance trading program of 1990; the same can be said for leaded gasoline and CFCs. Hence, there were no existing constituencies—in the private sector, the environmental advocacy community, or government—for the status quo approach, because there was no status quo approach. It is reasonable to be more optimistic about introducing market-based instruments for "new" problems, such as global climate change, than for highly regulated existing problems, such as abandoned hazardous waste sites.

Sixth, by the late 1980s, there already had been a perceptible shift of the political center toward a more favorable view of using markets to solve social problems. The Bush administration, which proposed the SO_2 allowance trading program and then championed it through an initially resistant democratic Congress, was (at least in its first two years) "moderate Republican" and phrases such as "fiscally responsible environmental protection" and "harnessing market forces to protect the environment" do sound like quintessential moderate Republican issues.[50] But beyond these issues, support for market-oriented solutions to various social problems had been increasing across the political spectrum for the previous fifteen years, as was evidenced by deliberations on deregulation of the airline, telecommunications, trucking, railroad, and banking industries. Indeed, by 1990, the concept (or at least the phrase) "market-based environmental policy" had evolved from being politically problematic to politically attractive.

Seventh, the adoption of the SO_2 allowance trading program for acid rain control, like any major innovation in public policy, can partly be attributed to a healthy dose of chance that placed specific persons in key positions—in this case, in the White House, EPA, Congress, and environmental organizations.[51] The result was what remains the golden era for market-based environmental strategies.

Conclusion

Some eighty years ago, economists first proposed the use of corrective taxes to internalize environmental and other externalities. Fifty years later, the portfolio of potential economic incentive instruments was expanded to include quantity-based mechanisms: tradable permits. Thus, economic incentive approaches to environmental protection are clearly not a new policy idea. Over the past two decades, they have held varying degrees of prominence in environmental policy discussions.

Market-based instruments now have moved to center stage, and policy debates look very different from the time when these ideas were characterized as "licenses to pollute" or dismissed as completely impractical. Market-based instruments are considered seriously for each and every environmental problem that is tackled, ranging from endangered species preservation (see Goldstein 1991; Bean 1997) to regional smog (U.S. EPA 1998) and what may be the greatest of environmental problems, the greenhouse effect and global climate change (see Chapter 5, Global Climate Policy; also see Hahn and Stavins 1995; Fisher and others 1996; Schmalensee 1996; Stavins 1997). It seems clear that market-based instruments, particularly tradable permit systems, will enjoy increasing acceptance in the years ahead.

No particular form of government intervention, no individual policy instrument—market-based or conventional—is appropriate for all environmental problems. Which instrument is best in any given situation depends on a variety of characteristics of the environmental problem as well as the social, political, and economic context in which it is being regulated. There is no policy panacea. Indeed, the real challenge for bureaucrats, elected officials, and other participants in the environmental policy process is analyzing and then selecting the best instrument for each situation that arises.

Notes

1. Although the discussion of goals typically precedes examination of alternative means for achieving goals, this is not necessarily the case. For example, both the Bush and Clinton administrations endorsed market-based methods for addressing global climate change before either had committed itself to specific greenhouse policy goals.

2. This section draws in part on Hockenstein and others (1997).

3. Another strain of literature—known as "free market environmentalism"—focuses on the role of private property rights in achieving environmental protection (Anderson 1991).

4. Under certain circumstances, substituting a market-based instrument for a command-and-control instrument can lower environmental quality, because command-and-control standards tend to lead to over-control (for more information, see Oates and others 1989).

5. Each source's marginal cost of pollution control is the additional or incremental cost for that source to achieve an additional unit of pollution reduction. If this marginal cost of control is not equal across sources, then the same aggregate level of pollution control could be achieved at lower overall cost by simply reallocating the pollution control burden among sources, so that low-cost controllers controlled more, and high-cost controllers controlled proportionately less. Additional savings could theoretically be achieved through such reallocations until marginal costs were identical for all sources (see Baumol and Oates 1988). Reference here is to *marginal abatement* cost, that is, marginal cost of *emission* reduction. Things become more complicated, but the general point holds with nonuniformly mixed pollutants, where the focus is on ambient concentration or exposure, not simply emissions (see Montgomery 1972; Tietenberg 1995).

6. Downing and White (1986), Malueg (1989), Milliman and Prince (1989), and Jung and others (1996) have presented theoretical analyses of the dynamic incentives of technological change under alternative policy instruments. The empirical literature is considerably thinner (see Jaffe and Stavins 1995).

7. For example, a pollution charge might take the form of a charge per unit of sulfur dioxide emissions, but not a charge per unit of electricity generated. The choice of whether to tax pollution quantities, activities preceding discharge, inputs to those activities, or actual damages will depend on trade-offs among costs of abatement, mitigation, damages, and program administration, including monitoring and enforcement. Pigou (1952) is generally credited with developing the idea of a corrective tax to discourage activities that generate externalities such as environmental pollution.

8. Thirty years ago, Thomas Crocker and John Dales independently developed the idea of using transferable discharge permits to allocate the pollution-control burden among firms or individuals (see Crocker 1966; Dales 1968). David Montgomery (1972) provided the first rigorous proof that a tradable permit system could, in theory, provide a cost-effective policy instrument for pollution control. A sizeable literature on tradable permits has followed. Early surveys of the literature were presented by Tietenberg (1980; 1985) and Hahn and Noll (1982). Much of the literature on tradable permits may actually be traced to Coase's (1960) treatment of negotiated solutions to externality problems.

9. Reference here is to so-called "cap-and-trade" programs, but—as discussed later—some programs, such as EPA's Emissions Trading Program, operate differently, as "credit programs." In this case, permits or credits are assigned only when a source reduces emissions below what is required by existing, source-specific limits.

10. In addition, the Energy Policy and Conservation Act of 1975 established a program of Corporate Average Fuel Economy (CAFE) standards for automobiles and light trucks. The standards require manufacturers to meet a minimum sales-weighted average fuel efficiency for their fleet of cars sold in the United States. A penalty is charged per car sold per unit of average fuel efficiency below the standard. The program operates much like an internal-firm tradable permit system or "bubble" scheme, because manufacturers can improve efficiency wherever it is cheapest within their fleets. Firms that do better than the standard can "bank" their surpluses and—in some cases—are permitted to borrow against their future rights. (For reviews of the literature on CAFE standards, with particular attention to the program's costs relative to "equivalent" gasoline taxes, see Crandall and others 1986 and Goldberg 1997.) Light trucks, which are defined by the federal government to include sport utility vehicles, face significantly weaker CAFE standards (see Bradsher 1997). Also, California has used a vehicle retirement program that operates much like a tradable-permit system to reduce mobile-source air emissions by removing the oldest and most polluting vehicles from the road (see Kling 1994; Alberini and others 1995; Tietenberg 1997). In addition, the Northeast and Middle Atlantic states have organized a NO_x permit trading program to control regional smog (Tietenberg 1997).

11. The bubble policy, which treats multiple emission points as if they were located within an imaginary bubble, allows existing sources to use credits to satisfy their control responsibilities; netting allows sources that are being modified or expanded to purchase credits, which allow them to avoid the need to meet the strict requirements for new sources.

12. The Emissions Trading Program has been evaluated by Tietenberg (1985) and by Foster and Hahn (1995). EPA's experience with tradable permit policies was assessed more broadly by Hahn (1989).

13. In each year of the program, more than 60% of the lead added to gasoline was associated with traded lead credits (see Hahn and Hester 1989a).

14. The program experienced some relatively minor implementation difficulties related to imported leaded fuel; it is not clear that a comparable command-and-control approach would have done better in terms of environmental quality (U.S. GAO 1986).

15. The Montreal Protocol called for a 50% reduction in the production of particular CFCs from 1986 levels by 1998. In addition, the protocol froze halon production and consumption at 1986 levels beginning in 1992.

16. In addition, there have been a very few international trades, but such trading is limited by the Montreal Protocol.

17. The CFC tax was enacted principally as a "windfall-profits tax," to prevent private industry from retaining scarcity rents created by the quantity restrictions (see Merrill and Rousso 1990).

18. As of 1992, no firms were producing CFCs up to their maximum allowable level, and permits could not be banked (carried forward). As a result, there was an excess supply of permits. However, it is possible that the excess would exist even in the absence of a tax, because firms reacted to changes in regulations and new policy initiatives that called for a more rapid phaseout of CFCs and halons.

19. See Clean Air Act Amendments of 1990, Public Law No. 101-549, 104 Statute 2399, 1990. Title IV is described by Ferrall (1991).

20. Utilities that install scrubbers receive bonus allowances for early cleanup. Also, specified utilities in Ohio, Indiana, and Illinois receive extra allowances during both phases of the program. All of these extra allowances are essentially compensation intended to benefit Midwestern plants that rely on high-sulfur coal. The political origins of this aspect of the program are discussed by Joskow and Schmalensee (1998).

21. Schmalensee and others (1998) and Stavins (1998) have assessed the program's performance.

22. Detailed case studies of the evolution of the use of economic incentives in the SCAQMD have been presented by the National Academy of Public Administration (1994), Thompson (1997), and Harrison (1999).

23. Johnson and Pekelney (1996) presented an early assessment of the program. Johnston (1994) provided a prospective critique. Additional analyses include those by Lents (1998) and by Klier and others (1997).

24. Effluent charges have been used more extensively in Europe than in the United States, but it is questionable whether the levels have been sufficient to affect behavior in significant ways. For a discussion of the economics and politics surrounding the taxation of sulfur dioxide, nitrous oxide, and carbon dioxide in the Scandinavian nations, the Netherlands, France, and Germany, see OECD (1993; 1995) and Cansier and Krumm (1999).

25. Russell (1988) further discussed this point.

26. Minnesota was the first state to implement deposit–refund legislation for car batteries in 1988. By 1991, ten states had such legislation: Arizona, Arkansas, Connecticut, Idaho, Maine, Michigan, Minnesota, New York, Rhode Island, and Washington. Deposits range from $5 to $10.

27. Requiring a sales receipt for a refund removes the incentive for the return of batteries that have already been purchased. Furthermore, given the extended life of most batteries, it may be unrealistic to expect consumers to maintain a receipt for many years.

28. See U.S. Congress. 1951. Independent Offices Appropriations Act of 1951, August 31, 1951, ch. 375, §501, 654 Stat. 290. 31 U.S.C. §9701.

29. See Public Law 99-499, Sec. 522(a), 1986.

30. See Public Law 101-239, Revenue Reconciliation Act of 1989.

31. See Sect. 9501 of Internal Revenue Code of 1954.

32. See U.S. Congress. 26 U.S.C. Sec. 4064, Gas Guzzler Tax. 1978.

33. See The Omnibus Budget Reconciliation Act of 1989 Sect. 7506: Excise Tax on the Sale of Chemicals which Deplete the Ozone Layer and of Products Containing Such Chemicals.

34. In March 1983, the Environmental Defense Fund (EDF) published a proposal calling for the Metropolitan Water District to finance the modernization of the Imperial Irrigation District's water system in exchange for use of conserved water (see Stavins 1983). In November 1988, after five years of negotiation, the two water giants agreed on a $230 million water conservation and transfer arrangement, much like the

EDF's original proposal to trade conservation investments for water (see Morris 1988).

35. The three primary arguments for restructuring are that the electricity industry is no longer a natural monopoly, because small generation technologies are now competitive with large centralized production; consumers will benefit from buying cheaper electricity from more efficient producers, who currently face significant barriers to entry; and the old system with cost-of-service pricing provides poor incentives for utilities to reduce costs. Brennan and others (1996) have given the historical background of electricity restructuring.

36. The Electric Consumers' Power to Choose Act (104 H.R. 3790 and 105 H.R. 655) is one example of such legislation. For a brief overview of the politics of electricity restructuring, see Kriz 1996.

37. There is considerable debate on this point, because in the short run, more electricity may be generated from old surplus capacity coal plants in the Midwest, increasing pollutant emissions. In any event, in the long run, competition will encourage a more rapid turnover of the capital stock (see Palmer and Burtraw 1997).

38. However, environmental advocates are very concerned that state public utility commissions (PUCs) will have much less influence than previously over the industry. In the past, PUCs encouraged "demand-side management" and supported the use of renewable forms of electricity generation through the investment approval process or by requiring full-cost pricing for generation. Several policies have been proposed to provide these functions in the new, more competitive environment (for example, a system of tradable "renewable energy credits," wherein each generator would need to hold credits for a certain percentage of their generation, and a tax on the transmission of electricity, used to subsidize renewable generation).

39. These incentives frequently are neither simple nor direct, because firms and individuals may choose to reduce their exposure to liability by taking out insurance. In this regard, see the earlier discussion in this chapter, Insurance Premium Taxes.

40. Tietenberg (1997) has produced a comprehensive review of information programs and their apparent efficacy. Morris and Scarlett (1996) have given an overview of international experience with eco-labels.

41. The national forests have approximately 380,000 miles of roads—roughly eight times the length of the interstate highway system. Although these roads do have recreational and other uses, they primarily serve as access for logging companies. They have been constructed either directly by the Forest Service or through a "purchaser road credit" system. The road credit system began with the 1964 Forest Roads and Trails Act, which allows the Forest Service to credit logging companies for their expenses in constructing the logging roads they need to access timber. Under this system, companies deduct road construction expenses directly from the amount they pay the Forest Service for the timber they extract. In 1996, direct outlays totaled $84 million and purchaser credits were valued at nearly $50 million (Bryan 1997).

42. On November 11, 1997, after a prolonged debate on below-cost timber sales, President Bill Clinton signed a spending bill that included provisions to continue subsidies for the construction of logging roads in national forests.

43. The Alliance to Save Energy study (1993) claims that end-use efficiency receives $1 for every $35 received by energy supply.

44. In the mid-1980s, the Environmental Defense Fund was the first environmental advocacy organization to aggressively welcome the use of market-based instruments (Krupp 1986).

45. There are some exceptions. Enron, for example, has attempted to use market-based instruments for its strategic benefit by becoming a leader in creating new markets for trading acid rain permits. Other firms have appointed environmental, health, and safety leaders who are familiar with a wide range of policy instruments—not only command-and-control approaches—and who bring a strategic focus to their company's pollution-control efforts (see Hockenstein and others 1997).

46. Stavins (1998) and Keohane and others (1999) have more thoroughly explored the answers to this question.

47. But understanding of market-based approaches among policymakers has increased somewhat. It has partly been due to increased understanding by their staffs, which is to some degree a function of the economics training that is now common in law schools and of the proliferation of schools of public policy (see Hahn and Stavins 1991).

48. However, the environmental advocacy community is by no means unanimous in its support for market-based instruments (see Seligman 1994).

49. When the memberships (and financial resources) of other environmental advocacy groups subsequently declined with the election of the environmentally friendly Clinton–Gore administration, the Environmental Defense Fund continued to prosper and grow (see Lowry 1993).

50. The Reagan administration enthusiastically embraced a market-oriented ideology but demonstrated no interest in using actual market-based policies in the environmental area.

51. In the White House, among the most active and influential enthusiasts of market-based environmental instruments were Counsel Boyden Gray and his Deputy John Schmitz, Domestic Policy Adviser Roger Porter, Council of Economic Advisers (CEA) Member Richard Schmalensee, CEA Senior Staff Economist Robert Hahn, and Office of Management and Budget Associate Director Robert Grady. At the Environmental Protection Agency, Administrator William Reilly—a "card-carrying environmentalist"—enjoyed valuable credibility with environmental advocacy groups, and Deputy Administrator Henry Habicht was a key, early supporter of market-based instruments. In the Congress, Senators Timothy Wirth and John Heinz provided high-profile bipartisan support for the SO_2 allowance trading system and, more broadly, for a wide variety of market-based instruments for environmental problems through their *Project 88* (Stavins 1988). Finally, in the environmental community, Environmental Defense Fund Executive Director Fred Krupp, Senior Economist Daniel Dudek, and Staff Attorney Joseph Goffman worked closely with the White House to develop the initial allowance trading proposal. Moreover, many individuals within the government supported market-based instruments as far back as the Carter administration, helping to lay the groundwork for what was to come.

References

Adar, Z., and J. M. Griffin. 1976. Uncertainty and the Choice of Pollution Control Instruments. *Journal of Environmental Economics and Management* 3:178–88.

Alberini, Anna, Winston Harrington, and Virginia McConnell. 1995. Determinants of Participation in Accelerated Vehicle Retirement Programs. *RAND Journal of Economics* 26:93–112.

Alliance to Save Energy. 1993. *Federal Energy Subsidies: Energy, Environmental, and Fiscal Impacts.* Lexington, MA: Alliance to Save Energy.

Anderson, Robert. 1997. *The U.S. Experience with Economic Incentives in Environmental Pollution Control Policy.* Washington, DC: Environmental Law Institute.

Anderson, Terry L. 1983. *Water Crisis: Ending the Policy Drought.* Washington, DC: Cato Institute.

Anderson, Terry L., and Donald R. Leal. 1991. *Free Market Environmentalism.* Boulder, CO: Westview Press.

Bailey, Elizabeth M. 1996. Allowance Trading Activity and State Regulatory Rulings: Evidence from the U.S. Acid Rain Program. MIT-CEEPR 96-002 WP. Cambridge, MA: Massachusetts Institute of Technology, Center for Energy and Environmental Policy Research.

Barthold, Thomas A. 1994. Issues in the Design of Environmental Excise Taxes. *Journal of Economic Perspectives* 8(1):133–51.

Baumol, William J., and Wallace E. Oates. 1988. *The Theory of Environmental Policy,* 2nd edition. New York: Cambridge University Press.

Bean, Michael J. 1997. Shelter from the Storm: Endangered Species and Landowners Alike Deserve a Safe Harbor. *The New Democrat,* March/April, 20–21.

Bellandi, R. L., and R. B. Hennigan. 1977. The Why and How of Transferable Development Rights. *Real Estate Review* 7:60–64.

Bohi, Douglas. 1994. Utilities and State Regulators Are Failing to Take Advantage of Emissions Allowance Trading. *The Electricity Journal* 7:20–27.

Bohm, Peter. 1981. *Deposit–Refund Systems: Theory and Applications to Environmental, Conservation, and Consumer Policy.* Baltimore, MD: Johns Hopkins University Press for Resources for the Future.

Bovenberg, A. Lans, and Lawrence H. Goulder. 1996. Optimal Environmental Taxation in the Presence of Other Taxes: General-Equilibrium Analyses. *American Economic Review* 86:985–1000.

Bovenberg, A. Lans, and R. de Mooij. 1994. Environmental Levies and Distortionary Taxation. *American Economic Review* 84:1085–9.

Bowes, Michael D., and John V. Krutilla. 1989. *Multiple-Use Management: The Economics of Public Forestlands.* Washington, DC: Resources for the Future.

Bradsher, Keith. 1997. Light Trucks Increase Profits but Foul Air More Than Cars. *New York Times,* November 30, A1, A38–A39.

Brennan, Timothy J., Karen L. Palmer, Raymond J. Kopp, Alan J. Krupnick, Vito Stagliano, and Dallas Burtraw. 1996. *A Shock to the System: Restructuring America's Electricity Industry.* Washington, DC: Resources for the Future.

Brotzman, Thomas. 1996. Opening the Floor to Emissions Trading. *Chemical Marketing Reporter*, May 27, SR8.
Bryan, Richard. 1997. Senator Richard Bryan's Statements before the U.S. Senate. Conference Report on Interior Appropriations Act. October 28. Washington, DC.
Burtraw, Dallas. 1996. The SO_2 Emissions Trading Program: Cost Savings without Allowance Trades. *Contemporary Economic Policy* 14:79–94.
Cansier, D., and R. Krumm. 1999. Air Pollution Taxation: An Empirical Survey. *Ecological Economics* 23:59–70.
Cason, Timothy N. 1995. An Experimental Investigation of the Seller Incentives in EPA's Emission Trading Auction. *American Economic Review* 85:905–22.
CFR (Code of Federal Regulations). 1995. 16 *C.F.R.*, Chapter 1, Federal Trade Commission, Part 305—Appliance Labeling Rule.
Coase, Ronald. 1960. The Problem of Social Cost. *Journal of Law and Economics* 3:1–44.
Crandall, Robert W., Howard K. Gruenspecht, Theodore E. Keeler, and Lester B. Lave. 1986. *Regulating the Automobile.* Washington, DC: The Brookings Institution.
Crocker, Thomas D. 1966. The Structuring of Atmospheric Pollution Control Systems. In *The Economics of Air Pollution*, edited by Harold Wolozin. New York: W. W. Norton.
Dales, John. 1968. *Pollution, Property, and Prices.* Toronto, Ontario: University Press.
Downing, Paul B., and Lawrence J. White. 1986. Innovation in Pollution Control. *Journal of Environmental Economics and Management* 13:18-27.
Easterbrook, Gregg. 1997. Greenhouse Common Sense: Why Global-Warming Economics Matters More than Science. *U.S. News and World Report*, December 1, 58–62.
Efaw, Fritz, and William N. Lanen. 1979. Impact of User Charges on Management of Household Solid Waste. Report prepared for the U.S. Environmental Protection Agency under Contract No. 68-3-2634. Princeton, NJ: Mathtech, Inc.
El-Ashry, Mohamed T., and Diana C. Gibbons. 1986. *Troubled Waters: New Policies for Managing Water in the American West.* Washington, DC: World Resources Institute.
Feldman, Richard D. 1991. Letter from Richard D. Feldman, U.S. Environmental Protection Agency. January 7.
FERC (Federal Energy Regulatory Commission). 1996. Order 888. April. Washington, DC.
Ferrall, Brian L. 1991. The Clean Air Act Amendments of 1990 and the Use of Market Forces to Control Sulfur Dioxide Emissions. *Harvard Journal on Legislation* 28:235–52.
Field, B. C., and J. M. Conrad. 1975. Economic Issues in Programs of Transferable Development Rights. *Land Economics* 51:331–40.
Fisher, B., S. Barrett, P. Bohm, B. Fisher, M. Kuroda, J. Mubazi, A. Shah, and R. Stavins. 1996. Policy Instruments to Combat Climate Change. In *Climate Change 1995: Economic and Social Dimensions of Climate Change,* edited by J. P. Bruce, H. Lee, and E. F. Haites. Intergovernmental Panel on Climate Change, Working Group III. Cambridge, U.K.: Cambridge University Press, 397–439.
Foster, Vivien, and Robert W. Hahn. 1995. Designing More Efficient Markets: Lessons from Los Angeles Smog Control. *Journal of Law and Economics* 38:19–48.

Frederick, Kenneth D., ed. 1986. *Scarce Water and Institutional Change.* Washington, DC: Resources for the Future.

Fullerton, Don, and Thomas C. Kinnaman. 1995. Garbage, Recycling, and Illicit Burning or Dumping. *Journal of Environmental Economics and Management* 29:78–92.

——. 1996. Household Responses to Pricing Garbage by the Bag. *American Economic Review* 86:971–84.

Fullerton, Don, and Gilbert Metcalf. 1997. Environmental Controls, Scarcity Rents, and Pre-Existing Distortions. National Bureau of Economic Research Working Paper No. 6091. July.

Fulton, William. 1996. The Big Green Bazaar. *Governing Magazine,* June, 38.

Goldberg, Penelopi K. 1997. The Effects of the Corporate Average Fuel Efficiency Standards. Working paper. Princeton, NJ: Department of Economics, Princeton University.

Goldstein, Jon B. 1997 The Prospects for Using Market Incentives to Conserve Biological Diversity. *Environmental Law* 21 (3, Part I). Northwestern School of Law, Portland, OR.

Goulder, Lawrence. 1995. Effects of Carbon Taxes in an Economy with Prior Tax Distortions: An Intertemporal General Equilibrium Analysis. *Journal of Environmental Economics and Management* 29:271–97.

——. 1995. Environmental Taxation and the Double Dividend: A Reader's Guide. *International Tax and Public Finance* 2:157–83.

Goulder, Lawrence, Ian Parry, and Dallas Burtraw. 1997. Revenue-Raising versus Other Approaches to Environmental Protection: The Critical Significance of Preexisting Tax Distortions. *RAND Journal of Economics* 28:708–31.

Hahn, Robert, and Roger Noll. 1982. Designing a Market for Tradeable Permits. In *Reform of Environmental Regulation,* edited by W. Magat. Cambridge, MA: Ballinger.

Hahn, Robert W., and Robert N. Stavins. 1992. Economic Incentives for Environmental Protection: Integrating Theory and Practice. *American Economic Review* 82(May):464–8.

——. 1995. Trading in Greenhouse Permits: A Critical Examination of Design and Implementation Issues. In *Shaping National Responses to Climate Change: A Post-Rio Policy Guide,* edited by Henry Lee. Cambridge, MA: Island Press, 177–217.

Hahn, Robert W. 1984. Market Power and Transferable Property Rights. *Quarterly Journal of Economics* 99:753–65.

——. 1989. Economic Prescriptions for Environmental Problems: How the Patient Followed the Doctor's Orders. *Journal of Economic Perspectives* 3:95–114.

Hahn, Robert W. 1990. Regulatory Constraints on Environmental Markets. *Journal of Public Economics* 42:149–75.

Hahn, Robert W., and Gordon L. Hester. 1989a. Marketable Permits: Lessons for Theory and Practice. *Ecology Law Quarterly* 16:361–406.

——. 1989b. Where Did All the Markets Go? An Analysis of EPA's Emissions Trading Program. *Yale Journal of Regulation* 6:109–53.

Hahn, Robert W., and Albert M. McGartland. 1989. Political Economy of Instrumental Choice: An Examination of the U.S. Role in Implementing the Montreal Protocol. *Northwestern University Law Review* 83:592–611.

Hahn, Robert W., and Robert N. Stavins. 1991. "Incentive-Based Environmental Regulation: A New Era from an Old Idea? *Ecology Law Quarterly* 18:1–42.

Harrison, David Jr. 1999. Turning Theory into Practice for Emissions Trading in the Los Angeles Air Basin. In *Pollution For Sale: Emissions Trading and Joint Implementation*, edited by S. Sorrell and J. Skea. London, U.K.: Edward Elgar.

Helfand, Gloria E. 1991. Standards versus Standards: The Effects of Different Pollution Restrictions. *American Economic Review* 81:622–34.

Hockenstein, Jeremy B., Robert N. Stavins, and Bradley W. Whitehead. 1997. Creating the Next Generation of Market-Based Environmental Tools. *Environment* 39(4):12–20, 30–33.

Hyde, William F. 1981. Timber Economics in the Rockies: Efficiency and Management Options. *Land Economics* 57:630–7.

International Institute for Sustainable Development. 1995. *Green Budget Reform: An International Casebook on Leading Practices.* London, U.K.: EarthScan.

Jaffe, Adam B., and Robert N. Stavins. 1995. Dynamic Incentives of Environmental Regulations: The Effects of Alternative Policy Instruments on Technology Diffusion. *Journal of Environmental Economics and Management* 29:S43–S63.

Jaffe, Adam B., Steven R. Peterson, Paul R. Portney, and Robert N. Stavins. 1995. Environmental Regulation and the Competitiveness of U.S. Manufacturing: What Does the Evidence Tell Us? *Journal of Economic Literature* 33:132–63.

Johnson, Scott Lee, and David M. Pekelney. 1996. Economic Assessment of the Regional Clean Air Incentives Market: A New Emissions Trading Program for Los Angeles. *Land Economics* 72:277–97.

Johnston, James L. 1994. Pollution Trading in La La Land. *Regulation* 3:44–54.

Jorgenson, Dale, and Peter Wilcoxen. 1994. The Economic Effects of a Carbon Tax. Paper presented to the IPCC Workshop on Policy Instruments and Their Implications, Tsukuba, Japan, January 17–20.

Joskow, Paul L., and Richard Schmalensee. 1998. The Political Economy of Market-Based Environmental Policy: The U.S. Acid Rain Program. *Journal of Law and Economics* 41:81–135.

Joskow, Paul L., Richard Schmalensee, and Elizabeth M. Bailey. 1988. The Market for Sulfur Dioxide Emissions. *American Economic Review* 88:69–85.

Jung, Chulho, Kerry Krutilla, and Roy Boyd. 1996. Incentives for Advanced Pollution Abatement Technology at the Industry Level: An Evaluation of Policy Alternatives. *Journal of Environmental Economics and Management* 30:95–111.

Kashmanian, R. 1986. Beyond Categorical Limits: The Case for Pollution Reduction through Trading. Paper presented at the 59th Annual Conference of the Water Pollution Control Federation.

Kelman, Steven. 1981. *What Price Incentives? Economists and the Environment.* Boston, MA: Auburn House.

Keohane, Nathaniel O., Richard L. Revesz, and Robert N. Stavins. 1999. The Positive Political Economy of Instrument Choice in Environmental Policy. *Environmental Economics and Public Policy*, edited by Paul Portney and Robert Schwab. London, U.K.: Edward Elgar, Ltd.

Kerr, Suzi, and David Maré. 1997. Efficient Regulation through Tradeable Permit Markets: The United States Lead Phasedown. Working Paper 96-06. January. College Park, MD: University of Maryland, Department of Agricultural and Resource Economics.

Klier, Thomas H., Richard H. Mattoon, and Michael A. Prager. 1997. A Mixed Bag: Assessment of Market Performance and Firm Trading Behaviour in the NO_x RECLAIM Programme. *Journal of Environmental Planning and Management* 40(6):751–74.

Kling, Catherine L. 1994. Emission Trading vs. Rigid Regulations in the Control of Vehicle Emissions. *Land Economics* 70:174–88.

Kriz, Margaret. 1996. A Jolt to the System. *National Journal*, August 3, 1631–6.

Krupp, Frederic. 1986. New Environmentalism Factors in Economic Needs. *Wall Street Journal*, November 20, 34.

Lave, Lester, and Howard Gruenspecht. 1991. Increasing the Efficiency and Effectiveness of Environmental Decisions: Benefit–Cost Analysis and Effluent Fees. *Journal of Air and Waste Management* 41(May):680–90.

Lents, James. 1998. The RECLAIM Program at Three Years. Working paper. April 28. Riverside, CA: University of California, Department of Science and Engineering.

Liroff, Richard A. 1986. *Reforming Air Pollution Regulations: The Toil and Trouble of EPA's Bubble.* Washington, DC: Conservation Foundation.

Lowry, Robert C. 1993. The Political Economy of Environmental Citizen Groups. Unpublished Ph.D. thesis. Cambridge, MA: Harvard University.

Macauley, Molly K., Michael D. Bowes, and Karen L. Palmer. 1992. *Using Economic Incentives to Regulate Toxic Substances.* Washington, DC: Resources for the Future.

MacDonnell, Lawrence J. 1990. *The Water Transfer Process as a Management Option for Meeting Changing Water Demands, Volume I.* April. Contracted, unpublished document. Submitted to the U.S. Geological Survey (Washington, DC) from the University of Colorado, Boulder, Natural Resources Law Centre.

Main, Jeremy. 1988. Here Comes the Big New Cleanup. *Fortune*, November, 102–18.

Malueg, David A. 1989. Emission Credit Trading and the Incentive to Adopt New Pollution Abatement Technology. *Journal of Environmental Economics and Management* 16:52–57.

McFarland, J. M. 1972. Economics of Solid Waste Management. In *Comprehensive Studies of Solid Waste Management, Final Report.* Report no. 72-3:41-106. Berkeley, CA: University of California, College of Engineering and School of Public Health, Sanitary Engineering Research Laboratory.

Menell, Peter. 1990. Beyond the Throwaway Society: An Incentive Approach to Regulating Municipal Solid Waste. *Ecology Law Quarterly* 17:655–739.

Merrill, Peter R., and Ada S. Rousso. 1990. Federal Environmental Taxation. Presented at the Eighty-third Annual Conference of the National Tax Association. November 13, San Francisco, CA.

Milliman, Scott R., and Raymond Prince. 1989. Firm Incentives to Promote Technological Change in Pollution Control. *Journal of Environmental Economics and Management* 17:247–65.

Mills, D.E. 1980. Transferable Development Rights Markets. *Journal of Urban Economics* 7:63–74.

Miranda, Marie Lynn, Jess W. Everett, Daniel Blume, and Barbeau A. Roy Jr. 1994. Market-Based Incentives and Residential Municipal Solid Waste. *Journal of Policy Analysis and Management* 13:681–98.

Misolek, W. S., and H. W. Elder. 1989. Exclusionary Manipulation of Markets for Pollution Rights. *Journal of Environmental Economics and Management* 16:156–66.

Montgomery, David. 1972. Markets in Licenses and Efficient Pollution Control Programs. *Journal of Economic Theory* 5:395–418.

Morandi, Larry. 1992. An Outside Perspective on Iowa's 1987 Groundwater Protection Act. Washington, DC: National Conference of State Legislatures.

Morris, Julian, and Lynn Scarlett. 1996. Buying Green: Consumers, Product Labels, and the Environment. Policy study no. 202. August. Los Angeles, CA: The Reason Foundation.

Morris, Willy. 1988. IID Approves State's First Water Swap with MWD. *Imperial Valley Press*, November 9.

National Academy of Public Administration. 1994. *The Environment Goes to Market: The Implementation of Economic Incentives for Pollution Control*, Chapter 2. July. National Academy of Public Administration.

Newell, Richard G., Adam B. Jaffe, and Robert N. Stavins. 1997. The Induced Innovation Hypothesis and Energy-Saving Technological Change. Working paper. October 18. Cambridge, MA: Harvard University, John F. Kennedy School of Government.

Nichols, Albert L. 1997. Lead in Gasoline. In *Economic Analyses at EPA: Assessing Regulatory Impact*, edited by Richard D. Morgenstern. Washington, DC: Resources for the Future, 49–86.

Oates, Wallace E., Paul R. Portney, and Albert M. McGartland. 1989. The Net Benefits of Incentive-Based Regulation: A Case Study of Environmental Standard Setting. *American Economic Review* 79:1233–43.

OECD (Organization for Economic Cooperation and Development). 1989. *Economic Instruments for Environmental Protection*. Paris, France: OECD.

——. 1991. *Environmental Policy: How to Apply Economic Instruments*. Paris, France: OECD.

——. 1993. *Taxation and the Environment, Complementary Policies*. Paris, France: OECD.

——. 1994a. *Evaluating Economic Incentives for Environmental Policy*. Paris, France: OECD

——. 1994b. *Managing the Environment—The Role of Economic Instruments*. Paris, France: OECD.

——. 1994c. *The Distributive Effects of Economic Instruments for Environmental Policy*. Paris, France: OECD.

——. 1995. *Environmental Taxation in OECD Countries*. Paris, France: OECD.

Palmer, Karen, and Dallas Burtraw. 1997. Electricity Restructuring and Regional Air Pollution. *Resource and Energy Economics* 19:139–74.

Parry, Ian. 1995. Pollution, Taxes, and Revenue Recycling. *Journal of Environmental Economics and Management* 29:64–77.

Peskin, Henry M. 1986. Nonpoint Pollution and National Responsibility. *Resources,* Spring, 10–11, 17.

Pigou, Arthur C. 1952. *The Economics of Welfare,* 4th edition. London, U.K.: MacMillan.

Porter, Richard. 1978. A Social Benefit–Cost Analysis of Mandatory Deposits on Beverage Containers. *Journal of Environmental Economics and Management* 5:351–75.

Porter, Richard. 1983. Michigan's Experience with Mandatory Deposits on Beverage Containers. *Land Economics* 59:177–94.

Reiling, Stephen D., and M. J. Kotchen. 1996. Lessons Learned from Past Research on Recreation Fees. In *Recreation Fees in the National Park Service: Issues, Policies and Guidelines for Future Action,* edited by A. L. Lundgren. Minnesota Extension Service Pub. No. BU-6767. St. Paul, MN: University of Minnesota, Department of Forest Resources, Cooperative Park Studies Unit.

Repetto, Robert, and Malcolm Gillis, eds. 1988. *Public Policies and the Misuse of Forest Resources.* New York: Cambridge University Press.

Repetto, Robert, Roger C. Dower, Robin Jenkins, and Jacqueline Geoghegan. 1992. *Green Fees: How a Tax Shift Can Work for the Environment and the Economy.* Washington DC: World Resources Institute.

Revesz, Richard L. 1997. *Foundations in Environmental Law and Policy.* New York: Oxford University Press.

Rose, Kenneth. 1997. Implementing an Emissions Trading Program in an Economically Regulated Industry: Lessons from the SO_2 Trading Program. In *Market-Based Approaches to Environmental Policy: Regulatory Innovations to the Fore,* edited by Richard F. Kosobud and Jennifer M. Zimmerman. New York: Van Nostrand Reinhold.

Russell, Clifford S. 1988. Economic Incentives in the Management of Hazardous Wastes. *Columbia Journal of Environmental Law* 13:257–74.

Schmalensee, Richard. 1996. Greenhouse Policy Architecture and Institutions. Paper prepared for the National Bureau of Economic Research Conference, Economics and Policy Issues in Global Warming: An Assessment of the Intergovernmental Panel Report, Snowmass, CO, July 23–24.

Schmalensee, Richard, Paul L. Joskow, A. Denny Ellerman, Juan Pablo Montero, and Elizabeth M. Bailey. 1998. An Interim Evaluation of Sulfur Dioxide Emissions Trading. *Journal of Economic Perspectives* 12(3, Summer):53–68.

Scodari, Paul, Leonard Shabman, and D. White. 1995. Commercial Wetland Mitigation Credit Markets: Theory and Practice. IWR Report 95-WMB-7. Alexandria, VA: U.S. Army Corps of Engineers, Institute for Water Resources, Water Resources Support Center.

Seligman, Daniel A. 1994. *Air Pollution Emissions Trading: Opportunity or Scam? A Guide for Activists.* San Francisco, CA: Sierra Club.

Shelby, Mike, Robert Shackleton, Malcolm Shealy, and Alex Cristofaro. 1997. *The Climate Change Implications of Eliminating U.S. Energy (and Related) Subsidies.* Washington, DC: U.S. EPA. October.

Sigman, Hilary A. 1995. A Comparison of Public Policies for Lead Recycling. *RAND Journal of Economics* 26(3, Autumn):452–78.

Skumatz, Lisa A. 1990. Volume-Based Rates in Solid Waste: Seattle's Experience. Report for the Seattle Solid Waste Utility. Seattle, WA: Seattle Solid Waste Utility.

Stavins, Robert N. 1983. *Trading Conservation Investments for Water.* March. Berkeley, CA: Environmental Defense Fund.

——. 1995. Transaction Costs and Tradeable Permits. *Journal of Environmental Economics and Management* 29:133–47.

——. 1996. Correlated Uncertainty and Policy Instrument Choice. *Journal of Environmental Economics and Management* 30:218–32.

——. 1997. Policy Instruments for Climate Change: How Can National Governments Address a Global Problem? *The University of Chicago Legal Forum,* Volume 1997. Chicago, IL: University of Chicago School of Law, 293–329.

——. 1998. What Can We Learn from the Grand Policy Experiment? Lessons from SO_2 Allowance Trading. *Journal of Economic Perspectives* 12(3, Summer):69–88.

Stavins, Robert N., ed. 1988. *Project 88: Harnessing Market Forces to Protect Our Environment.* December. Sponsored by Senator Timothy E. Wirth, Colorado, and Senator John Heinz, Pennsylvania. Washington, DC.

——. 1991. *Project 88: Round II Incentives for Action: Designing Market-Based Environmental Strategies.* May. Sponsored by Senator Timothy E. Wirth, Colorado, and Senator John Heinz, Pennsylvania. Washington, DC.

Stavins, Robert N., and Bradley W. Whitehead. 1992. Pollution Charges for Environmental Protection: A Policy Link between Energy and Environment. *Annual Review of Energy and the Environment* 17:187–210.

Stevens, B. J. 1978. Scale, Market Structure, and the Cost of Refuse Collection. *The Review of Economics and Statistics* 40:438–48.

Thompson, Dale B. 1997. The Political Economy of the RECLAIM Emissions Market for Southern California. March. Working paper. Charlottesville, VA: University of Virginia.

Tietenberg, Tom. 1980. Transferable Discharge Permits and the Control of Stationary Source Air Pollution: A Survey and Synthesis. *Land Economics* 56:391–416.

——. 1985. *Emissions Trading: An Exercise in Reforming Pollution Policy.* Washington, DC: Resources for the Future.

——. 1997. Information Strategies for Pollution Control. Paper presented at the Eighth Annual Conference, European Association of Environmental and Resource Economists, Tilburg, the Netherlands, June 26–28.

——. 1997. Tradeable Permits and the Control of Air Pollution in the United States. Paper prepared for the 10th Anniversary Jubilee Edition of *Zeitschrift Fürangewandte Umweltforschung.*

Tietenberg, Tom H. 1995. Tradeable Permits for Pollution Control When Emission Location Matters: What Have We Learned? *Environmental and Resource Economics* 5:95–113.

Tisato, P. 1994. Pollution Standards vs. Charges under Uncertainty. *Environmental and Resource Economics* 4:295–304.

Tripp, James T. B., and Daniel J. Dudek. 1989. Institutional Guidelines for Designing Successful Transferable Rights Programs. *Yale Journal of Regulation* 6:369–91.

U.S. Congressional Budget Office. 1990. *Reducing the Deficit: Spending and Revenue Options.* February. Washington, DC: U.S. Government Printing Office (GPO).

U.S. Department of State. 1997. *U.S. Climate Action Report.* Publication 10496. Washington, DC: U.S. Government Printing Office (GPO).

U.S. EPA (Environmental Protection Agency). 1982. *Regulation of Fuel and Fuel Additives.* Proposed rule 38,078-90 (August), final rule 49,322-24 (October). Washington, DC: U.S. EPA.

——. 1984. *Case Studies on the Trading of Effluent Loads, Dillon Reservoir.* Final report. Washington, DC: U.S. EPA, Office of Policy Analysis.

——. 1985. *Costs and Benefits of Reducing Lead in Gasoline, Final Regulatory Impact Analysis.* February. Washington, DC: U.S. EPA, Office of Policy Analysis.

——. 1986. *Emissions Trading Policy Statement.* Final policy statement. *Federal Register* 51:43814.

——. 1990. *Environmental Investments: The Cost of a Clean Environment.* Report of the administrator to Congress. December. Washington, DC: U.S. EPA.

——. 1991. *Economic Incentives, Options for Environmental Protection.* Document P-2001. Washington, DC: U.S. EPA.

——. 1992. *States' Efforts to Promote Lead-Acid Battery Recycling.* Washington, DC: U.S. EPA.

——. 1992. *The United States Experience with Economic Incentives to Control Environmental Pollution.* EPA-230-R-92-001. Washington, DC: U.S. EPA.

——. 1998. EPA Proposes Emissions Trading Program to Help Protect Eastern U.S. from Smog. Press release. April 29. Washington, DC: U.S. EPA.

U.S. GAO (General Accounting Office). 1986. *Vehicle Emissions: EPA Program to Assist Leaded-Gasoline Producers Needs Prompt Improvement.* GAO/RCED-86-182. August. Washington, DC: U.S. GAO.

——. 1990. *Solid Waste: Trade-offs Involved in Beverage Container Deposit Legislation.* GAO/RCED-91-25. Washington, DC: U.S. GAO.

U.S. OTA (Office of Technology Assessment). 1992. *Building Energy Efficiency.* Washington, DC: U.S. OTA.

——. 1995. *Environmental Policy Tools: A User's Guide.* Washington, DC: U.S. OTA.

Voigt, Paul C., and Leon E. Danielson. 1996. Wetlands Mitigation Banking Systems: A Means of Compensating for Wetlands Impacts. Working paper AREP96-2. Raleigh, NC: North Carolina State University, Department of Agricultural and Resource Economics, Applied Resource Economics and Policy Group.

Wahl, Richard W. 1989. *Markets for Federal Water: Subsidies, Property Rights, and the Bureau of Reclamation.* Washington, DC: Resources for the Future.

Weitzman, Martin L. 1974. Prices vs. Quantities. *Review of Economic Studies* 41:477–91.

Wertz, Kenneth L. 1976. Economic Factors Influencing Households' Production of Refuse. *Journal of Environmental Economics and Management* 2:263–72.

four

Air Pollution Policy

Paul R. Portney*

As the United States enters the twenty-first century, the Clean Air Act exemplifies—perhaps better than any other environmental statute—the strengths and weaknesses of pollution control regulation here. Arguably the most important of all environmental laws, the Clean Air Act affects the very air we breathe, the rate at which our cities grow (and sometimes shrink), the quality of our visits to national parks and wilderness areas, and the prices we pay for virtually everything we buy. In fact, with the Federal Food, Drug and Cosmetic Act (administered by the Food and Drug Administration), the Clean Air Act may even be the most significant of *all* the regulatory statutes aimed at advancing environmental quality, safety, and health. Finally and most importantly, in one very important dimension the Clean Air Act has been an almost unqualified success: *air quality has improved significantly in virtually every metropolitan area around the country since 1970.*

As the Clean Air Act has aged, frustrations have developed with its basic structure. From the vantage point of the year 2000, in fact—and in spite of the success of the Act—these frustrations seem greater than at any point in its past. Nevertheless, with but a few exceptions, the changes made to the act in 1977 and 1990 have never amounted to too much more than tinkering with the basic framework laid out in 1970.

*Deidre Farrell and Pam Jagger performed invaluable research assistance and helped draft parts of this chapter.

The purpose of this chapter is to review and analyze the fundamentals of that framework. First, I present a history and background of air pollution control efforts in the United States leading up to the two major restructurings (in 1970 and 1990). Next, I outline the major features of the new approaches represented by the 1970 and the 1990 amendments to the Clean Air Act, emphasizing the substantial shift in responsibility from state and local governments to the federal government that took place in 1970. Then, I discuss the very favorable trends in air quality since 1970, emphasizing that these changes may not be exclusively the result of legislation passed since 1970—a most important point to grasp. I subsequently discuss results from a recent and very comprehensive economic evaluation of the Clean Air Act. Evidence is presented on the costs and benefits of historical air pollution control efforts, as well as the likely costs and benefits over the next decade. Where few or no data are available, the pros and cons of the Clean Air Act are discussed in qualitative terms. So as not to slight the most recent developments in clear air policy, I review the changes since the 1990 amendments. Finally, I summarize my judgments and make recommendations for improving the Clean Air Act.

Air Pollution Control before 1970

Neither air pollution nor efforts to control it are recent phenomena.[1] The earliest air pollution problems probably resulted from volcanic eruptions, the decomposition of organic matter, other natural sources, and the first fires set by primitive people. Although even earlier examples no doubt exist, one of the first "modern" pollution control ordinances can be traced to thirteenth-century England, where King Edward I banned the burning of certain highly polluting coals in London (Davies 1970).

Closer to home and nearer in time, air pollution was first combated in the United States through nuisance or trespass suits brought in the courts (Stern 1982, 44) during the industrial revolution. The very first air pollution statutes in the United States, designed to control smoke and soot from furnaces and locomotives, were passed by the cities of Chicago and Cincinnati in 1881. Within thirty years, county governments had begun to pass their own pollution control laws, and in 1952, Oregon became the first state to enact meaningful ordinances for combating foul air. Other states followed, generally with legislation aimed at smoke and particulate matter; only California took on the air quality problems associated with motor vehicle exhausts. Table 4-1 shows the growth over time of city, county, and state laws protecting air quality.

In 1955, the federal government entered the picture for the first time with the passage of the Air Pollution Control Act. This law, which did little

Table 4-1. Development of U.S. Municipal, County, and State Air Pollution Control Legislation, 1880–1980.

	Number of jurisdictions with statutes		
Year	*Municipal*	*County*	*State*
1880	—	—	—
1890	2	—	—
1900	5	—	—
1910	23	—	—
1920	40	1	—
1930	51	2	—
1940	52	3	—
1950	80	2	—
1960	84	17	8
1970	107	81	50
1980	81	142	50

Notes: "Municipal" includes city-county agencies; "county" includes multicounty agencies.
Source: Adapted from Stern 1982.

more than authorize federal funds to assist the states in paying for air pollution research and in training technical and managerial personnel, was prompted by the states' agitation over dealing with what they thought was a national problem. Congress extended the Air Pollution Control Act in 1959 and again in 1962. Prior to both extensions, more ambitious powers were contemplated for the federal government, but they were scrapped in favor of merely continuing appropriations for research, technical assistance, and training for state air pollution control agencies.

The mid-1960s saw the passage of three new acts that foreshadowed the later, more radical restructuring of federal air pollution control efforts. In 1963, Congress passed the original Clean Air Act. This law provided permanent federal support for air pollution research, continued and increased federal assistance to the states for the development of their pollution control agencies, and—in an important departure from previous legislation—introduced a mechanism through which the federal government could assist the states when cross-boundary air pollution problems arose. In 1965, the Motor Vehicle Air Pollution Control Act was passed. Most important among its provisions was permission (not compulsion) for the secretary of the federal Department of Health, Education, and Welfare (DHEW) to establish emissions standards for new motor vehicles. This legislation marked the real beginning of an active federal role in controlling air pollution from mobile sources. Nevertheless, it stopped short of imposing vehicle emissions standards, something only California had done by that time. However, also in 1965, Congress amended the Clean Air Act for the first time and directed DHEW to set the first federal emissions standards for motor vehicles.

In 1967, Congress passed the Air Quality Act. This law—the temporal and philosophical precursor of the 1970 changes—again provided funds for the states, this time to help them plan as well as implement their pollution control strategies. More important, the act required states to establish air quality control regions (now known as AQCRs), geographic areas that share common air quality concerns in much the same way so-called watersheds deal with water pollution problems in common. The act also directed DHEW to investigate and publish information about the adverse health effects associated with numerous common air pollutants so that the states could then set air quality standards for them. Moreover, DHEW (through its National Center for Air Pollution Control) was to identify viable pollution control techniques so that each state could get on with the business of regulating polluters to attain the air quality standards that each was to have established.

This review of legislation up to 1970 is by necessity both brief and selective, but it conveys the overall evolution of federal policy before 1970. What began in 1955 as simple grants-in-aid to state and local governments gradually evolved to the point where those recipients were given very specific responsibilities by the federal government for combating air pollution problems. The changes made in 1970 were, in a sense, the culmination of an increasing federal role. It was not until the middle-to-late 1990s that the pendulum began to swing back to the states.

The New Direction in Air Pollution Control Policy

Congressional impatience with the pace of activity under the pre-1970 approach, coupled with increasing public activism (exemplified by Earth Day in spring 1970), led inexorably to the Clean Air Act Amendments of 1970 (hereafter referred to simply as the Clean Air Act, the 1970 amendments, or CAA). A few perceived problems were associated with the previous approach. For example, DHEW had been slow in issuing the guidance documents that detailed the adverse health effects associated with common air pollutants; where the guidance documents had been prepared, states either had failed to set air quality standards or were slow in developing implementation plans that showed how they would meet the standards; and the automobile manufacturers, who had done little in the 1960s to inspire confidence in their commitment to pollution control, appeared capable of wriggling out of the emissions standards set by DHEW.[2] These factors and others combined to produce a major overhaul in federal air pollution policy.

It is impossible to neatly summarize all the important features of the Clean Air Act and its amendments. Indeed, the act as currently amended contains about 100 complicated sections and is 173 pages long. Nevertheless, the basic structure is comprehensible. The simplest way to think of it is in terms of the two major kinds of responsibilities it creates: The first has to do with the goals of air pollution control, and the second with the means of attaining them.

Goals

As a result of the Clean Air Act, the federal government—specifically, the administrator of the then newly created U.S. Environmental Protection Agency (EPA)—was charged with establishing the preeminent environmental objectives of federal air quality policy, the National Ambient Air Quality Standards (NAAQSs). These standards, which were to be set by the states under the old approach, represented the maximum permissible concentrations of the common air pollutants (which DHEW had been studying at the time the Clean Air Act was amended in 1970). Primary standards would protect human health; secondary standards would be established if the health-based standards were insufficient to protect exposed materials, agricultural products, forests, or other nonhealth values.

Congress made several important decisions about the primary standards. First, they were to be uniform across the country—that is, each of the AQCRs that had been established under the earlier legislation (or subsequently would be created) had to reduce ambient (that is, outdoor) air pollution concentrations at least to the level of the NAAQSs. The states could elect to impose stricter standards if they wished; however, no area could elect to have air more polluted than that called for in the NAAQSs.[3] Second, Congress directed that the primary standards be set by EPA at levels that would "provide an adequate margin of safety" to protect the public "from any known or anticipated adverse effects associated with such air pollutant[s] in the ambient air."

Both of these features are very important. National uniformity of standards means that areas where control costs are high or control benefits are low cannot respond by setting less stringent air quality standards to account for the situation; rather, such areas are locked into the same goals as other parts of the country. The margin-of-safety provision has become controversial because it embraces what is known as the "threshold model" of illness related to air pollution. So, by calling for a margin of safety in setting primary standards, Congress implicitly signaled its belief that "safe" levels existed and could be identified for the common air pollutants. After safe levels were identified for a particular pollutant, each NAAQS would be set at a level below that threshold, thus providing the putative margin of safety. This margin-of-safety concept is illustrated in Figure 4-1, which shows air pollution concentrations graphed against excess sick days (above some natural background of occasional illness), a hypothetical measure of the health of the population. Line segment *AA′*, which illustrates one possible relationship, indicates that for pollution concentrations between *O* and *A*, no excess adverse health effects occur. Beyond point *A*, the threshold concentration, increases in pollution imply more illness. Thus, according to the threshold approach, the NAAQS for this hypothetical pollutant might be set at *OB*, providing a margin of safety equal to *BA*.

Two problems have arisen with this approach. The first is apparent because a growing body of physiological and toxicological evidence suggests that for the air pollutants that require NAAQSs, no safe levels may exist—that is, some individuals may be so sensitive to pollution that almost any positive concentration may increase their risk of illness or discomfort (see Friedman 1981). In the intervening years, studies have repeatedly failed to produce evidence of a clear health effects threshold for any of the criteria pollutants—something EPA acknowledged in 1997 when it proposed new standards for ozone and fine particulates. Therefore, the true dose–response curve that links pollution to illness may look more like *OC* in the figure. Flat at low ranges, this curve nevertheless indicates that even very low pollution concen-

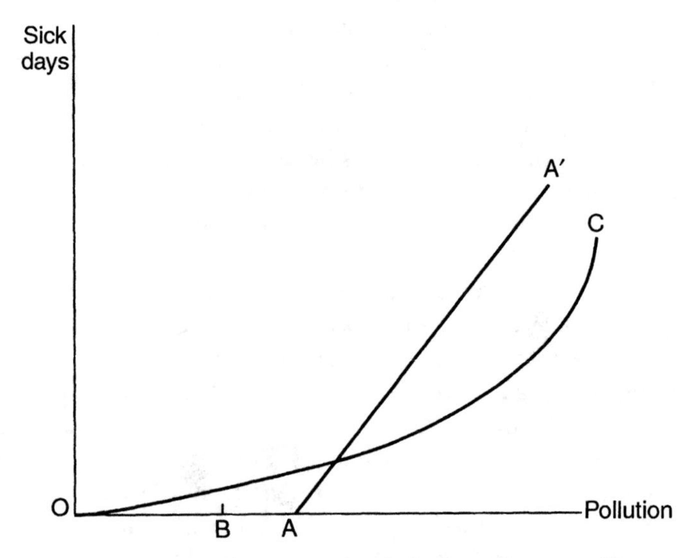

Figure 4-1. Dose–response functions that link air pollution to illness.

trations may result in some excess risk of illness. This leads to the following conclusion: If no threshold exists, then the only concentration that provides the statutorily required margin of safety is zero. Yet, setting permissible concentrations at zero would imply an end to all fossil fuel combustion and to commercial and industrial activity—indeed, to modern society as we know it.[4] Because Congress clearly did not contemplate such extremes, the EPA administrator is caught between the apparent mandate of the law and the realities of science and economics.

The second problem with the standard-setting process is more practical. Under current interpretations, the NAAQSs are to be set without regard to the costs of attainment (much like the zero-risk approach discussed in Chapter 2, The Evolution of Federal Regulation). Yet, even if threshold concentrations could be identified, might not the costs of meeting standards set at those levels—say, as opposed to slightly less protective standards—be so great that the weaker standards would be preferred? Even if the answer were yes, such a decision appears to be prohibited under the Clean Air Act.

This latter issue was brought to the fore in national debate in 1997 when EPA promulgated a revision of the NAAQS for ozone because respiratory illness had been observed at concentrations below the current standard. No evidence supported the existence of a health effects threshold, yet EPA was faced with a legal mandate to revise the standard in light of current epidemiological and toxicological literature. The EPA administrator, caught in a bind, issued the following statement: "EPA recognized that since there is no discernible threshold below which no adverse health effects occur, no level would eliminate all risk. Thus, a zero-risk standard is not possible, *nor is it required by the Clean Air Act.* The selected … level is based on the judgment that at this level the public health will be protected with an adequate margin of safety" [emphasis added]. In setting a nonzero standard, the administrator interpreted the law in a way that did not guarantee every individual zero risk from air pollution, thus reflecting the practical impossibility of doing so. In May 1999, a three-judge panel of the U.S. Court of Appeals overturned EPA's 1997 air quality standards on the basis that the agency provided no "intelligible principle" upon which it based those standards. This decision, subsequently upheld by the full appeals court, goes right to the heart of standard setting under Section 109.

With the possible exception of emissions standards for all new sources (discussed in the next section), ambient standards were the most important part of the Clean Air Act as amended in 1970. In fact, they remained the focus of a substantial amount of U.S. air pollution control activity in the 1990 amendments. The NAAQSs for the six common air pollutants as they were in the year 2000 are presented in Table 4-2.

Table 4-2. National Ambient Air Quality Standards (NAAQSs) in Effect in 2000.

	Primary (health related)		*Secondary (welfare related)*	
Pollutant	*Type of average*	*Standard level of concentration*[a]	*Type of average*	*Standard level of concentration*
CO[b]	8 hour	9 ppm (10 mg/m^3)	No secondary standard	
	1 hour	35 ppm (40 mg/m^3)	No secondary standard	
Lead	Maximum quarterly average	1.5 μg/m^3	Same as primary standard	
NO_x	Annual arithmetic mean	0.053 ppm (100 μg/m^3)	Same as primary standard	
Ozone[c]	Maximum daily 1-hour average	0.12 ppm (235 μg/m^3)	Same as primary standard	
PM_{10}[d]	Annual arithmetic mean	50 μg/m^3	Same as primary standard	
	24 hour	150 μg/m^3	Same as primary standard	
SO_2	Annual arithmetic mean	0.03 ppm (80 μg/m^3)	3 hour[b]	0.50 ppm (1,300 μg/m^3)
	24 hour[b]	0.14 ppm (365 μg/m^3)		

Note: CO = carbon monoxide; NO_x = nitrogen oxides; PM_{10} = particulate matter less than 10 micrometers in diameter; and SO_2 = sulfur dioxide..

[a] Values in parentheses are approximately equivalent concentrations.

[b] Not to be exceeded more than once per year.

[c] The standard is attained when the expected number of days per calendar year with maximum hourly average concentrations >0.12 ppm is ≤1, as determined according to Appendix H of the Ozone NAAQS.

[d] Particulate standards use PM_{10} as the indicator pollutant. The annual standard is attained when the expected annual arithmetic mean concentration is ≤50 μg/m^3; the 24-hour standard is attained when the expected number of days per calendar year >150 μg/m^3 is ≤1, as determined according to Appendix K of the Particulate Matter NAAQS.

Source: U.S. EPA 1998.

In 1970, Congress had expressed concern about less common air pollutants that might pose serious threats to health; the 1990 amendments outlined a strategy for controlling them. Rather than call for maximum ambient concentrations to be set, as with the NAAQSs, Congress directed the EPA administrator in 1970 to identify any such pollutants (now referred to as toxic or hazardous air pollutants, or HAPs) and propose discharge standards for the major sources that emit them. However, in wording quite similar to that concerning the common air pollutants, the EPA administrator was directed by Congress in 1970 to set these emissions standards so that the concentrations remaining in the air after controls were applied would be so low as to provide "an ample margin of safety" against adverse health effects. In essence, a similar environmental goal was established for both common pollutants and HAPs. In the 1990 amendments to the Clean Air Act, Congress went further. It directed the EPA administrator to determine Maximum Available Control Technology (MACT) for the major industrial facilities that emit one or more of the 189 substances presumed to be toxic and to issue regulations requiring sources to install this control equipment. EPA has been busily at work since 1990 issuing these standards.

What about those parts of the country where air quality was already better than the national air quality standards established by EPA? Would new industrial and other growth be allowed there until air quality had deteriorated to the level of the national standards? Although the 1970 amendments were silent on this subject, a landmark court case soon established that the Clean Air Act was to be implemented in such a way as to preserve air quality in clean regions as well as enhance it in polluted areas (see National Academy of Sciences 1981). In 1977, Congress amended the Clean Air Act again to formally declare the prevention of significant deterioration (PSD) in clean areas as an additional goal. The 1977 amendments established three classes of "already clean" areas. In Class I areas (which include national parks, forests, and wilderness areas, and other areas that states elect to include), very little additional deterioration in air quality is permitted, even if current concentrations are far below the NAAQSs. Somewhat more pollution is permitted in Class II areas (which make up most of the remaining clean air regions). In Class III areas, air quality is permitted to deteriorate up to but not beyond the level of the NAAQSs.

The 1977 amendments also established another goal of air pollution control policy: the protection and enhancement of visibility in the national parks and federal wilderness areas. Thus, even where air pollution was not threatening health, it was to be reduced if it impaired a visitor's view of, say, the Grand Canyon or the Yosemite Valley.

Such were the environmental goals Congress established through the Clean Air Act as amended in 1970, 1977, and 1990. As we turn our attention to the means of attaining them, it will become clear that the adoption of advanced technologies was another goal of the act, quite distinct from the quest to meet ambient environmental standards.

Emissions Standards as a Means of Attainment

Although Congress gave the federal government the sole responsibility for setting goals under the Clean Air Act, the authority for accomplishing those goals was divided. The federal government appears to have been given the upper hand here; nevertheless, some very important powers were reserved for lower levels of government.

Mobile Sources. On the unassailable logic that fifty different sets of state standards would wreak havoc with motor vehicle manufacturers, and in an attempt to spur carmakers to take serious action in pollution control, the Clean Air Act of 1970 mandated emissions standards for cars, trucks, and buses to be handled at the federal level.[5] In fact, Congress took the unusual step of writing directly into the Clean Air Act the emissions reductions that cars (and later trucks) would have to achieve, as well as the schedule for their accomplishment. (In contrast, in all but one other part of the act, the issuance of specific, numerical discharge standards was delegated to EPA or to the states.) The vehicle standards that Congress adopted called for a 90% reduction in average hydrocarbon and carbon monoxide emissions by 1975 (measured from the already controlled levels existing at the time) and the equivalent of an 82% reduction in the nitrogen oxides emitted by cars (measured from then-uncontrolled levels). With the 1990 amendments, EPA was required to issue emissions standards for diesel buses and was given the power to regulate the emissions of any nonroad engines (such as motorboats and lawnmowers) that contribute to urban air pollution.

The 1990 amendments to the Clean Air Act introduced a new wrinkle. Specifically, areas in violation of the NAAQS for ozone were categorized according to the severity of the air pollution problem. This situation set the stage for introducing a pyramid of control measures, arranged such that the cleanest of the nonattainment areas would have the fewest new measures to implement; each successively more polluted area would have more measures to implement than the last.

The 1990 amendments also targeted gasoline formulation, with certain areas required to offer only "cleaner" gasoline for sale to motorists. This approach allowed a level of flexibility in implementing control measures that previously had been impossible, because fuels with different attributes could easily be manufactured and shipped where needed. For example, specially formulated fuel has been provided to Denver, Colorado, throughout the winter to target the serious carbon monoxide problem associated with that city's unique winter weather patterns, which are attributable to its unique geography.

The 1990 amendments also required the more polluted regions of the country to introduce vapor recovery systems on the nozzles of gasoline pumps and gave California—the state that for years had had the most serious air quality problems—the power to require carmakers to manufacture and sell a certain umber of "zero emission" vehicles if they wanted to sell cars in that state. These technological innovations were combined with requirements for more rigorous vehicle emissions tests for residents of polluted areas. Residents with vehicles that failed the more-stringent emissions tests were obligated to have their automobiles repaired and retested. Also, in keeping with regulations to reduce pollutants from automobiles, centrally managed vehicle fleets were required to purchase "clean fuel" vehicles that ran on alternative fuels (such as compressed natural gas) in many areas. To combat the number of miles driven by the work forces of large employers in polluted regions, EPA attempted to enact rules that would require employers to force employees into or compensate them for carpooling. Considerable resistance on the part of both employers and employees, however, has forced EPA to back down on this particular issue.

Stationary sources. Factories, power plants, refineries, and other large industrial facilities are known collectively as stationary sources. Were emissions limits on these sources intended to be keyed to local air quality or—like the controls on cars and other vehicles—to be uniform across the country? A little bit of both, as it turns out.

Without a doubt, the most important provision introduced in the 1970 Clean Air Act amendments relating to stationary sources was that pertaining to newly constructed (or substantially modified) plants. In Section 111 of the act, Congress gave EPA the power to set binding emissions standards for all new sources of the common air pollutants. These standards are known as new source performance standards (NSPSs) and have two important characteristics. First, EPA must set NSPSs on the basis of the "best technological system of continuous emission reduction" (in other words, these standards must be technology-based, determined by the state of the art in pollution control at the time the standards are set). Second, the NSPSs must be affordable to affected parties. This latter requirement does not introduce balancing into the NSPS process, because the controls are assumed to be worthwhile as long as the source can afford them. But the requirement does introduce, at least crudely, some economic considerations into the standard-setting process. The NSPSs initially were meant to be applied uniformly throughout the

United States, without regard for air quality in the area where a plant was being built.

Over time, some variation based on local conditions has been introduced into the new source performance standards. Congress decided that in areas where at least one of the NAAQSs was being violated (called *nonattainment areas*), new sources had to meet even stricter emissions standards than those called for by the NSPSs. In 1977, Congress amended the Clean Air Act to require new sources that wished to locate in nonattainment areas to install technology consistent with the lowest achievable emissions rate (LAER).[6] Furthermore, in most PSD areas, where air quality is already better than that called for by the NAAQSs, major new sources were required to install the best available control technologies (BACTs).

Readers confused by the distinctions among NSPSs, LAERs, and BACTs should not despair of their analytic capabilities. Defining and distinguishing these levels of technological control for new sources has tied EPA, the courts, and the regulated community in knots. As a practical matter, however, if EPA has defined an NSPS for a particular kind of new source, that standard often is considered satisfactory as a LAER or a BACT in nonattainment or PSD areas, respectively, even though LAERs and BACTs, in principle, are supposed to be more strict than NSPS emissions limits.

With EPA authorized to establish more or less uniform emissions standards for all new plants, the only unfinished business in the Clean Air Act concerned the controls to be imposed on *existing* sources, that is, those already in operation in 1970. Here, in an important departure from the overall shift toward federal authority, the states were given the responsibility for establishing these emissions standards. Specifically, the Clean Air Act mandated that each state prepare a state implementation plan (SIP) that would demonstrate how existing sources would be controlled. According to the CAA, all parts of the country were to be in compliance with the NAAQSs by 1975; this deadline was extended many times before the 1990 amendments. Under the 1990 amendments, the last area is not required to come into compliance with the CAA until 2010. EPA did retain the authority to reject a SIP if it was inadequate to ensure attainment and, in the limit, the authority to step in and impose federal controls to meet the NAAQSs.

This division of responsibility gave the states and local governments primary control over existing sources, which was important to them from a political standpoint. When EPA imposes strict federal standards for all new plants to meet, the construction of new plants may be slowed, and opportunities for new jobs may be reduced. But the losers from such a policy are difficult to identify (because it cannot be determined who would have gotten the new jobs had a plant been built) and hence politically weak. However, stringent controls imposed on existing plants can result in their closure and thus endanger the jobs of current workers. Because the latter scenario could bring considerable political pressure to bear on local officials, the states wanted—and were granted—the responsibility for regulating existing local sources.

Local control of existing local sources has had another important effect. Suppose pollution in one jurisdiction could be transported by prevailing winds to another jurisdiction (as we know to be the case). In such an instance, the first (pollution-exporting) region could impose weak controls on its local sources, thereby enhancing its economic position, with little fear of suffering from the resulting pollution. On the other hand, the second (pollution-receiving) area might be controlling the sources within its jurisdiction quite stringently, only to suffer from "secondhand" pollution. Although the 1970 amendments included language that, in principle, gave EPA the power to act in such cases, it proved to be inadequate in practice. The problem was treated somewhat more thoroughly in the 1990 amendments, but the remedies have yet to be fully implemented and thus have not demonstrated their full potential.

In particular, the 1990 amendments acknowledge the problem known as "pollution transport" explicitly and provide for the establishment of multistate air quality management regions. One such management region is named in the 1990 amendments; the Northeast Ozone Transport Region (OTR) is made up of eleven northeastern states. These states are collaborating to develop strategies to reduce air pollution in that region; however, the partnership is far from perfect, and numerous lawsuits have resulted as one state has attempted to compel another to impose regulations not directly beneficial to the population of the offending state. One possible solution to this problem is to require the state that benefits from emissions reductions elsewhere to pay some of the polluting state's control costs. The development of such a system is currently being considered for the northeastern states for nitrogen oxide emissions.

In an exception to the rule of "federal controls on new sources, but state controls on existing sources," Congress took on coal-fired power plants in 1990. Specifically, under Title IV of the 1990 amendments, Congress directed EPA to come up with a system to reduce emissions of sulfur dioxide from these plants by nearly 50% over a ten-year period and to reduce emissions of nitrogen oxides as well. In a most welcome departure from the past, EPA was instructed *not* to mandate specific types of control equipment to effect these reductions. Rather, the reductions in sulfur dioxide emissions were to result from an initial set of required cutbacks from more than 100 coal-burning power plants (and, subsequently, from an even larger group), with these plants having the flexibility to determine how they would meet their emissions reduction goal.

Not only could these plants shift to lower sulfur coal, wash their coal in advance of burning it, or invest in energy conservation to reduce emissions; for the first time, they were given another option—buy emissions reductions from other power plants that were willing and able to reduce their emissions by more than they were required to do. This program has resulted in the creation of a nationwide market for emissions reductions, or *allowances*, a market which has brought about greater reductions in sulfur dioxide emissions than called for under Title IV—and at much lower cost than would have been required under the technology-based approach of the past.

Finally, the 1990 amendments require stationary sources to obtain operating permits that explicitly describe the regulations to which the source is subject as well as the means by which it plans to comply with all applicable laws. Furthermore, it requires a source to submit results of any mandatory air quality monitoring it undertakes to the permitting authority on a regular basis. Sources that operate without a permit and without exemption are in violation of the law.

Area Sources. The 1990 amendments add one additional and important category of pollution sources. *Area sources* are the collection of pollution sources that are not mobile and that are not large (such as factories or power plants). Area sources run the gamut from dry cleaners to printing shops to auto paint shops.

Whereas in the early days of pollution regulation, this sector may have represented only a tiny fraction of the total air pollution generated, two factors have raised its profile and attracted the attention of regulators. First, as the large individual point sources that historically generated the greatest emissions have reduced their pollution, total emissions (and the fraction of total emissions that could be attributed to point sources) decreased. In contrast, emissions from area sources have not remained constant but grown as the economy has grown, leading to an even faster increase in the fraction of total emissions generated by these sources.

Given the nature of area sources, the method of choice for regulation is through consumer product manufacturing guidelines. Provisions in the CAA that govern both the NAAQSs and hazardous air pollutants require EPA to consider the contribution of area sources to total emissions of the criteria pollutants, particularly volatile organic compounds, and to identify strategies for reducing emissions. EPA also must identify the thirty HAPs that "present the greatest risk to public health from air pollution in urban areas" and develop a strategy to achieve a 75% reduction in cancer risk from those sources. Currently, EPA is considerably behind schedule in issuing this particular set of guidelines.

One final important responsibility was reserved for state and local governments in the 1970 amendments: monitoring for and enforcing compliance with both the environmental goals and the individual source discharge standards called for in the Clean Air Act. This provision was in part a simple continuation of the pre-1970 policy under which localities began to monitor air pollutants, particularly sulfur dioxide and total suspended particulate matter; monies were appropriated under the 1970 Amendments for continued financial assistance to local air pollution control agencies. The policy of local monitoring and enforcement also served a political purpose: It gave local areas the flexibility to coax local sources into compliance. In practice, some recalcitrant polluting sources have thumbed their noses at local enforcement efforts; for this reason, EPA can—and has, from time to time—backed up local governments with federal enforcement authority.

Summary

Such is the structure that emerged in the Clean Air Act. The federal government sets the air quality goals that all parts of the country are to meet and also limits degradation in the already clean areas. Emissions standards for motor vehicles, all new stationary sources of the common air pollutants, and sources of hazardous air pollutants also are set in Washington (by Congress for cars, by EPA for all other sources), as are the characteristics of fuels in most places. Against the backdrop of these new source standards, individual states impose appropriate limits on existing sources to ensure that the national ambient air quality standards are met. Finally, the states are required to monitor and enforce the entire system of emissions limitations.

Accomplishments since 1970

By far, the most important measure of the success of the Clean Air Act amendments of 1970 and 1990 is what has happened to air quality since 1970. Because it took most of the early and middle part of the 1970s to get pollution controls in place on both stationary and mobile sources, and because the breadth and quality of air pollution monitoring data improved toward the end of that same decade, data are presented here for the period 1979–1998 (the last year for which complete air quality data are available). The six panels in Figure 4-2 illustrate the trends in ambient (or outdoor) concentrations of the six pollutants for which NAAQSs have been set.

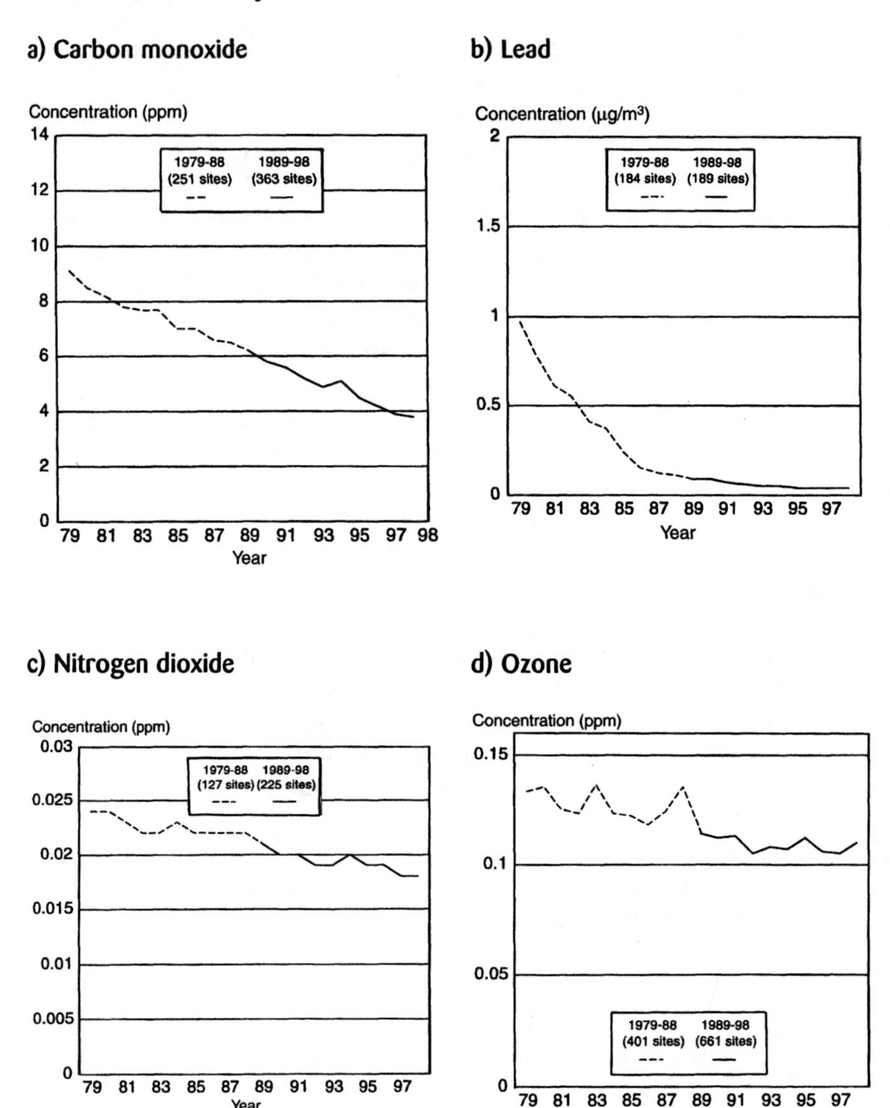

Figure 4-2. National air quality trends, 1987–96 (continues on next page).

Source: U.S. EPA 1998.

e) Particulate matter

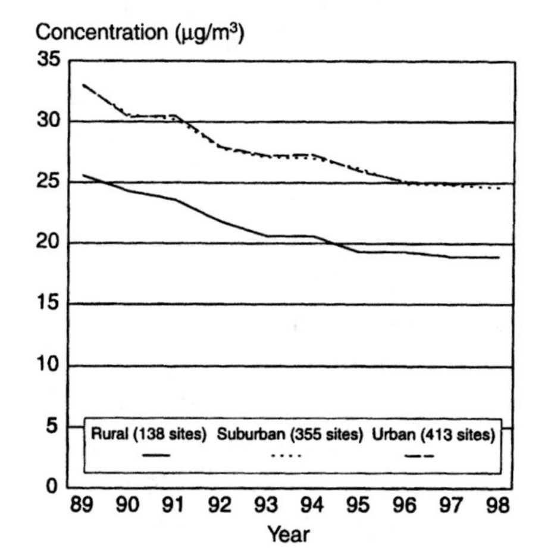

f) Sulfur dioxide

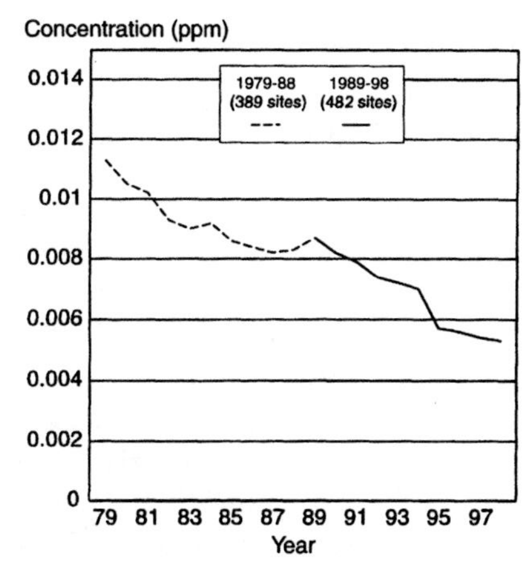

Figure 4-2. (continued).

The story they tell is a most encouraging one.

- Concentrations of carbon monoxide have fallen 58% during that twenty-year period.
- Concentrations of nitrogen dioxide are down 25% over the same period.
- Ambient levels of ozone (smog) are down 17%, despite a great increase over the period in both the number of cars on the road and the number of vehicle miles traveled each year.
- Concentrations of particulate matter of 10 microns or less (referred to as PM_{10}) are down by a quarter (measured only since 1989, since that is the first year a nationwide monitoring network was in place).
- Ambient levels of sulfur dioxide have fallen by 53%.
- Airborne lead concentrations are down by a spectacular 96% since 1979.

It is especially encouraging to see the fall in airborne concentrations of lead over this twenty-year period, because these reductions have been accompanied by a corresponding reduction in the levels of lead measured in children's blood. Since blood–lead levels themselves are inversely correlated with children's IQ, the twenty-year decline is of considerable significance. Even if one restricts one's attention to the last ten years, the improvements in air quality are impressive.

If one is willing to consider air pollution monitoring data from a smaller and perhaps less reliable network of monitors, it is possible to show even longer term trends. Goklany (1999) has assembled such trends, one of which—that for sulfur dioxide—is shown in Figure 4-3. Since 1962, according to these data, ambient concentrations have fallen by nearly 75%; note that this trend had begun prior to 1970, a point to which I return below.

Attainment Status

Despite fairly steady reductions in average ambient concentrations, the NAAQSs still are violated much more often than is permitted under the Clean Air Act. The problem of widespread nonattainment, as it has come to be called, is one of the most vexing in current air pollution policy.

Under the Clean Air Act, no more than one violation per year is permitted in a metropolitan statistical area for each of the one-hour, eight-hour, and twenty-four–hour standards (technically, the one-hour ozone standard cannot be violated more than three times during any three-year period). However, violations are quite common in many metropolitan areas. The

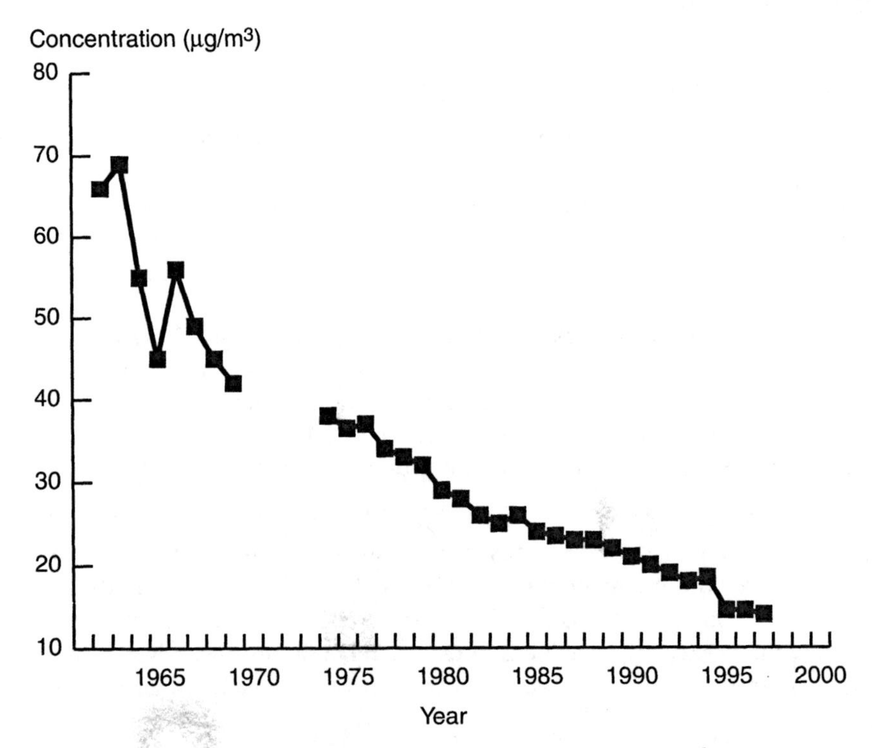

Figure 4-3. Ambient sulfur dioxide concentrations, mean annual average, 1962–1997.

Source: Adapted from Goklany 1999, 57.

ozone standard is most often violated in the United States; according to EPA, at the time of this writing, thirty-two areas (with a combined population of 93 million people) witnessed two or more days per year during which the one-hour ozone standard was exceeded in 1999 (down from 101 million in the early 1990s). Interestingly, Los Angeles is no longer the definitive example of urban air pollution. In 1998, while it experienced 46 days on which the ozone standard was exceeded (down from 149 days in 1989), it trailed Bakersfield, California (75 days), Fresno, California (67 days), and Riverside/San Bernardino, California (94 days). Table 4-3 shows the extent of nonattainment for ozone and other criteria air pollutants as of the end of 1999. The total number of nonattainment areas is 168, a significant decrease from the 293 nonattainment areas observed in the early 1990s.

Beyond the frequency of violations, the seriousness of the nonattainment problem is difficult to assess. As is too often the case, more data are needed. Among the questions needing answers are the following:

- In areas where persistent violations of one or more of the NAAQSs occur, how serious are those violations? (In some places—Riverside, California, for instance—ambient ozone concentrations are sometimes nearly double the maximum permissible one-hour level of 0.12 ppm; elsewhere, the standards are often barely exceeded, giving less cause for concern.)
- In areas where multiple violations are the rule, do they all occur at the same monitoring station(s) or do they occur more randomly? Any day on which the reading at any one of the monitoring stations exceeds the standard is counted as a day of violation; thus, if each of ten stations in a metropolitan area exceeds the relevant ambient standard only once per year but on different days, that area will be classified as having ten yearly violations. Presumably, we would be more concerned about persistent violations at any one site, especially if it is located in a heavily populated area.

Table 4-3. Nonattainment Areas for Ozone and Other Criteria Air Pollutants.

Criteria air pollutant	*Original number of areas*	*Number of areas in 1999*	*1999 population (thousands)*
Carbon monoxide	43	20	33,230
Lead	12	8	1,116
Nitrogen dioxide	1	0	0
Ozone	101	32	92,505
PM_{10}	85	77	29,880
Sulfur dioxide	51	31	4,371

The most difficult question about the nonattainment problem is one we can barely attempt to answer here: What improvements in human health would result from eliminating most or all violations of the NAAQSs? Unfortunately, no unambiguous answer to this question is possible given any amount of space, because of the great uncertainty and conflicting evidence about the effects of any of the criteria air pollutants on human health.[7] Nevertheless, several generalizations seem warranted.

Fairly steady improvements in urban air quality all around the United States have eliminated many of the more immediate or acute threats to health that are associated with air pollution. It is highly unlikely, for instance, that the United States will ever again experience episodes that involve the criteria pollutants serious enough to trigger significant premature mortality such as occurred in Donora, Pennsylvania, in 1948, when a temperature inversion trapped a thick blanket of smoke over the city for an extended period. Yet, ambient concentrations are clearly high enough in some areas during certain periods that even healthy individuals experience considerable discomfort and acute adverse health effects (smog in the Houston or Los Angeles areas is a definitive example). Of the pollutants that pose the most acute threats to health at existing concentrations, sulfates—particularly acidic sulfates—appear to be the most harmful.

Although existing air pollution concentrations may be sufficient to cause or at least exacerbate chronic respiratory or other illnesses, it is difficult to establish the likelihood or magnitude of such effects (see National Academy of Sciences 1985). In a sense, we are victims of our own successes. Pollution levels are low enough in most places now that their effects on chronic illness often are obscured by other causes, foremost among which are smoking and certain occupational exposures to harmful substances. However, the reduction in ambient lead concentrations during the past two decades in all likelihood has had significant favorable effects on chronic illnesses, especially among children.

International Experience

For purposes of comparison with U.S. air quality, Figure 4-4 presents data on ambient concentrations of suspended particulate matter from cities around the world. For the period 1990–95, at least, urban air quality was quite variable worldwide. In three cities in Canada and two in Japan, particulate concentrations averaged about 40–50 $\mu g/m^3$, whereas levels were more than seven times higher in Tehran, Iran, and six times that in Calcutta, India. Cities in the United States or Canada that seem polluted would be considered extraordinarily clean in many other parts of the world.

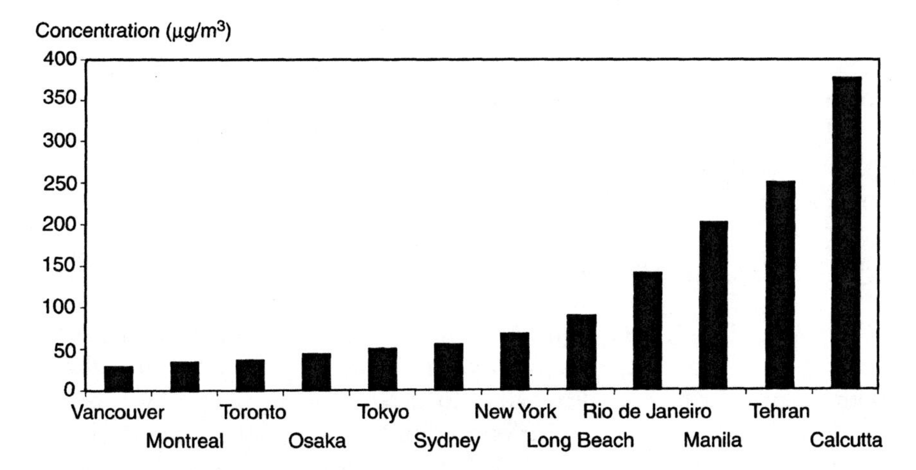

Figure 4-4. Suspended particulate matter in cities.

Notes: Data are most recently available for the period 1990–95; most data are for 1995. United States data are for 1990.

Sources: World Bank 1998; U.N. Environment Program 1990.

Linking Policies to Accomplishments

If the data on ambient concentrations can be taken at face value, cannot one automatically conclude that the Clean Air Act has been a smashing success? After all, pollution levels in the air have fallen since the act has been in effect. The answer is, not necessarily, as the following examples may demonstrate.

Suppose (purely hypothetically, of course) that simultaneously, all U.S. pollution control requirements were removed and the U.S. and world economies entered a severe recession. Pollutant emissions might fall dramatically (and with them, ambient concentrations) because the production of automobiles, steel, chemicals, gasoline, and other products of "dirty" industries would be curtailed during the economic slump. Yet, no one would argue that the lifting of pollution controls caused air quality to improve (or at least we would hope not!). Obviously, other factors would be at work. Another example would be a short-term or even seasonal temperature inversion in a particular area (as in the cases of Donora and Denver, mentioned earlier), which might trap pollutants and cause ambient concentrations to increase even though pollution controls are actually reducing emissions.

My point is that both pollutant emissions and ambient concentrations depend on a number of factors. These factors include the stringency of environmental regulations, to be sure. But they also include other, nonregulatory factors such as the general level of business activity; the rate and type of economic growth in particular areas; the prices of coal, oil, wood, gasoline, and other fuels used in home furnaces, cars, and industrial boilers; and even the weather.[8] Before passing judgment on the efficacy of the CAA, these other factors somehow must be netted out. This is no easy task—even though the isolation of variables is a common problem in the social sciences—because controlled experiments in which these other factors can be held constant generally are not possible.

For the Record: Before 1970

One final piece of evidence to consider in reviewing the Clean Air Act and its possible effects is the trend in air quality before the act, especially before the 1970 amendments. If air quality had been deteriorating prior to 1970 but then began to improve, some contribution on the part of the amendments passed that year might be suggested. Although the air quality data extending back into the 1960s are less reliable than today's, they do tell an interesting story. (See Goklany 1999 for an in-depth investigation of this issue.)

According to data from EPA (U.S. EPA 1981), average ambient concentrations of total suspended particulates fell about 22% between 1960 and 1970 (as monitored at ninety-five sites around the United States). During the period 1966–1971, annual average ambient sulfur dioxide concentrations fell by an even larger margin, 50% (based on data from thirty-one sites). Although we must be leery of trends based on so few sites, these data are important because they suggest that air quality was improving as fast or faster before the Clean Air Act as it has since 1970.

These data also call into question one of the fundamental premises behind the act: that states and local governments never would impose the controls necessary to achieve healthful air. Although some of the pre-1970 improvements were no doubt due to economic factors of the kind discussed earlier (such as the relocation of some industrial activity), they also were hastened by state and local ordinances that regulated the incineration of garbage and the burning of coal or high-sulfur fuel oil in residential, commercial, and industrial furnaces. Whether local governments acting alone could have continued to make progress in the fight against air pollution is arguable; certainly, it would have been difficult for them to deal with those pollutants that emanated from smaller and more diffuse sources. Nevertheless, their accomplishments prior to 1970 should not be ignored.

Thirty Years of Air Quality Regulation

Despite somewhat conflicting evidence, the following conclusions about the U.S. air quality experience since 1970 seem warranted. First, air quality in the United States appears to have improved quite significantly in most places since 1970. True, part of this improvement may be due to the closure of some large sources of pollutants or their relocation to remote, unmonitored regions. It is also true that some metropolitan areas—particularly in southern California—still are experiencing far too many days on which some NAAQSs are being violated. Nevertheless, most people in large metropolitan areas experience much better air quality than they did even ten years ago.

Linking these changes to the 1970, 1977, and 1990 amendments to the Clean Air Act is more problematic. Other factors (for example, the economy, fuel prices, and weather) also affect emissions and/or ambient concentrations. The fact that at least some measures of air quality were improving at an impressive rate before 1970 suggests that factors in addition to the CAA are at work behind the improvements since 1970.

Nevertheless, it seems indisputable that the Clean Air Act and its amendments have played an important role in improving U.S. air quality and in preventing its further degradation. For this not to be so, one would have to believe that all the investments in stationary and mobile source pollution control have been for naught. Yet, tons and tons of scrubber sludge, flyash, and other solid wastes attest to the at least partial efficacy of air pollution control equipment on factories and power plants. Similarly, although there can be no disputing the deterioration of catalytic converters and other automobile pollution controls over time, it seems clear that the CAA provisions for dealing with mobile sources have had a positive effect on air quality.[9]

Even where air quality has deteriorated since 1970, it probably is fair to say that the deterioration would have been worse had it not been for the controls imposed by the CAA on stationary, and especially mobile, sources. Ultimately, of course, this is the true measure of any regulatory or other government intervention: How different does the world look *with* the program in question from how it would have looked *without* it? Because other things generally do change, this question is not the same as asking what has happened since the program was put into effect.

Although my overall judgment about the role of the Clean Air Act in improving air quality may be comforting, much more needs to be known before the act can be declared a success or a failure. The ultimate judgment necessitates comparing the improvements in air quality associated with the act (and the resulting enhancement of human welfare) with the sacrifices that society has been forced to make to obtain them. In other words, we must compare the benefits and costs associated with CAA and its 1990 amendments. We now turn to these and related matters.

An Economic Evaluation of the Clean Air Act

In the first edition of this book, it was difficult to say much definitively about the costs and benefits associated with the 1970 amendments to the CAA. Valiant efforts had been made during the period 1970–1990 to pull together fragmentary information about improvements in air quality, the effects of pollution on human health, ecosystems, and visibility, as well as the compliance costs arising from the regulatory program initiated in 1970 (see Freeman 1982 for perhaps the best of these efforts). All such efforts were hampered by the cost—and the complexity—of even attempting a thorough analysis of pros and cons.

In the 1990 amendments to the CAA, and driven partly by frustration over *not* knowing whether the CAA was doing more good than harm, Congress stepped into the breach. In Section 812 of the 1990 amendments, Congress directed EPA to undertake the first comprehensive benefit–cost analysis of the nation's federal air quality regulatory programs. Specifically, Congress asked for one assessment of the benefits and costs during the period 1970–1990 (dubbed the "retrospective" study, intended to inform debate regarding the 1970 amendments to the CAA) and another assessment for the period 1990–2010 (naturally, the "prospective" study, which was intended to provide information about the 1990 amendments). In 1997, EPA completed the retrospective analysis (U.S. EPA 1997); in November of 1999, it reported to Congress on the findings of the prospective study (U.S. EPA 1999).

It is impossible here to summarize comprehensively the findings from these two studies. Each cost millions of dollars and runs to hundreds of pages. Nevertheless, it is possible to provide a general overview of the approach in the two studies, identify key assumptions, and highlight the important findings.

The fundamental question that EPA addressed in its retrospective analysis is this: What would the United States have looked like in the absence of the regulatory programs put in place beginning in 1970, and how does that compare to the situation we actually observed? For the prospective study, EPA asked: What would be expected to happen to air quality during 1990–2010 with the regulations that are scheduled to be implemented, and how would that differ if no additional regulations were put into place during that period?

The Retrospective Study: Benefits

To answer these questions with respect to benefits, EPA—and the contractors on whom they relied—first made assumptions about the growth in pollutant emissions that would have occurred during 1970–1990 had there been no federal controls imposed on tailpipe emissions from cars, nor any required pollution reductions from electric power plants, steel mills, petroleum refineries, chemical factories and other large stationary sources of air pollution. Needless to say, this is no easy task. In some cases, it is probably fair to assume that emissions that had been going up prior to 1970 would have continued to increase thereafter; in other cases, the natural ebb and flow of industry might have resulted in emission decreases even in the absence of federal controls. In any event, EPA laid out two clear alternatives on air pollutant emissions—a "control" and a "no-control" scenario. By 1990, sulfur dioxide emissions were assumed to be 40% lower in the control than in the no-control scenario, with the differences for nitrogen oxides, volatile organic compounds (VOCs), and carbon monoxide being 30%, 45%, and 50% respectively.

Next, these two very different scenarios on emissions had to be converted into alternative pictures concerning ambient air quality—one picture of air quality in the United States with the 1970 amendments to the CAA and another under the assumption that no regulations had been issued. To accomplish this task, the agency used several types of computer models. One type allows researchers to predict the formation in the air of ozone (smog) and other pollutants based on assumptions about emissions, temperature, and time of day. (Smog forms when VOCs combine with nitrogen oxides in the presence of sunlight, but will not form in their absence.) Another type of model combines data on pollutant formation with observed wind and weather patterns to predict how air pollution will be transported from one region to another. Both types of models are extremely complex; gluttons for punishment may wish to pore over the technical appendices in both the retrospective and prospective studies for details!

Briefly, average nationwide ambient concentrations of sulfur dioxide were modeled to be 40% lower with the CAA regulations than without, while those for nitrogen oxides and carbon monoxide were 30% and 50% lower. Ozone concentrations nationwide were predicted to be 15% less in the control scenario. Especially in this latter case, it is tempting to ask why the change in ozone concentrations is so small, given that emissions of the VOCs and nitrogen oxides that combine to form it were predicted to have fallen by much more. The answer is, first, that there are natural sources of VOCs—trees, for example—and, second, that the complex reactions in ozone chemistry are nonlinear; that is, a 10% reduction in precursor emissions produces a less than 10% reduction in airborne concentrations of ozone.

The next step in the studies is, in many ways, the most important. In order to determine the benefits of air quality regulation, the changes that the regulations are assumed to have produced (that is, the difference in air qual-

Table 4-4. Criteria Pollutant Health Benefits: Distributions of 1990 Incidences of Avoided Health Effects for 48-State Population.[a]

			Annual effects avoided (in thousands of incidents reduced)[b]			
Endpoint	*Pollutant(s)*	*Affected population*	*5th percentile*	*Mean*	*95th percentile*	*Unit*
Premature mortality	PM_{10}[c]	Age 30 and over	112	184	257	cases
	Lead	All	7	22	54	cases
Chronic bronchitis	PM_{10}	All	498	674	886	cases
Lost IQ points	Lead	Children	7,440	10,400	13,000	points
IQ less than 70	Lead	Children	31	45	60	cases
Hypertension	Lead	Men 20–74	9,740	12,600	15,600	cases
Coronary heart disease	Lead	Ages 40–74	0	22	64	cases
Atherothrombotic brain infarction	Lead	Ages 40–74	0	4	15	cases
Initial cerebrovascular accident	Lead	Ages 40–74	0	6	19	cases
Hospital admissions						
All respiratory	PM_{10} and ozone	All	75	89	103	cases
Chronic obstructive pulmonary disease and pneumonia	PM_{10} and ozone	Over age 65	52	62	72	cases
Ischemic heart disease	PM_{10}	Over age 65	7	19	31	cases
Congestive heart failure	PM_{10} and carbon monoxide	Age 65 and over	28	39	50	cases

continued on next page

ity between the control and the no-control scenarios) must be translated into improvements in human health, reduced damage to agricultural crops and exposed materials, and improvements to visibility and other things that people value. (Presumably, we value clean air because it will make us and our children healthier and because reduced pollution means more aesthetically pleasing vistas, less corrosion to buildings and other exposed materials, and less damage to forests and agricultural crops, among other things.)

For health benefits, which always form the lion's share of air pollution control benefits, EPA relied on many epidemiological analyses that linked air pollution concentrations to various adverse human health effects. In some cases, so-called clinical studies were used to predict health improvements associated with the control scenario. Table 4-4 shows all the categories for

Table 4-4. (continued).

Endpoint	Pollutant(s)	Affected population	Annual effects avoided (in thousands of incidents reduced)[b] 5th percentile	Mean	95th percentile	Unit
Other respiratory-related ailments						
Shortness of breath	PM_{10}	Children	14,800	68,000	133,000	days
Acute bronchitis	PM_{10}	Children	0	8,700	21,600	cases
Upper and lower respiratory symptoms	PM_{10}	Children	5,400	9,500	13,400	cases
Any of 19 acute symptoms	PM_{10} and ozone	Ages 18–65	15,400	130,000	244,000	cases
Asthma attacks	PM_{10} and ozone	Asthmatics	170	850	1,520	cases
Increase in respiratory illness	Nitrogen dioxide	All	4,840	9,800	14,000	cases
Any symptom	Sulfur dioxide	Asthmatics	26	264	706	cases
Restricted activity and work-loss days						
Minor restricted activity days	PM_{10} and ozone	Ages 18–65	107,000	125,000	143,000	days
Work-loss days	PM_{10}	Ages 18–65	19,400	22,600	25,600	days

[a] The following additional human welfare effects were quantified directly in economic terms: household soiling damage, visibility impairment, decreased worker productivity, and agricultural yield changes.

[b] The 5th and 95th percentile outcomes represent the lower and upper bounds, respectively, of the 90% credible interval for each effect as estimated by uncertainty modeling. The mean is the arithmetic average of all estimates derived by the uncertainty modeling.

[c] In this analysis, PM_{10} is used as a proxy pollutant for all nonlead criteria pollutants that may contribute to premature mortality.

Source: U.S. EPA 1997, Table ES-1.

which predictions of improved health were made. To take several examples from the table, there were 184,000 fewer premature deaths among citizens thirty years of age or older in 1990 on account of the reductions in fine particulate matter with diameters of 10 microns or less (referred to as PM_{10}). EPA estimates that there is a 5% chance that the true number of fatalities avoided is as small as 112,000 and a 5% chance that the number is as large as 257,000. Under the heading "other respiratory-related ailments," the table shows that EPA finds 8.7 million fewer cases of acute bronchitis in the control as opposed to the no-control scenario.

Taking these predictions of adverse health effects avoided as given for the moment, the next step is to translate them into dollar terms so that the benefits might be compared with costs. It should go without saying that this engenders great controversy. At least part of this controversy is unnecessary. Those unfamiliar with benefit–cost analysis sometimes assume that economists make up the values to be assigned to the prevention of adverse health effects, substituting their judgment for the preferences of others. This is not the case at all. Rather, a very important part of environmental economics (and a most intriguing one, too) involves looking at a variety of decisions made by ordinary people that provide clues as to the values that these people attach to improvements in their own health.

Here is an example. Suppose that reducing ambient ozone levels brings with it a reduction in the number of asthma attacks experienced by the population. In order to place a dollar value on this benefit, economists look at several things. First, how much is out-of-pocket spending reduced for asthma medicines, doctor visits, and perhaps even hospitalizations in severe episodes? Second, what are the savings in lost income that asthma sufferers would have incurred had air quality not improved? While these two measures added together probably still underestimate the benefits of reduced ozone levels (after all, people would pay to prevent an asthma attack even if they incurred no out-of-pocket losses, simply because of the unpleasantness of being sick), they may provide a good lower boundary. These same two measures (medical costs avoided plus lost income) can also be used to value other diseases related to air pollution, such as acute and chronic respiratory disease, heart ailments, hypertension, and eye irritation.

But what about the value of preventing a premature death? How does one put a value on something so important? We begin by noting that we are never preventing the death of someone whose identity we know. That is, the question is not, "How much should we spend to extend Paul's life [or the life of another individual] by seven years?" Rather, what epidemiology can do is give us an estimate of how many fewer people in a population of 270 million people can be expected to die if, say, fine particle concentrations are reduced by *x*% nationwide. With that understanding, the relevant question becomes "How much will the average person pay for an air quality improvement that will reduce his or her chances of dying in a given year by, say, one in one hundred thousand (or 0.00001)?"

It may be surprising to learn that there are data that help us answer this question. For many years, economists have been analyzing data from the labor market, looking particularly at a wide variety of occupations that pose discernible annual fatality risks. In particular (and using powerful statistical techniques), researchers have looked to see how much employees have to be paid each year to accept riskier jobs, holding constant other dimensions. While these studies have found a wide range of answers, they tend to conclude that workers are willing to sacrifice between $300 to $700 in additional annual pay for a job that carries with it a reduced annual fatality risk of one in 10,000. This means that 10,000 such workers would be willing collectively to sacrifice $3–$7 million to prevent one sure death. We refer to this as the *value of a statistical life*, and some number within this range is almost always used to value one case of premature mortality prevented.

Table 4-5 shows the values that EPA used to convert its estimates of adverse health effects prevented in the control scenario into dollar values. As shown, EPA used a value of $4.8 million per case of premature mortality prevented—more or less in the middle of the range described above. According to EPA, preventing one case of acute bronchitis is worth $45 (again, using data on medical costs and lost income avoided). The table also indicates the value that EPA attached to improving visibility—$14 per unit—which it derived from questionnaires in which individuals are asked their willingness to pay for improved air quality and clarity.

Combining both estimates of physical effects with values per effect, Table 4-6 shows the total benefits for the period 1970–1990 that EPA attributes to the 1970 amendments to the Clean Air Act. If one were to add up the numbers in the "mean" column, they would total $22.2 *trillion*, or about $1.1 trillion annually.

The Retrospective Study: Costs

What about the costs associated with twenty years worth of regulatory controls? How do they compare to the benefits? In its retrospective analysis, EPA uses a variety of approaches to determine the cost of cleaner air during the period 1970–1990. One approach involves looking at all the expenditures made by regulated parties during those two decades. (Actually, because virtually no money was spent on environmental compliance until 1973, the estimates are actually for the seventeen-year period 1973–1990). To make such estimates, EPA relied heavily upon surveys of manufacturing firms conducted during much of that period by the Bureau of Economic Analysis within the Department of Commerce. These surveys elicited information annually on spending for pollution control by regulated firms. Using this approach, EPA found that expenditures for pollution control came to $628 billion during 1973–1990.

However, regulation also can involve costs that do not show up as out-of-pocket expenditures. For instance, waiting in line to have my car inspected imposes a cost on me over and above what I pay the gas station to conduct the emissions test. Similarly, if emissions control equipment on my new car imposes a fuel economy penalty on me, I have incurred a cost even though I

Table 4-5. Unit Valuation of Health and Welfare Effects.

Endpoint	*Pollutant(s)*	*Mean estimate valuation ($1990)*
Mortality	PM and lead	$4,800,000 per case
Chronic bronchitis	PM	$260,000 per case
IQ changes		
Lost IQ points	Lead	$3,000 per IQ point
IQ < 70	Lead	$42,000 per case
Hypertension	Lead	$680 per case
Strokes[a]		
Males	Lead	$200,000 per case
Females	Lead	$150,000 per case
Coronary heart disease	Lead	$52,000 per case
Hospital admissions		
Ischemic heart disease	PM	$10,300 per case
Congestive heart failure	PM	$8,300 per case
Chronic obstructive pulmonary disease	PM and ozone	$8,100 per case
Pneumonia	PM and ozone	$7,900 per case
All respiratory	PM and ozone	$6,100 per case
Respiratory illness and symptoms		
Acute bronchitis	PM	$45 per case
Acute asthma	PM and ozone	$32 per case
Acute respiratory symptoms	PM, ozone, nitrogen dioxide, and sulfur dioxide	$18 per case
Upper respiratory symptoms	PM	$19 per case
Lower respiratory symptoms	PM	$12 per case
Shortness of breath	PM	$5.3 per day
Work-loss days	PM	$83 per day
Mild restricted activity days	PM and ozone	$38 per day
Welfare effects		
Visibility	DeciView[b]	$14 per unit change in DeciView
Household soiling	PM	$2.50 per household per PM_{10} change
Decreased worker productivity	Ozone	$1[c]
Agriculture (new surplus)	Ozone	Estimated change in economic surplus

[a] Strokes include atherothrombotic brain infarctions and cerebrovascular accidents; both are estimated to have the same monetary value.

[b] A DeciView is a measure of clearer air.

[c] Decreased worker productivity is valued as change in daily wages: $1 per worker per 10% decrease in ozone.

Source: U.S. EPA 1997.

Table 4-6: Present Value of 1970–90 Monetized Benefits by Endpoint Category for 48-State Population.

		Present value ($1990 billions, discounted to 1990 at 5%)		
Endpoint	*Pollutant(s)*	*5th percentile*	*Mean*	*95th percentile*
Mortality	Particulate matter	$2,369	$16,632	$40,597
	Lead	$121	$1,339	$3,910
Chronic bronchitis	Particulate matter	$409	$3,313	$10,401
IQ (lost IQ points and children with IQ < 70)	Lead	$271	$399	$551
Hypertension	Lead	$77	$98	$120
Hospital admissions	Particulate matter, ozone, lead, and carbon monoxide	$27	$57	$120
Respiratory-related symptoms, restricted	Particulate matter, ozone, nitrogen dioxide, and sulfur dioxide	$123	$182	$261
Activity and decreased productivity	Particulate matter	$6	$74	$192
Visibility	Particulate matter	$38	$54	$71
Agriculture (net surplus)	Ozone	$11	$23	$35

Source: U.S. EPA 1997.

make no upfront environmental expenditure that shows up neatly in some account. In order to see how such *opportunity costs* might affect its cost estimate, EPA calculated costs in a slightly different way, one that ostensibly took better account of such things. Using this approach, EPA concluded that air quality regulation in the control scenario cost $523 billion during the period 1973–1990. (This was lower than the total for expenditures, according to EPA, because air quality regulation actually ended up saving automobile buyers and owners money they would otherwise have spent for tuneups and maintenance). Using still a third approach, one that is more consistent with the approach used to estimate benefits, EPA found that the regulations imposed during the period 1973–1990 reduced the well-being of consumers by $569 billion. (Their well-being was reduced because prices were higher, job opportunities fewer, and returns to capital lower than they would have been in the absence of air quality regulation.)

You might well ask, "Which is the right cost measure to use?" The truth is that cost estimation is much more difficult—and imprecise—than economists sometimes let on. There is not very much difference among these three approaches. They range from a low of $523 billion (or $31 billion per year over the seventeen-year period) to as much as $628 billion (or $37 billion annually), and we might as well acknowledge that this is as close as we are likely to come without a much more ambitious push to better understand the "con" side of air quality regulation.

The Prospective Study

For the period 1990–2010, I will say much less about EPA's study of the benefits and costs associated with the 1990 amendments to the Clean Air Act, principally because the methodologies used were the same as in the retrospective analysis. There is one difference, however. Instead of comparing actual emissions during the period 1970–1990 with assumed emissions under a no-control scenario, the prospective study bases benefit and cost calculations on differences between two *assumed* futures—one with additional CAA controls in place and another in which it is assumed that the 1990 amendments never happened. Other than that, however, the same epidemiological studies, air quality models, and other staples of the retrospective analysis were employed.

According to EPA, by the year 2010, the 1990 amendments will have resulted in emissions of VOCs that are 35% lower in the control than in the no-control case. Nitrogen oxide emissions will be lower by nearly 40%, according to EPA, with smaller emissions reductions expected for carbon monoxide and PM_{10}. Using air quality models, these expected emissions reductions were mapped into improvements in ambient air quality and then—using epidemiological and other analyses—into improvements in human health, visibility, agricultural output, and other benefit categories.

To estimate costs of control for the period 1990–2010, EPA made assumptions about the types of pollution control equipment that manufacturing firms and electricity-generating plants would have to install, the added costs incurred to either manufacture or purchase cleaner fuels, the costs of more rigorous vehicle inspection and maintenance programs, and the costs to automakers for producing low- or zero-emitting vehicles in California and possibly other states.

Table 4-7 shows the results of EPA's assessment of the benefits and costs of the 1990 amendments to the Clean Air Act. As the table indicates, in the year 2000, those amendments are costing the nation about $19 billion (in $1990). By the year 2010, the annual price tag will have grown to $27 billion. According to EPA, the benefits of the 1990 amendments compare more than favorably with the costs. EPA's "central" (or most likely) estimate of benefits

Table 4-7. Summary Comparison of Annual Estimates of Benefits and Costs

	2000 ($1990 millions)	*2010 ($1990 millions)*
Monetized direct costs		
Central	$19,000	$27,000
Monetized direct benefits		
Low	$16,000	$26,000
Central	$71,000	$110,000
High	$160,000	$270,000
Net benefits		
Low	($3,000)	($1,000)
Central	$52,000	$83,000
High	$140,000	$240,000
Benefit–cost ratio		
Low	less than 1:1	less than 1:1
Central	4:1	4:1
High	more than 8:1	more than 10:1

Source: U.S. EPA 1999, iii.

is $71 billion annually in 2000 and $110 billion by 2010 (again expressed in $1990). In terms of net benefits—that is, benefits minus costs—the 1990 amendments will be generating a favorable surplus of $52 billion in the year 2000 using the central estimate of benefits, a surplus that EPA expects to rise to $83 billion by 2010. Again, using the central estimate of benefits, the benefit–cost ratio is four to one in the years 2000 and 2010.

Bottom Line Assessment and Caveats

Taking EPA's estimate of air pollution control benefits as given and regardless of the cost measure used, it appears that the benefits associated with the 1970 amendments to the Clean Air Act have been a smashing success. Even using the high estimate for costs—$37 billion annually from 1973–1990, annual benefits of $1.1 trillion exceed costs by a margin of nearly thirty to one. While this difference is much, much larger than that found in previous benefit–cost assessments of the 1970 amendments (Freeman 1982, for instance), these findings are consistent with the earlier studies in finding a favorable balance of benefits against costs. Likewise, EPA's 1999 study suggests that the 1990 amendments to the Clean Air Act also do well in a benefit–cost assessment. The benefit–cost ratio is slimmer for the latter changes (4:1 as contrasted to 30:1 for the 1970 amendments, at least for the "central estimate"), though net benefits are positive and large in all but the "low" benefit case.

A few caveats are in order at this point. While none is likely to alter the conclusions above, they are worth bearing in mind. First, in one respect, the findings of EPA's retrospective analysis strain credulity. In the year 1990, gross domestic product (GDP) in the United States was slightly more than $5.7 trillion. Yet, according to EPA, benefits that same year resulting from the 1970 amendments to the CAA came to $1.1 trillion. Interpreted literally, this means that U.S. citizens would have been willing to give up one-fifth of their incomes in 1990 rather than be deprived of the advantages of clean air. While there can be no denying the importance of good air quality, it seems unlikely to me, at least, that people would sacrifice this much to have it.

The reasons for EPA's benefits estimate being so large call into question other assumptions of the retrospective analysis. Fully 80% of the total dollar benefits that EPA ascribes to the 1970 amendments are due to the value of predicted reductions in premature mortality (see Table 4-4). This dollar sum is the product of the reduction in the number of deaths prevented each year (which EPA calculates to be in excess of 200,000 annually in the retrospective study) and the value per death prevented ($4.8 million). Both numbers have raised eyebrows.

The number of premature deaths prevented was calculated using an epidemiological study that compared death rates across cities, using as possible explanatory variables both the ambient concentrations of fine particles in the air as well as a host of other possible determinants, including age, education, income, and many other variables (Pope and others 1995). While this study was carefully done, it seems implausible (to me, at least) that nearly one-tenth of all deaths in the United States each year could be due to exposure to fine particles in the air, especially when previous efforts to apportion total mortality have attributed no more than 2% of deaths to *all* environmental causes (Doll and Peto 1981). It is possible, though by no means assured, that future work in epidemiology could result in a downward adjustment in the number of deaths prevented through the control of fine particles in the air.

There is also some controversy about the value that EPA attaches to each premature death prevented. As indicated above, the $4.8 million value per-death-averted comes from occupational studies linking fatality risk to wage rates. But a death in the workplace is almost always immediate. Since the average age of the workers in the populations surveyed was about forty-three, preventing an occupational death saves about thirty to thirty-five "life years," given current life expectancies. However, it is highly unlikely that those who die prematurely from air-pollution-related diseases are losing this many life years; rather, they may die at age seventy rather than live to be seventy-eight.

Thus, the question arises whether it is appropriate to use the same value of a statistical life to value the loss of, say, seven life years as it would be to value the loss of thirty life-years. Most people, I believe, would respond negatively to this question and, thus, would hold that EPA's benefit total may be too high.

Suppose, though, that we were to arbitrarily reduce the number of deaths avoided by half and also value each death prevented by $1 million rather than $4.8. This would still result in benefits that exceeded costs for the 1970 amendments by a factor of at least 3:1. Such smaller, but still positive, results would not necessarily be the case, however, for the prospective study that assessed the benefits and costs associated with the 1990 amendments to the CAA. Since the benefit–cost ratio in the "central" case is 4:1, any significant change in the predicted number of premature deaths avoided or any significant downward revision in the value of prolonging a life for a relatively small number of life-years could "flip" the results of the benefit–cost analysis into an unfavorable balance.

As it is careful to point out, however, in a number of categories EPA chose not to make a monetary estimate of pollution control benefits, including some types of health effects. For such categories, either techniques were not available to make quantitative estimates of adverse physical effects avoided as a result of reduced pollution or EPA deemed it too difficult to assign dollar values to these effects. These categories include such effects as the reduced damage to lakes, streams, and rivers from reduced airborne deposition of pollutants into the water; possible increases in forest productivity; and the adverse health effects avoided through the control of hazardous air pollutants. If these benefit categories were included, clearly the benefit total would go up.

The best we can do, it would seem, is to use common sense in conjunction with the results of studies like these. Prior to 1970, there were no federal controls on the tailpipe emissions of cars and no federal regulations governing emissions from power plants, factories, refineries, and other industrial facilities. Perhaps not surprisingly, air quality (though improving along some dimensions) was quite poor by today's standards. It would be surprising to me if the 1970 amendments did not look good on a benefit–cost basis in view of the quite significant improvements in air quality that followed their passage. On the other hand, by 1990, the incremental cost of further emission reductions had grown significantly—during the previous two decades we had exhausted most of the inexpensive opportunities to cut emissions ("picked the low-hanging fruit," to use the vernacular). Similarly, having addressed the most obvious threats to health during the 1970s and 1980s—such as lead in gasoline and gross emissions from very dirty plants—it makes sense that the *incremental* benefits associated with further emissions reductions might be of less value. Thus, we would expect future proposals to reduce air pollution emissions to be closer calls on benefit–cost grounds than their earlier brethren.

Cost-Effectiveness Analysis and Air Pollution Control

Because of difficulties in identifying and valuing the favorable effects of air pollution control (particularly reduced mortality and morbidity), benefit–cost comparisons always will be controversial. Indeed, such comparisons generally do not play much of a role in establishing the goals of air pollution control or other environmental policies. These goals, as embodied in legislation, usually are the result of political compromises between environmental and industry groups, state and local officials, the executive branch and members of Congress, and other interested and politically active parties.

Even so, difficulty and controversy do not negate the potential contributions of analytical methods to air pollution policy. These methods can play a large role in helping society meet predetermined environmental or other goals as inexpensively as possible by using an approach referred to as *cost-effectiveness analysis.* Such analyses of air pollution policy have come to much less ambiguous conclusions than those that attempt to compare costs and benefits. Specifically, a very large body of research has demonstrated clearly that existing limits on pollutant emissions—and perhaps current air quality goals—could be met at a fraction of what is now being spent.

The studies that reach this conclusion generally use a common method. Typically, they search for possible variations in the cost of controlling the last ton(s) of air pollutants removed at regulated sources. Where variations exist, cost savings are possible. For instance, if a coal-fired power plant is spending $2,000 per ton to remove the last several tons of sulfur dioxide, but a nearby smelter could remove an additional ton for $1,000, total control costs could be reduced by $1,000 by allowing the power plant to emit an additional ton (thereby saving $2,000) while requiring the smelter to remove an additional ton (for $1,000). Emissions remain the same, but control costs fall. If this process were carried to its logical conclusion, total control costs would be minimized when all regulated sources faced the same marginal cost of control on the last unit of pollution removed. The same logic suggests that controls on multiple sources (or stacks) at one plant could be manipulated to minimize control costs for that plant.

In part to determine the savings possible through more cost-effective air quality regulation, Tietenberg (1985) surveyed the most important of these studies. His findings are summarized in Table 4-8; the column labeled "CAC

Table 4-8. Empirical Studies of Air Pollution Control.

Study and year	*Pollutant(s)*	*Geographic area*	*CAC benchmark*	*Assumed pollutant type*	*Ratio of CAC cost to least cost*
Atkinson and Lewis (1974)	Particulates	St. Louis metropolitan area	SIP regulations	NMA	6.00[a]
Roach and others (1981)	Sulfur dioxide	Four Corners in UT, CO, AZ, and NM	SIP regulations	NMA	4.25
Hahn and Noll (1982)	Sulfates	Los Angeles	California emission standards	NMA	1.07
Krupnick (1983)	Nitrogen dioxide	Baltimore	Proposed RACT regulations	NMA	5.96[b]
Seskin, Anderson, and Reid (1983)	Nitrogen dioxide	Chicago	Proposed RACT regulations	NMA	14.4[b]
McGartland (1984)	Particulates	Baltimore	SIP regulations	NMA	4.18
Spofford (1984)	Sulfur dioxide	Lower Delaware Valley	Uniform percentage reduction	NMA	1.78
	Particulates			NMA	22.0
Harrison (1983)	Airport noise	United States	Mandatory retrofit	UMA	1.72[c]
Maloney and Yandle (1984)	Hydrocarbons	All domestic du Pont plants	Uniform percentage reduction	UMA	4.15[d]
Palmer, Mooz, Quinn, and Wolf (1980)	Chlorofluorocarbon emissions from nonaerosol applications	United States	Proposed emission standards	UMA	1.96

Abbreviations: CAC = command and control, the traditional regulatory approach; SIP = state implementation plan; RACT = reasonably available control technologies, a set of standards imposed on existing sources in nonattainment areas; NMA = nonuniformly mixed assimilative; and UMA = uniformly mixed assimilative.

[a] Based on 40 g/m^3 at worst receptor.

[b] Based on a short-term, one-hour average of 250 g/m^3.

[c] Because it is a benefit-cost study instead of a cost-effectiveness study, the Harrison comparison of the CAC approach with the least-cost allocation involves different benefit levels. Specifically, the benefit levels associated with the least-cost allocation are only 82% of those associated with the CAC allocation. To produce cost estimates based on more comparable benefits, as a first approximation, the least-cost allocation was divided by 0.82 and the resulting number was compared with the CAC cost.

[d] Basedon an 85% reduction of emissions from all sources.

Source: Tietenberg 1985.

benchmark" indicates the regulatory approach (command-and-control) against which cost savings were measured. Thus, in the study by Roach and others, actual SIP regulations on sulfur dioxide emissions in four southwestern states were compared to a least-cost solution. Similarly, Maryland SIP regulations formed the basis for McGartland's study of the control of total suspended particulates in Baltimore, although he focused on ambient air quality goals rather than emissions goals. The ratios in the right-hand column show how much more expensive the traditional (command-and-control) regulatory approach is when compared to the most cost-effective set of controls: the range is from a low of 7% more expensive in a study of sulfur dioxide control in the Los Angeles basin up to 2,100% more expensive in a study of hypothetical nitrogen oxide regulations in Chicago. It appears from Tietenberg's survey that traditional air quality regulation may be, on average, about three or four times more expensive than the most cost-effective approach, if the studies Tietenberg surveyed are representative of real-world experience.

Tietenberg cited several reasons why actual savings may be less than what was suggested by the studies he reviewed. The most important reason is that pollution control equipment was already in place at many facilities because of existing regulations and this equipment could not be dismantled and reinstalled at sources that would be controlled under a cost-minimizing approach without incurring some expense (Tietenberg 1985, 48).

A problem also could arise under a least-cost approach if one source in an area were allowed to increase its pollution substantially in exchange for reductions from several other sources spread throughout that area. In such a case, even though aggregate pollution might remain the same, it would become more heavily concentrated at one location. For this reason, many cost-minimizing schemes include provisions to prevent "hot-spot" problems. These provisions would reduce the cost savings possible from reallocating the control burden. In fact, the New York state legislature passed a bill in May 2000 that would prohibit utilities in that state from selling emissions allowances to power plants in states upwind of New York. The concern is that such trades would result in increased acidic deposition in the lakes of the Adirondack region of the state, despite evidence to the contrary (see Burtraw and Mansur 1999).

These and other qualifications notwithstanding, the studies that Tietenberg reviewed give some idea of the kind of savings that would be possible through more selective regulation. Suppose, for example, that command-and-control regulation were 25% more expensive than more cost-effective rules (rather than 200% to 300%, as suggested in Table 4-8). If the estimates of annual air pollution control costs discussed above are at all accurate, it would have been possible to save $11 billion in 1990 alone through wiser air quality regulation, with aggregate emissions held constant. (By 2000, according to EPA, the 1970 amendments were costing $26 billion annually while the 1990 amendments were generating an addition $19 billion; 25% of this is slightly more than $11 billion). The possibility of saving society this much money *without compromising air quality* is what attracts economists to alternative forms of air quality regulation.

Recent unpublished data supplied by EPA reinforce this conclusion. These data show variations in control costs among different sources of nitrogen oxides, VOCs, particulate matter, and ammonia that range from zero dollars per ton removed to as much as $4 million for VOCs, $10,000 for nitrogen oxides, $30,000 for particulate matter, and $55,000 for ammonia. These data give further credence to the view that pollution taxes or emissions trading could result in dramatic savings in pollution control costs if substituted for technology-based standards.

Other Issues in Air Quality Regulation

A handful of other issues in air quality regulation have quite interesting economic ramifications but do not fit neatly within the topics discussed above. A brief discussion of each concludes this chapter.

Role of Costs in Setting Air Quality Standards.

Above, I made it quite clear that the Clean Air Act as amended in 1970 has been consistently read and interpreted as directing the EPA administrator to set NAAQSs on the basis of health evidence alone. On several occasions in the past thirty years, business groups have challenged this interpretation, claiming that, surely, Congress intended for the administrator to consider also how much it would cost to comply with each of the possible alternative standards under consideration. Each time that EPA's decisionmaking was challenged in court, a court of appeals ruled, in effect, that if Congress had intended for costs to be considered, then that is what Section 109 of the CAA would have said. In other words, courts of appeals have ruled that silence on the issue of costs in setting air quality standards (as opposed, say, to emissions standards from individuals sources—where costs can be considered) is tantamount to a prohibition.

The Supreme Court has always refused to hear an appeal of the decisions of the lower courts on this issue. Until the year 2000, that is. The Supreme Court announced in mid-2000 that it would hear an appeal by EPA of the decision by an appeals court in 1999 that overturned EPA's attempt to revise the NAAQS for ozone and to establish a new standard for fine particulate matter. At about the same time the Supreme Court made this announcement, it dropped a bombshell, saying that the court would also consider whether costs could be one of the factors taken into account when setting air quality standards—something everyone thought had been decided once and for all a long time ago.

If the Supreme Court says that Section 109 of the Clean Air Act *can* be interpreted to mean that costs may be considered in setting air quality standards, would this mean a revolutionary change in the nature of our air quality regulations? While many regulated parties hope and perhaps believe that the answer to this question is "yes," it seems unlikely to me. First, the Supreme Court would never conclude that Section 109 implies that the standards should be set on the basis of a benefit–cost analysis alone—there is simply no way to read Section 109 and arrive at this conclusion. The best that critics of the section as currently interpreted could hope for is a ruling that costs are one of the factors, along with health, ecological, and other adverse impacts, that may be considered by the EPA administrator in setting air quality standards.

Second, even if this victory is won, there is the question as to how costs would be balanced against other factors in standard setting—factors such as cases of premature mortality prevented, reduced incidence of acute and chronic illnesses, increased agricultural output, and so on. Imagine a scale that had perched, on one side, higher prices to consumers, some job losses due to regulation, and lower returns to all those owning stock in a regulated company and, on the other side, reduced risks of premature mortality and acute and chronic morbidity, not to mention improved visibility. It is by no means clear to which side that imaginary scale would tilt. The virtue of benefit–cost analysis is that, if carefully done, it can illuminate these trade-offs and even simplify them by expressing some (perhaps many) benefits and costs in a common metric—dollars.

But even when all benefits and costs are expressed in dollars terms, there are usually ranges—and sometimes wide ranges—around the individual elements and the totals. Thus, the EPA administrator could always say "I have looked at both sides of the ledger—with some costs and some benefits being expressed in dollar terms, and others not expressible in this form—and concluded that it is in the best interests of the country to set the standard at this level [whatever level was selected]." While it is unlikely to be the case that the administrator could justify *any* standard she or he wanted to choose, economic analysis has sufficient uncertainties and latitude that regulations might not be very different if the Supreme Court comes up with an interpretation of Section 109 that departs from previous decisions by lower courts.

One aspect of the bottom line, however, is easy to describe. Should the Supreme Court expand the role of costs in setting future NAAQSs, more money will be spent and more attention devoted to the costs—as well as the benefits—of air quality regulation in the years ahead. Because trade-offs will be allowed explicitly, moreover, there ought to be a healthy debate about the compromises that are and are not acceptable to the public.

At the same time, it should be acknowledged that there is a downside to a ruling by the Supreme Court that costs may be considered in standard setting: There will be much more litigation surrounding the establishment of the NAAQSs. Businesses will appeal final standards, suing on the grounds that costs were underestimated and, therefore, standards should be made less stringent. So, too, will environmental advocates appeal; they will sue on the grounds that EPA gave too much weight to economic impacts in its decisionmaking and too little weight to the health considerations that used to be the sole basis for standard setting. This litigation will drive up the cost of rulemaking and delay the implementation of whatever standards are eventually decided upon—not an appealing prospect!

On the other hand, it seems disingenuous to have a law that has been interpreted to prohibit costs from being considered in setting the NAAQSs when, in fact, virtually everyone knows that costs do—and should—get factored into decisionmaking anyway. Why not acknowledge openly that difficult trade-offs have to be made in setting air quality—and, indeed, all other environmental standards—and make those trade-offs open and explicit?

Impediments to Emissions Trading

In several places in this chapter, I pointed out how much money could be saved through regulatory programs that capped the total amount of a pollutant that could be emitted, allocated this total amount among existing sources (creating "rights" to discharge pollutants), and then allowed these rights to be bought and sold in open markets. As I point out in Chapter 2, sources holding rights might choose to sell them if they can reduce pollution inexpensively, while those finding it very inexpensive to cut emissions might be on the buyer's side of the market. It has long been argued that emissions trading programs have the potential to reduce the costs of pollution control considerably, while holding overall emissions to some desired level, by taking advantage of the variations in the marginal costs of control among sources. The success of the trading programs that have been put in place for sulfur dioxide and lead (see Chapter 3) proves that the theory translates well into the real world.

These programs have an Achilles heel, however, and opposition to emissions trading has begun to develop around it. In a nutshell, the issue is this:

even if *total* annual emissions of a particular pollutant are reduced, it is possible under a trading regime for emissions in one or more areas to increase. Suppose, for example, that when nationwide emissions of sulfur dioxide were reduced by 50% under the 1990 amendments, sources in one part of the country had bought up most of the allowances that were created (because they found it very expensive to reduce their emissions). Emissions in this area would have gone up even though the national total was falling. As one might imagine, such a situation would not be pleasing to those in the now more-polluted area.

I quickly note that this has *not* happened under sulfur dioxide trading and it would be unlikely to occur under trading programs for nitrogen dioxide, VOCs, or other air pollutants. If the reduction in overall emissions is significant, it is overwhelmingly likely that sources everywhere will reduce relative to their preregulation baseline—even if their reduction would be smaller under trading than it would be if every source had to reduce its emissions by the same percentage. But the very fact that an increase in local emissions in some places is a possibility is a source of great concern to some citizens, particularly minority and low-income groups that feel they regularly get the short end of the stick. (This concern has given rise to what is known as the environmental justice movement, an affiliation of individuals and groups with some political clout.)

How might the concerns of these citizens be addressed? One way is to observe that emissions trading under Title IV of the 1990 amendments to the CAA—the first really ambitious application anywhere in the world—resulted in significant reductions in emissions everywhere. No hot spots where emissions increased developed as a result of trading. This will almost always be the case for any application where the overall level of emissions is to be reduced by a significant amount.

If we wish to go a step further, however, it might be necessary in certain cases to guarantee that no areas will experience emissions increases *relative to their preprogram baselines.* The emphasized phrase is important. If we were to guarantee that no areas would have greater emissions under emissions trading than they would if all areas were required to meet the same percentage reductions, we would lose a great deal of the cost savings that emissions trading makes possible. But it should be possible to guarantee no net increase in emissions under a trading program without losing much or any of the efficiency gains that such a system would promise.

Another solution would be to focus on ambient concentrations on the pollutants being traded. Rather than guarantee that emissions will not be higher under trading than they would be under a more expensive regulatory approach, we could guarantee concerned citizens that air quality will not be allowed to violate some predetermined standards. Since air quality is overwhelmingly likely to improve as a result of cap-and-trade programs, this should not be a difficult condition to meet. Because of the very great savings in air pollution control costs that trading makes possible, it is essential that we think creatively about ways to meet the concerns of those who fear the implementation of such programs.

Long-Range Transport and Environmental Federalism

A persistent problem in air quality management in the United States has to do with the fact that, once emitted, air pollutants don't stand still. As might be expected, at least some pollutants are carried long distances by the wind, usually from west to east. (Certain pollutants, like carbon monoxide, are quite chemically reactive and hence dissipate before they can be transported elsewhere.) This means that one city, state, or region of the country may be regulating the air pollution sources within its boundaries quite stringently and yet still find itself in violation of one or more air quality standard because of the pollution "air mailed" to it by another jurisdiction.

Under the Clean Air Act, there has always been authority to deal with such problems, but this authority has seldom been used because it is difficult to pinpoint exactly who is responsible for which pollution violations. Downwind areas always claim to be doing as much as they can to control local sources of pollution while insisting that neighboring (or perhaps even distant) states are causing all their problems. Upwind states, quite predictably, say that this is just an excuse on the part of the "downwinders" to avoid imposing hard and costly pollution control measures on local industries and motorists. Both sides are occasionally right.

In the mid-1990s, and motivated by a concern about smog affecting large parts of the eastern United States, EPA initiated an ambitious and creative experiment. It created and empowered a large (thirty-seven states) organization called the Ozone Transport and Assessment Group (or OTAG) and charged the group with two important responsibilities. The first was to make use of the very best air pollution chemistry and modeling to determine the overall emissions reductions that were needed to address widespread violations of the ozone standard in the eastern United States. The second and much more challenging assignment was to forge recommendations about how the overall emissions reductions ought to be apportioned among the many participating states. The group was quite skillfully lead by the head of the Illinois state Environmental Protection Agency, and it invited participation by business and environmental groups, state and local environmental officials, academics, and interested citizens. OTAG concluded its deliberations and made recommendations to the U.S. EPA in 1998, which the latter then acted upon.

What's important here is not so much the specific recommendations made by OTAG as the precedent that may have been set. Increasingly, the air pollution problems we face in the United States are associated not so much with large, stationary sources (the exception being coal-fired power plants) as with smaller businesses (such as drycleaners and auto paint shops) and, especially, the cars that we drive. In other words, because we have concentrated for nearly thirty years on controlling emissions from steel mills, petroleum refineries, chemical plants, municipal incinerators, and other large sources, the remaining pollution comes increasingly from more disperse sources.

From a public policy standpoint, this is significant because the federal government may have lost whatever advantage it once may have had as the "regulator of first resort." In other words, it was probably appropriate for the federal government to have taken on controlling the large factories and other major sources of air pollution in the 1970s (though this could have been done in a much more efficient way than was chosen). But other issues—namely, dealing with smaller area sources, such as drycleaners, and figuring out how to reduce traffic congestion and take other measures to alleviate the car-related air pollution problems of metropolitan areas—are almost surely best left to individual cities, states, or regional authorities. While the federal government should still play an important role in this (mandating emissions standards for new cars, trucks and buses, for instance), more authority has to devolve to lower levels of government. And if this is so, then, *how much* regulatory authority ought to devolve to the states and metropolises? Currently, they are the principal enforcement arm of the federal EPA and do virtually all of the monitoring for both air quality and compliance with individual source discharge standards. Should they be given still more authority?

To take an extreme example, one pollutant for which EPA sets a uniform national air quality standard—carbon monoxide—does not travel long distances at all. Should states or metropolitan areas be allowed to decide for themselves what the maximum permissible ambient concentration of carbon monoxide should be? Perhaps not, since cars are the principal source of carbon monoxide and it would not make sense for automakers to have to meet fifty different state standards, much less a different standard for each and every urban area. What about pollutants that pose regional problems, such as the smog addressed by OTAG? Should we allow regions to determine not only how they will meet the national standard, but perhaps also elect to adopt a standard that is less strict than the current NAAQS? This is an extraordinarily controversial subject, one that is even hotter than the appropriate role for benefit–cost balancing in setting the NAAQSs. But as state environmental authorities become more sophisticated, it becomes possible to contemplate an evolution in air quality regulation in which regional, state, and municipal officials play a larger and more important role.

Regulation in the Information Age

Among the dramatic changes to the U.S. economy in the last thirty (or ten) years, services have become a much larger share of GDP—and manufacturing itself has altered. In industries such as pharmaceutical, microprocessor, and telecommunications equipment manufacturing, to name but a few, product cycles have gotten shorter and shorter. That is, going from the conception of an innovation, to its adoption by a company, to its manufacture and delivery to market (not to mention the time in which it becomes obsolete) often takes no more than eighteen months.

This has important albeit subtle implications for air quality regulation. It is currently the case that the process to permit a new manufacturing facility, or even modify the production process at an existing facility, can take years, especially if there exists local opposition. Needless to say, when a variation on a product will become obsolete within eighteen months, no one will bother applying for the permit change necessary to manufacture that product if the decision to grant the permit will take forever. While every EPA administrator has paid lip service to this problem, and while each has had his or her preferred fix ("one-stop-shopping" for all necessary permits has been a recurrent theme in these efforts), the problem still exists and may get worse because of new permitting requirements imposed by the 1990 amendments.

It is absolutely essential that we find a way to continue to protect air quality—using permits or some other mechanism—while at the same time ensuring that firms on the cutting edge of innovation are not hamstrung by the glacial pace of regulation. One possibility might be to exempt firms with outstanding environmental records from many if not all of permitting requirements, especially if the change they are seeking will have minimal effects on emissions. This would not only enhance their innovative potential, but would at the same time create a powerful incentive for them to excel in their environmental performance.

Notes

1. This section draws heavily on Stern 1982.
2. For an interesting history of motor vehicle pollution control, see White 1982.
3. The 1990 amendments made a change that created the appearance of a departure from this approach (see below). But it remains the case that all parts of the

country must stay focused on bringing air quality at least to the level of the NAAQS, if not better.

4. Even this step would not reduce risks to zero, because almost all the common air pollutants also are emitted from natural sources—for example, windblown dust, volcanoes, sea spray, and the decomposition of organic matter.

5. For an extended discussion of these standards, see White 1982, 15–19.

6. New sources that wish to locate in nonattainment areas also are required to offset any emissions remaining after achieving the lowest achievable emissions rate by securing reductions from sources already in the area.

7. EPA's Office of Research and Development is required to periodically review the clinical, toxicological, and epidemiological evidence concerning the health effects associated with the criteria air pollutants. These reviews are issued in the form of huge documents that purport to summarize and interpret the studies. A cursory reading of any of these reports ("criteria documents") will illustrate the ambiguities inherent in the studies' findings.

8. Given constant pollutant emissions in a particular area, ambient ozone concentrations can vary significantly, depending on the amount of sunlight and the strength of the prevailing winds. Similarly, particulate concentrations will be lower during rainy periods than during dry spells, even if particulate emissions are constant.

9. For more details about the emissions record of in-use vehicles over time, see White 1982, 29–36, 56.

References

Burtraw, Dallas, and Erin Mansur. 1999. Environmental Effects of SO_2 Trading and Banking. *Environmental Science and Technology* 33(20):3489–3494.

Davies, J. Clarence. 1970. *The Politics of Pollution.* New York: Pegasus Press.

Doll, R., and R. Peto. 1981. *The Causes of Cancer.* Oxford: Oxford University Press.

Freeman, A. Myrick, III. 1982. *Air and Water Pollution Control: A Benefit–Cost Assessment.* New York: Wiley.

Friedman, Robert D. 1981. *Sensitive Populations and Environmental Standards.* Washington, DC: The Conservation Foundation.

Goklany, Indur. 1999. *Clearing the Air: The Real Story of the War on Air Pollution.* Washington, DC: Cato Institute.

National Academy of Sciences. 1981. *On Prevention of Significant Deterioration.* Washington, DC: National Academy Press.

———. 1985. *Epidemiology and Air Pollution.* Washington, DC: National Academy Press.

Pope, C.A., and others. 1995. Particulate Air Pollution as a Predictor of Mortality in a Prospective Study of U.S. Adults. *American Journal of Critical Care Medicine* 151: 669–674.

Stern, Arthur C. 1982. History of Air Pollution Legislation in the United States. *Journal of the Air Pollution Control Association* 32(1): 44–61.

Tietenberg, T.H. 1985. *Emissions Trading: An Exercise in Reforming Pollution Policy.* Washington, DC: Resources for the Future.

U.N. (United Nations) Environment Program. 1990. *Environmental Data Report, Concentration of Suspended Particulate Matter (SPM) at Selected GEMS/Air Sites (1990 or latest).* http://www.nihs.go.jp/GINC/asia/gis/spm/spmworld.htm (accessed June 30, 2000).

U.S. EPA (Environmental Protection Agency). 1981. *1980 Ambient Assessment—Air Portion.* February. EPA-4501 4-81-014. Washington, DC: U.S. EPA

———. 1997. *The Benefits and Costs of the Clean Air Act, 1970 to 1990.* October. Washington, DC: U.S. EPA.

———. 1998. *National Air Quality and Emissions and Trend Report, 1996.* January. Research Triangle Park, NC: U.S. EPA, Office of Air Quality Planning and Standards.

———. 1999. *The Benefits and Costs of the Clean Air Act Amendments of 1990.* EPA 410-R-99-001. Washington, DC: U.S. EPA.

White, Lawrence J. 1982. *The Regulation of Air Pollutant Emissions from Motor Vehicles.* Washington, DC: American Enterprise Institute.

World Bank. 1998. *World Development Indicators.* Washington, DC: World Bank.

five

Climate Change Policy

Jason F. Shogren and Michael A. Toman*

Having risen from relative obscurity as few as ten years ago, climate change now looms large among environmental policy issues. Its scope is global; the potential environmental and economic impacts are ubiquitous; the potential restrictions on human choices touch the most basic goals of people in all nations; and the sheer scope of the potential response—a significant shift away from using fossil fuels as the primary energy source in the modern economy—is daunting. The magnitude of these changes has motivated experts the world over to study the natural and socioeconomic effects of climate change as well as policy options for slowing climate change and reducing its risks. The various options serve as fodder for often testy negotiations within and among nations about how and when to mitigate climate change, who should take action, and who should bear the costs.

In this chapter, we explore the economics of climate change policy. We examine the risks that climate change poses for society, the benefits of pro-

*The authors are very grateful to Larry Goulder for his extensive and constructive comments on an earlier draft of this paper, and to comments and advice provided by Sally Kane, Rob Stavins, Jonathan Wiener, and numerous other colleagues. Emily Aronow, Marina Cazorla, Sarah Cline, and Jennifer Lee provided very capable research assistance in the preparation of the paper, and Kay Murphy provided excellent assistance in the preparation of the manuscript. Special thanks are due to Joel Darmstadter for his work in compiling the data for Tables 5-1 and 5-2. As usual, we take full responsibility for the contents of the paper.

tection against the effects of climate change, and the costs of alternative protection policies. We organize our discussion around three broad themes: why costs and benefits matter in assessing climate change policies, as does the uncertainty surrounding them; why well-designed, cost-effective climate policies are essential in addressing the threat of climate change; and why a coherent architecture of international agreements is key to successful policy implementation.

We consider first the state of knowledge about climate change and its effects, and developments in international and U.S. domestic policy. Next, we elaborate the first theme, stressing the importance of considering costs and benefits as well as uncertainty. Elaborating on the second theme, we discuss how to design climate change policies. Then, we consider the challenges to securing effective and economically sound international responses. We conclude the chapter with a summary of key policy lessons and gaps in knowledge.

A Brief Overview of Climate Change

Scientific Background

Life on Earth is possible partly because some gases such as carbon dioxide (CO_2) and water vapor, which naturally occur in Earth's atmosphere, trap heat—like a greenhouse. CO_2 released from use of fossil fuels (coal, oil, and natural gas) is the most plentiful human-created greenhouse gas (GHG). Other gases—including methane (CH_4),[1] chlorofluorocarbons (CFCs; now banned) and their substitutes currently in use, and nitrous oxides associated with fertilizer use—are emitted in lower volumes than CO_2 but trap more heat. Global GHG inventories are hard to calculate reliably. Tables 5-1 and 5-2 list U.S. emissions sources and energy consumption levels, respectively, as of 1997. U.S. carbon emissions were roughly 25% of the global total in 1996 (U.S. EPA 1999).

Human-made GHGs work against us when they trap too much sunlight and block outward radiation. Scientists worry that the accumulation of these gases in the atmosphere has changed and will continue to change the climate. Potential climate risks include more severe weather patterns; hobbled ecosystems, with less biodiversity; changes in patterns of drought and flood, with less potable water; inundation of coastal areas from rising sea levels; and a greater spread of infectious diseases such as malaria, yellow fever, and cholera. On the plus side, climate change might benefit agriculture and forestry in certain locations by increasing productivity as a result of longer growing

TABLE 5-1. U.S. Sources of Greenhouse Gas Emissions, 1997.

	Carbon equivalents (million metric tons)			
Sector	CO_2	CH_4	*Other*	*Total*
Energy	1,466	58	29	1,553
Other	22	122	117	261
Total	1,488	180	146	1,814

Source: U.S. EPA 1999.

TABLE 5-2. U.S. Energy Consumption and CO_2 Emissions, 1997.

Source or emitter	*Energy consumption (quads)*	*CO_2 emissions (million tons of carbon)*
Energy source		
Fossil	80.5	1,466
Coal	20.9	533
Natural gas	22.6	319
Petroleum	37.0	613
Nonfossil	13.7	NA
Nuclear	6.7	NA
Hydro	3.9	NA
Other	3.1	NA
Total	94.2	1,466
Sector		
Fossil	80.5	1,466
Electric power	22.3	532
Industry	21.9	307
Transportation	24.7	446
Residential/commercial	10.9	168
Nonfossil	13.7	NA
Electric power	11.2	NA
Other	2.5	NA
Total	94.2	1,466

Notes: NA = Not applicable. CO_2 emissions include small unallocable amounts emitted in U.S. territories (not shown separately) and exclude small amounts attributable to nonfossil (biogenic) resources. For the purpose of this presentation, the electric power sector is treated as a consumer of energy sources and emitter of CO_2. An alternative treatment would bypass the power sector and ascribe its energy use to ultimate consumers of electricity.

Source: U.S. DOE 1998.

seasons and increased fertilization. Although climate change is not the same as day-to-day or even year-to-year fluctuations in the weather, the nature of these fluctuations could be altered by climate change.

Climate change is a historical fact, as illustrated by the ice ages. Part of the controversy today is the extent to which human activities are responsible for changes in the climate system. While acknowledging the many uncertainties about the precise nature and strength of the link between human activities and climate change, many scientists argue that the evidence points to an effect from people emitting too much CO_2 and other GHGs into the atmosphere.

Scientists reach this conclusion by looking at two trends. First, global surface temperature data show that Earth has warmed 0.5 °C (1 °F) over the past 100 years. At the same time, atmospheric concentrations of GHGs such as CO_2 have increased by about 30% over the past 200 years. Scientists attempt to capture the interactions of a complex dynamic climate system and human activities that put additional GHGs in the atmosphere by developing complicated computer models to simulate how future climate conditions might change with, for example, double the preindustrial concentration of GHGs in the atmosphere. Critics of these efforts stress that correlation and causation should not be confused. They also question the current ability to separate human-made changes from natural variability.

Although the causation between human actions and higher temperatures continues to be debated, the Intergovernmental Panel on Climate Change (IPPC) concluded in its Second Assessment Report that "the balance of evidence suggests that there is a discernible human influence on global climate" (IPCC 1996a). (This phrase has generated some controversy in its own right. The many uncertainties are characterized in Chapter 8 of the same report.) A recent report by the National Research Council (NRC 2000) found that evidence for a human contribution is rising. At the same time, however, the report found that scientists were becoming *less* confident in current quantitative forecasts of climate change.[2]

GHGs remain in the atmosphere for tens or hundreds of years. GHG concentrations reflect long-term emissions; changes in any one year's emissions have a trivial effect on current overall concentrations. Even significant reductions in emissions made today will not be evident in atmospheric concentrations for decades or more. In addition, the major GHG emitters change over time. The industrialized world currently accounts for the largest portion of emissions. However, by the middle of the twenty-first century, developing countries with growing population and wealth probably will generate the largest share of emissions. Both of these factors affect climate policy design.

Potential Physical and Socioeconomic Consequences

The risk of climate change depends on the physical and socioeconomic implications of a changing climate. Climate change might have several effects:

- Reduced productivity of natural resources that humans use or extract from the natural environment (for example, lower agricultural yields, smaller timber harvests, and scarcer water resources).
- Damage to human-built environments (for example, coastal flooding from rising sea levels, incursion of salt water into drinking water systems, and damages from increased storms and floods).
- Risks to life and limb (for example, more deaths from heat waves, storms, and contaminated water, and increased incidence of tropical diseases).
- Damage to less managed resources such as the natural conditions conducive to different landscapes, wilderness areas, natural habitats for scarce species, and biodiversity. For example, rising sea levels could inundate coastal wetlands, and increased inland aridity could destroy prairie wetlands.

All of these damages are posited to result from changes in long-term GHG concentrations in the atmosphere. Very rapid rates of climate change could exacerbate the damage. The adverse effects of climate change most likely will take decades or longer to materialize, however. Moreover, the odds that these events will come to pass are uncertain and not well understood. Numerical estimates of physical impacts are few, and confidence intervals are even harder to come by. The rise in sea level as a result of polar ice melting, for instance, is perhaps the best understood, and the current predicted range of change is still broad. For example, scenarios presented by the IPCC (1996a) indicate possible increases in sea level of less than 20 cm to almost 100 cm by 2100 as a result of a doubling of Earth's atmospheric GHG concentrations. The uncertainty in these estimates stems from not knowing how temperature will respond to increased GHG concentrations and how oceans and ice caps will respond to temperature change. The risks of catastrophic effects such as shifts in the Gulf Stream and the sudden collapse of polar ice caps are even harder to gauge.

Unknown physical risks are compounded by uncertain socioeconomic consequences. Cost estimates of potential impacts on market goods and services such as agricultural outputs can be made with some confidence, at least in developed countries. But cost estimates for nonmarket goods such as human and ecosystem health give rise to serious debate.

Moreover, existing estimates apply almost exclusively to industrial countries such as the United States. Less is known about the adverse socioeconomic consequences for poorer societies, even though these societies arguably are more vulnerable to climate change. Economic growth in developing countries presumably will lessen some of their vulnerability—for example, threats related to agricultural yields and basic sanitation services would decline. But economic growth in the long term could be imperiled in those regions whose economies depend on natural and ecological resources that would be adversely affected by climate change. Aggregate statistics mask considerable regional variation: some areas probably will benefit from climate change while others lose (IPCC 1998).

In weighing the consequences of climate change, it is important to keep in mind that humans adapt to risk to lower their losses. In general, the ability to adapt contributes to lowering the net risk of climate change more in situations where the human control over relevant natural systems and infrastructure is greater. Humans have more capacity to adapt in agricultural activities than in wilderness preservation, for example. The potential to adapt also depends on a society's wealth and the presence of various kinds of social infrastructure, such as educational and public health systems. As a result, richer countries probably will face less of a threat to human health from climate change than poorer societies that have less infrastructure. For additional discussion about adaptation possibilities, see Rosenberg 1993; Smith and others 1996; Pielke 1998; Sohngen and Mendelsohn 1999; Kane and Shogren 2000.

Policymakers must address the perceived risks of climate change in the population, not only the risks indicated in scientific assessments. So far, climate change does not appear to be that salient an issue in the minds of many Americans. Although the public's understanding of the issue seems to be increasing, the topic still is not well understood; and so far, no dramatic climate change event has hit the media to give the issue a permanent place at the front of public attention.[3] Even if climate change becomes a more prominent issue in the United States, people may disregard the issue because they believe that the probability that severe results will come to pass is very low, so immediate action on their part is not required. For further general discussion of risk perception issues, see Lichtenstein and others 1978; Camerer and Kunreuther 1989; Viscusi 1992; Crocker and Shogren 1997.

In constructing a viable and effective risk-reducing climate policy, policymakers must address hazy estimates of the risks, the benefits from taking action, and the potential for adaptation against the uncertain but also consequential cost of reducing GHGs. Costs of mitigation matter, as do costs of climate change itself. One must consider the consequences of committing resources to reducing climate change risks that could otherwise be used to meet other human interests, just as one must weigh the consequences of different climatic changes.

We now consider how policymakers worldwide and in the United States have responded in the policy domain so far.

A Chronology of Policy and Institutional Responses

International Developments

Figure 5-1 summarizes some milestones in the evolution of global climate policy. The negotiation of the 1992 United Nations Framework Convention on Climate Change (UNFCCC 1999a) was a watershed in that process. Article 2 of the convention states that the objective is to stabilize GHG concentrations within a time frame that would prevent "dangerous" human damage to the climate system. Article 3 states that precautionary risk reduction should be guided by equity across time and wealth levels, as expressed in the concept of "common but differentiated responsibilities." Article 4 states that nations should cooperate to improve human adaptation and mitigation of climate change through financial support and low-emission technologies. Articles 3 and 4 also refer to the use of cost-effective response measures. In concert with the Framework Convention, advanced industrialized countries pledged to reduce emissions to 1990 levels by 2000. However, this pledge was not a legally binding international agreement.

The 1997 Kyoto Protocol of the Framework Convention (UNFCCC 1999b) was the next major milestone. The protocol states that the industrialized "Annex B" countries (known in the 1992 convention document as "Annex I" countries) agreed to legally binding reductions in net GHG emissions that would average about 5% below 1990 levels by 2008–12.[4] This target was negotiated in a context wherein it was clear that almost all advanced industrialized countries would not achieve the pledged reduction of emissions to 1990 levels by 2000. Given expected business-as-usual emissions growth between 1990 and 2010, the actual emissions reductions needed for compliance are substantial (on the order of one-third of what otherwise would prevail in the United States, for example). No numerical targets for the emissions of developing countries were set in the protocol. In other words, the approach taken was "deep then broad"—a few countries are to make significant cuts early, with the hope of increased participation from other countries later—rather than the "broad then deep" strategy promoted by many critics of the Kyoto Protocol. For a critique of the Kyoto Protocol's deep-then-broad character, see Jacoby and others 1998; Shogren 1999.

The Kyoto Protocol includes several flexibility mechanisms that allow nations some latitude as to how they will meet the targets and timetables. The details of how these mechanisms would operate was largely left for

1979	■ First World Climate Conference
1990	■ First Assessment Report of the IPCC; initial evidence that human activities might be affecting climate, but significant uncertainty
1990	■ Second World Climate Conference; agreement to negotiate a "framework treaty"
1992	■ UNFCCC established at the UNCED (also known as the Earth Summit) in Rio de Janeiro, Brazil
	■ Annex I developed countries pledge to return emissions to 1990 levels by 2000
	■ United States ratifies UNFCCC later in the year
1993	■ Clinton administration publishes its Climate Change Action Plan, a collection of largely voluntary emission-reduction programs
1995	■ IPCC Second Assessment Report completed (published in 1996); stronger conviction expressed that human activities could be adversely affecting climate
1995	■ Berlin Mandate developed at the first COP (COP1) to the UNFCCC
	■ Agreement to negotiate *legally binding* targets and timetables to limit emissions in Annex I countries
1997	■ COP3 held in Kyoto Japan, leading to the Kyoto Protocol
	■ Annex I/Annex B countries agree to binding emission reductions averaging 5% below 1990 levels by 2008–12, with "flexibility mechanisms" (including emissions trading) for compliance; no commitments for emission limitation by developing countries
1997	■ U.S. Senate passes Byrd–Hagel resolution, 95 to 0, stating that the United States should accept no climate agreement that did not demand comparable sacrifices of all participants and calling for the administration to justify any proposed ratification of the Kyoto Protocol with analysis of benefits and costs
1998	■ COP4 held in Buenos Aires, Argentina; emphasis on operationalizing the "flexibility mechanisms" of the Kyoto Protocol
	■ IPCC Third Assessment begins
1999	■ COP5 held in Bonn, Germany; continued emphasis on operationalizing the flexibility mechanisms

FIGURE 5-1. Summary of Key Milestones in Climate Policy, 1979–99.

Notes: IPCC = Intergovernmental Panel on Climate Change; UNFCCC = United Nations Framework Convention on Climate Change; UNCED = United Nations Conference on Environment and Development; COP = Conference of Parties.

Source: UNFCCC 1999c.

future negotiations. Individual Annex B countries are free to achieve their targets through any credible domestic policies they wish—domestic policies need not be coordinated. The protocol also provides for international "where flexibility" in which nations can reduce emissions through different forms of international trading of emissions quotas. The protocol further provides flexibility in that emissions targets can be met by reducing any of six different gases—not only CO_2—and via carbon sequestration through sinks such as forests. Non-CO_2 gas concentrations are compared with CO_2 concentrations by means of global warming potential equivalency factors that reflect the heat-trapping properties of different gases in the atmosphere.[5]

Post–Kyoto Protocol meetings continued the international debate, especially about the technical, legal, and moral foundations of the proposed flexibility mechanisms. This debate revealed sharp differences in opinion between the United States and some other industrialized countries versus the European Union and many developing countries. At issue was the extent to which reliance on international emissions trading could substitute for, or could only complement, domestic efforts to reduce energy use and CO_2 emissions. Although differences are beginning to be worked out, the ultimate fate of the flexibility mechanisms—and of the protocol itself—remains to be seen.

U.S. Developments

A few key events trace out the broad outlines of U.S. climate change policy debates from the mid-1980s to the mid-1990s. The Reagan and Bush administrations were skeptical about the need for substantial reductions in carbon emissions. They advocated "actions which will have broad-ranging benefits" such as eliminating CFCs and other stratospheric ozone–depleting substances that are also GHGs (under the Montreal Protocol); implementing various pollution-control measures that also would promote energy efficiency (under the Clean Air Act); increasing forest sinks; encouraging energy efficiency in buildings, appliances, and lighting; and increasing the use of renewable and nonfossil sources of energy.[6]

The Clinton administration was more enthusiastic about GHG control policies, at least in rhetoric. They initially embraced voluntary technology-based measures promulgated in the 1993 Climate Change Action Plan (Clinton and Gore 1993). This approach was based on a firm conviction that substantial progress toward reducing GHG emissions could be achieved without adverse economic consequences; to the contrary, the administration touted the economic benefits of cleaner and more climate-friendly technology.

However, by 1996, it was clear that the United States would fail to achieve the goal of reducing emissions to 1990 levels by 2000. Reasons cited for the failure included less program funding than anticipated from a Republican Congressional majority and, more important, overoptimistic goals and a misspecified baseline (for example, oil prices did not rise as projected).

In 1996, the administration announced its willingness to accept legally binding, long-term emissions reductions goals in the international negotiations without spelling out which goals or policies it would support. As negotiations proceeded toward the 1997 Kyoto Protocol, the U.S. Senate passed by a vote of 95 to 0 a nonbinding resolution offered by Senators Robert C. Byrd and Chuck Hagel in the summer of 1997. The Byrd–Hagel resolution stated that the United States should accept no climate agreement that did not demand comparable sacrifices of all participants. The resolution was stimulated by concern about the effects of a climate agreement on the U.S. economy, but it conflicted with the idea of common but differentiated responsibilities for developed and developing countries espoused by the Climate Convention (UNFCCC 1999a), which the United States already had ratified. The resolution also required the administration to provide an economic justification of any climate change policy regime—a demonstration that the prospective benefits were worth the costs.

From 1996 through 2000, the Clinton administration pursued a public campaign to increase awareness of climate change risks. It also engaged in efforts to estimate the costs to the U.S. economy of GHG limitations while repeating much of its earlier rhetoric about the economic benefits of GHG reduction. President Bill Clinton stated that the Kyoto Protocol would not be sent to the Senate for ratification until policymakers settled disputes about policies for flexibility in the means of compliance, costs of compliance, and "meaningful participation" by developing countries. Acrimonious debate erupted sporadically between the administration and Congress as well as among various nongovernmental stakeholders about budgetary priorities related to climate change and the consequences of climate policies for the U.S. economy.

Numerous studies were produced in and out of government to help fuel the controversy. Among the most noteworthy and controversial was a July 1998 report from the Clinton administration, prepared by the President's Council of Economic Advisers (CEA 1998). This report stated that under the most favorable policy circumstances, the effects on energy prices and the costs to the United States of meeting the Kyoto Protocol emissions target could be extremely small. It states that these costs are "likely to be modest if those reductions are undertaken in an efficient manner employing the [various international] flexibility measures for emissions trading."

By modest, the administration report means an annual GDP decrease of less than 0.5%—roughly, $10 billion dollars; no expected negative effect on the trade deficit; increases in gasoline prices of only about $0.05 a gallon; lower electricity rates; and no "significant aggregate employment effect."[7] A critical assumption in the administration scenario was a very high degree of success in implementing the Kyoto Protocol flexibility mechanisms, especially emissions trading with developing countries and the former Soviet Union. Essentially, a broad-and-deep baseline was built into the administration's cost estimates.

Critics labeled the report as unduly optimistic and out of step with mainstream economic analyses. These critics also savaged a study by a consortium of U.S. national laboratories (Interlaboratory Working Group 1997) in which it was concluded that large low-cost energy savings were possible in the United States. The critics claimed that the authors inadequately recognized the barriers to capturing these savings and the policies needed to reap them. Nevertheless, this idea continues to hold sway in the arguments of the Kyoto Protocol's advocates.

Economic studies that produced high cost estimates had their own set of critics. For example, a 1998 study by the Energy Information Administration (EIA 1998), which suggested that costs to meet the Kyoto Protocol with domestic policies would be very high, was criticized for assuming too little flexibility in energy use and for overstating the negative effects of energy price increases on the economy. A number of industry-sponsored studies that indicated high costs also were criticized, for similar reasons. These studies tended to show substantial costs to the United States even with efficient domestic emission control policies: The trade deficit would increase by tens of billions of dollars, gasoline prices would increase by more than one-third, electricity prices would nearly double, and millions of U.S. jobs would disappear.

Other observers found moderate costs; that is, the Kyoto Protocol would not be painless but would not destroy national economies either. To better understand the context of these estimates, we next examine the assessment of costs and benefits of GHG control.

Evaluating the Costs and Benefits of Climate Change Risk Mitigation

The Importance of Considering the Costs and Benefits of Policy Intervention[8]

Although uncertain, climate change risks are real and need to be better understood so as to avoid unwanted consequences. Many observers characterize responding to the risks of climate change as taking out insurance;

nations try to reduce the odds of adverse events occurring through mitigation and to reduce the severity of negative consequences by increasing the capacity for adaptation once climate change occurs. The insurance analogy underscores both the uncertainty that permeates how society and policymakers evaluate the issue and the need to respond to the risks in a timely way.

Responding effectively to climate change risks requires society to consider the potential costs and benefits of various actions as well as inaction. By costs we mean the opportunity costs of GHG mitigation or adaptation—what society must forgo to pursue climate policy. Benefits are the gains from reducing climate change risks by lowering emissions or by enhancing the capacity for adaptation. An assessment of benefits and costs gives policymakers information they need to make educated decisions in setting the stringency of a mitigation policy (for example, how much GHG abatement to undertake and when to do it) and deciding how much adaptation infrastructure to create.

It is important to consider the costs and the benefits of climate change policies because all resources—human, physical, and natural—are scarce. Policymakers must consider the benefits not obtained when resources are devoted to reducing climate change risks, just as they must consider the climate change risks incurred or avoided from different kinds and degrees of policy response. *Marginal* benefits and costs reveal the gain from an incremental investment of time, talent, and other resources into mitigating climate risks, and the other opportunities forgone by using these resources for climate change risk mitigation. It is not a question of *whether* to address climate change but *how much* to address it.

Critics object to a benefit–cost approach to climate change policy assessment on several grounds. Their arguments include the following:

- The damages due to climate change, and thus the benefits of climate policies to mitigate these damages, are uncertain and thus inherently difficult to quantify given the current state of knowledge. Climate change also could cause large-scale, irreversible effects that are hard to address in a simple benefit–cost framework. Therefore, the estimated benefits of action are biased downward.
- Climate mitigation costs are uncertain and could escalate rapidly from too-aggressive emission control policies. Proponents of this view are indicating a concern about the risk of underestimating mitigation costs.
- Climate change involves substantial equity issues—both among current societies and between current and future generations—that are questions of morality, not economic efficiency. Policymakers should be concerned with more than benefit–cost analysis in judging the merits of climate policies.

As these arguments indicate, some critics worry that economic benefit–cost analysis gives short shrift to the need for climate protection, whereas others are concerned that the results of the analysis will call for unwarranted expensive mitigation.

Both groups of critics have proposed alternative criteria for evaluating climate policies. These can be seen as different methods of weighing the benefits and costs of policies given uncertainties, risks of irreversibility, the desire to avoid risk, and distributional concerns. For example, under the "precautionary principle," which seeks to avoid "undue" harm to the climate system, cost considerations are absent or secondary. Typically, the idea is that climate change beyond a certain level simply involves too much risk, if one considers the distribution of benefits and costs over generations.

"Knee of the cost curve" analysis, in contrast, seeks to limit emission reductions to a point at which marginal costs increase rapidly. Benefit estimation is set aside in this approach because of uncertainty. The approach implicitly assumes that the marginal damages from climate change (which are the flip side of marginal benefits from climate change mitigation) do not increase much as climate change proceeds and that costs could escalate rapidly from a poor choice of emissions target.

The benefit–cost approach can address both uncertainty and irreversibility. We do not mean to imply that estimates in practice are always the best or that how one evaluates and acts on highly uncertain assessments will not be open to philosophical debate.[9] But it is fundamentally inaccurate to see analysis of economic benefits and costs from climate change policies as inherently biased because of uncertainty and irreversibility. Nor should benefit–cost analysis be seen as concerned only with market values accruing to developed countries. One great achievement in environmental economics over the past forty years has been a clear demonstration of the importance of nonmarket benefits, which include benefits related to the development aspirations of poorer countries. These values can be given importance equal to that of market values in policy debates.

Our advocacy that benefits and costs be *considered* when judging climate change policies does not mean we advocate a simple, one-dimensional benefit–cost test for climate change policies. In practice, decisionmakers can, will, and should bring to the fore important considerations about the equity and fairness of climate change policies across space and time. Decisionmakers also will bring their own judgments about the relevance, credibility, and robustness of benefit and cost information and about the appropriate degree of climate change and other risks that society should bear. Our argument in favor of considering both benefits and costs is that policy deliberations will be better informed if good economic analysis is provided. For additional discussion, see Toman 1998.

The alternative decision criteria advanced by critics also are problematic in practice. The definition of "undue" is usually heuristic or vague. The approach is equivalent to assuming a sharp spike, or peak, in damages caused by climate change beyond the proposed threshold. It may be the case, but not enough evidence yet exists to assume this property (let alone to indicate at what level of climate change such a spike would occur). With knee of the curve analysis, on the other hand, benefits are ignored so there is no assurance of a sound decision either.

Benefits and costs are unavoidable. How their impacts are assessed is what differentiates one approach from another. We maintain throughout this discussion that the assessment and weighing of costs and benefits is an inherent part of any policy decision.

Long-Term Equity and Fairness Issues

The fairness of climate change policies to today's societies and to future generations continues to be at the core of policy debates. These issues go beyond what economic benefit–cost analysis can resolve, though such analysis can help illustrate the possible distributional impacts of different climate policies. In this section we focus on intergenerational equity issues. Contemporaneous equity issues are addressed in a later section on the architecture of international agreements.

Advocates of more aggressive GHG abatement point to the potential adverse consequences of less aggressive abatement policies for the well-being of future generations as a moral rationale for their stance. They assert that conventional discounting—even at relatively low rates—may be inequitable to future generations by leaving them with unacceptable climate damages or high costs from the need to abate future emissions very quickly (see Howarth 1996; 1998). Critics also have argued that conventional discounting underestimates costs in the face of persistent income differences between rich and poor countries. Essentially, the argument is that because developing countries probably will not close the income gap over the next several decades, and because people in those countries attach higher incremental value to additional well-being than people in rich countries, the effective discount rate used to evaluate reductions in future damages from climate change should be lower than that applied to richer countries.

Supporters of the conventional approach to discounting on grounds of economic efficiency argue just as vehemently that any evaluation of costs and benefits over time that understates the opportunity cost of forgone investment is a bad bargain for future generations because it distorts the distribution of investment resources over time. These supporters of standard discounting also argue that future generations are likely in any event to be better off than the present generation is, casting doubt on the basic premise of the critics' concerns. For additional discussion of this view, see Schelling 1995; Weitzman 1999.

Experts attempting to address this complex mixture of issues increasingly recognize the need to distinguish principles of equity and efficiency, even though there is as yet no consensus on the practical implications for climate policy. We can start with the observation that anything society's decisionmakers do today—abating GHGs, investing in new seed varieties, expanding health and education facilities, and so on—should be evaluated in a way that reflects the real opportunity cost, that is, the options forgone both today and over the long term. This answer responds to the critics who fear a misallocation of investment resources if climate policies are not treated similarly to other uses of society's scarce resources.

Moreover, as Weitzman (1998, 1999) pointed out, long-term uncertainty about the future growth of the economy provides a rationale for low discount rates on grounds of efficiency, not equity. The basic argument is that if everything goes well in the future, then the economy will be productive, the rate of return on investment will remain high, and the opportunity cost of displacing investment with policy today likewise also will be high. However, if things do not go so well and the rate of return on capital is low because of climate change or some other phenomenon, then the opportunity cost of our current investment in climate change mitigation versus other activities also will be low.

But economic efficiency only means a lack of waste given some initial distribution of resources. Specifically how much climate change mitigation to undertake is a different question, one that refers to the distribution of resources across generations. The answer to this question depends on how concerned members of the current generation are about the future generally, how much they think climate change might imperil the well-being of their descendants, and the options at their disposal to mitigate unwelcome impacts on future generations. For example, one could be very concerned about the well-being of the future but also believe that other investments—such as health and education—would do more to enhance the well-being of future generations. Not surprisingly, experts and policymakers do not agree on these points. For additional discussion, see the papers in Portney and Weyant 1999.

Estimating Benefits and Costs: Integrated Assessment Models

Analyzing the benefits and costs of climate change mitigation requires understanding biophysical and economic systems as well as the interactions between them. Integrated assessment (IA) modeling combines the key elements of biophysical and economic systems into one integrated system (Figure 5-2). IA models strip down the laws of nature and human behavior to their essentials to depict how more GHGs in the atmosphere raise temperature and how an increase in temperature induces economic losses. The models also contain enough detail about the drivers of energy use and interactions between energy and economy that one can determine the economic costs of different constraints on CO_2 emissions.[10]

Researchers use IA models to simulate a path of carbon reductions over time that would maximize the present value of avoided damages (that is, the benefits of a particular climate policy) less mitigation costs. As noted above, considerable controversy surrounds this procedure.

A striking finding of many IA models is the apparent desirability of imposing only limited GHG controls over the next 20 or 30 years. According to the estimates in most IA models, the costs of sharply reducing GHG concentrations today are too high relative to the modest benefits the reductions

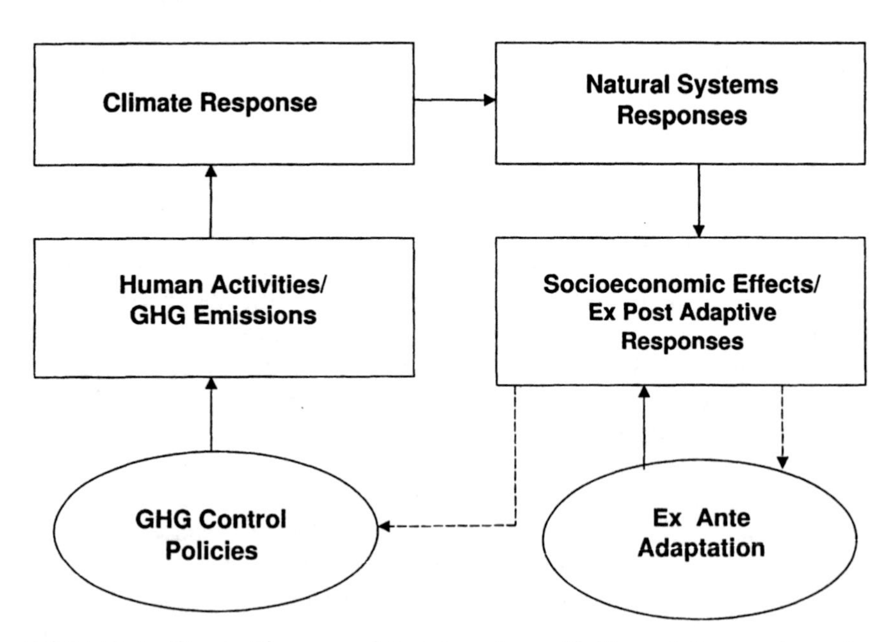

FIGURE 5-2. Climate Change and Its Interaction with Natural, Economic, and Social Processes.

Note: Key components of an integrated assessment model are illustrated. Solid lines represent physical changes; broken lines represent policy changes.

Source: Darmstadter and Toman 1993.

are projected to bring. The benefit of reducing GHG concentrations in the near term is estimated in many studies to be on the order of $5–25 per ton of carbon (see, for example, Nordhaus 1998; Tol 1999). Only after GHG concentrations have increased considerably do the impacts warrant more effort to taper off emissions, according to the models.

Even more striking is the finding of many IA models that emissions should rise well into this century (see Manne 1996). In comparison, the models indicate that policies pushing for substantial near-term control, such as the Kyoto Protocol, involve too much cost, too soon, relative to their projected benefits. Critics react to these findings along the lines noted earlier. Specifically, they argue that IA models inadequately address several important elements of climate change risks: uncertainty, irreversibility, and risk of catastrophe. Assessing the weight of these criticisms requires us to explore the influences on the economic benefits and costs of climate protection.

Influences on the Benefits

The IPCC Second Assessment Report (IPCC 1996b; 1996c) concluded that climate change could pose some serious risks. The IPCC presented results of studies showing that the damaging effects of a doubling of GHG concentrations in the atmosphere could cost on the order of 1.0–1.5% of GDP for developed countries and 2.0–9.0% of GDP for developing countries (Pearce and others 1996; see also Frankhauser and others 1998). Reducing such losses is the benefit of protecting against the negative effects of climate change.

Several factors affect the potential magnitude of the benefits. One is the potential scale and timing of damages avoided. Although IA models differ greatly in detail, most have economic damage representations calibrated to produce damages resulting from a doubling of atmospheric GHG concentrations roughly of the same order as the IPCC Second Assessment. This point is worth keeping in mind when evaluating the results. The models increasingly contain separate damage functions for different regions (see Nordhaus and Yang 1996; Tol 1995). Generally, the effects in developing countries are presumed to be worse than in the developed world, again as in the IPCC Second Assessment. For the most part, these costs would be incurred decades into the future. Consequently, the present value of the costs would be relatively low today.

Assumptions about adaptation also affect estimates of potential benefits. Some critics of the earlier IPCC estimates argue that damages likely will be lower than predicted because expected temperature increases from a doubling of atmospheric GHG concentrations probably will be less than projected, ecosystems seem to be more resilient over the long term than the estimates suggest, human beings can adapt more than was supposed, and damages are not likely to increase proportionally with GDP (see, for example, Mendelsohn 1999; Mendelsohn and Neumann 1999). The implication is that the optimal path for GHG control (in a present value sense) should be even less aggressive than the IA results indicate. These new assessments remain controversial.[11]

A third factor affects benefits: Damage costs not only are uncertain but also involve a chance of a catastrophe (see Cline 1992; Yohe 1993; Tol 1995; Pizer 1999; Roughgarden and Schneider 1999). However, a general finding from IA models is that GHG reductions should be gradual, even if damages are larger than conventionally assumed. A risk of catastrophe provides a rationale for more aggressive early actions to reduce GHG concentrations; however, the risk has to be very large to rationalize near-term actions as aggressive as those envisioned in the Kyoto Protocol in a present-value IA framework (see Peck and Teisberg 1993; 1996; Manne 1996; Gjerde and others 1999; Pizer 1999; a survey of experts in Nordhaus 1994b on climate change risks provides much of the grist for this ongoing debate). Part of the reason for this finding is that the outcome with the lowest cost also is the most likely to occur. IA models also do not incorporate risk-averse attitudes, which would provide a stronger rationale for avoiding large costs. Moreover, discounting in the models reduces the effective impact of all but the most catastrophic costs after a few decades.

Irreversibility of GHG emissions is yet another factor influencing the benefits of GHG abatement. Because GHG emissions persist in the atmosphere for decades, even centuries, the resulting long-term damages strengthen the rationale for early and aggressive GHG control (see Narain and Fisher 1999). Moreover, given that some damage costs from adjusting to a changed climate depend on the *rate* of climate change, immediate action also might be valuable. To date, however, the importance of this factor has not been conclusively demonstrated; the gradual abatement policies implied by the IA models do not seem likely to increase the speed of further climate change that much.

Finally, policies that reduce CO_2 also can yield ancillary benefits in terms of local environmental quality improvement—such as reduced human health threats or damage to water bodies from nitrogen deposition. The magnitudes of these ancillary effects remain fairly uncertain. They are lower to the extent that more environmental improvement would occur anyway, in the absence of GHG policy. They also depend on how GHG policies are implemented (for example, a new boiler performance mandate that encouraged extending the lives of older, dirtier boilers would detract from the environment). For additional discussion, see Burtraw and others 1999; Lutter and Shogren 1999.

Influences on the Costs

Estimates of the cost of mitigating GHG emissions vary widely. Some studies suggest that the United States could meet its Kyoto Protocol target at negligible cost; other studies claim that the United States would lose at least 1–2% of its GDP each year. A study by the Energy Modeling Forum helped explain the range of results in assessing the costs to meet the Kyoto Protocol policy targets (Weyant and Hill 1999). For example, the carbon price (carbon tax or emissions permit price) needed to achieve the Kyoto Protocol emissions target in the United States with domestic policies alone ranges from about $70 per metric ton of carbon to more than $400 per ton (in 1990 dollars) across the models. The corresponding GDP losses in 2010 range from less than 0.2% to 2.0% relative to baseline.[12] Carbon prices are put in perspective by relating them to prices for common forms of energy, as listed in Table 5-3.

The results reported by Weyant and Hill (1999) and previous assessments of GHG control costs (Hourcade and others 1996a; 1996b) reflect different views about three key assumptions that drive the estimated costs of climate policy: stringency of the abatement policy, flexibility of policy instruments, and possibilities for development and diffusion of new technology. First, as one would expect, the greater the degree of CO_2 reduction required (because the target is ambitious, baseline emissions are high, or both), the greater the cost.

Costs of GHG control depend on the speed of control as well as its scale. Wigley and others (1996) showed that most long-term target GHG concentrations could be achieved at substantially lower present value costs if abatement were increased gradually over time, rather than rapidly, as envisaged under the Kyoto Protocol. Subsequent elaboration of this idea has shown that, in principle, cost savings well in excess of 50% could be achieved by using a cost-effective strategy for meeting a long-term concentration target versus an alternative path that mandates more aggressive early reductions

TABLE 5-3. Implications of a Carbon Tax for Gasoline and Coal Prices.

	Price ($US)		
Commodity	*1997 U.S. average*	*With $100/ton carbon tax*	*With $400/ton carbon tax*
Bituminous coal	26.16	87.94	273.28
Motor gasoline	1.29	1.53	2.26

Notes: Coal price is national average annual delivered price per ton to electric utilities; gasoline price is national average annual retail price per gallon.

Sources: U.S. DOE 1999a,b.

(see Manne and Richels 1997). These cost savings come about not only because costs that come later are discounted more but also because less existing capital becomes obsolete prematurely. Kolstad (1996) points out that an irreversibility problem is associated with premature commitment to a form and scale of low-emissions capital, just as irreversibility is associated with climate change. The former irreversibility implies lower costs with a slower approach to mitigation.

Another important factor in assessing the costs of CO_2 control is the capacity and willingness of consumers and firms to substitute alternatives for existing high-carbon technologies. Substitution undertaken depends partly on the technological ease of substituting capital and technological inputs for energy inputs and partly on the cost of lower-carbon alternatives. Some engineering studies suggest that 20–25% of existing carbon emissions could be eliminated at low or negligible cost if people switched to new technologies such as compact fluorescent light bulbs, improved thermal insulation, efficient heating and cooling systems, and energy-efficient appliances (see IPCC 1996b, 1996c; NAS 1991; OTA 1991; Interlaboratory Working Group 1997). Economists counter that the choice of energy technology offers no free lunch. Even if new technologies are available, many people are unwilling to experiment with new devices at current prices. Factors other than energy efficiency also matter to consumers, such as quality, features, and the time and effort required to learn about a new technology and how it works. People behave as if their time horizons are short, perhaps reflecting their uncertainty about future energy prices and the reliability of the technology.

In addition, the unit cost of GHG control in the future may be lower than in the present, as a consequence of presumed continuation in trends toward greater energy efficiency in developed and developing countries (as well as some increased scarcity of fossil fuels). These trends will be enhanced by policies that provide economic incentives for GHG-reducing innovation. Kolstad (1996) argued that the cost associated with premature commitment to irreversible long-lived investments in low-emissions technologies is likely to be more important in practice than climatic irreversibility, at least over the medium term. The reason is that sunk investments cannot be undone if climate change turns out to be less serious than might be expected, whereas society can accelerate GHG control if it learns that the danger is greater than estimated. The strength of this point depends in part on how irreversible low-GHG investment is and on the costs of irreversible climate change (Narain and Fisher 1999).

Other analysts have argued that without early action to reduce GHG emissions, markets for low-emission technologies would not develop and societies would lock in to continued use of fossil fuel–intensive energy systems

(Grubb and others 1995; Grubb 1997; Ha-Duong and others 1997). When knowledge is gained through basic research and development (R&D), the optimal time path moves in the direction of maintaining current emissions levels and increasing future reductions to take advantage of accumulated knowledge (Goulder and Mathai 2000). However, when knowledge is gained through "learning by doing" there is a stronger case for earlier action.[13]

Still another important factor is the flexibility and cost-effectiveness of the policy instruments imposed, both domestically and internationally . For example, Weyant and Hill (1999) showed that the flexibility to pursue CO_2 reductions anywhere in the Annex I group of countries through some form of international emissions trading system could lower U.S. costs to meet the Kyoto Protocol target by roughly 30–50%. Less quantitative analysis has been done of alternative domestic policies. Nevertheless, it can be presumed from studies of the costs of abating other pollutants that cost-effective policies will lower the cost of GHG abatement, perhaps significantly. In contrast, constraints on the use of cost-effective policies—for example, the imposition of rigid technology mandates in lieu of more flexible performance standards—will raise costs, perhaps considerably.[14] This factor often is neglected in analyses of domestic abatement activity that consider only the use of cost-effective policies such as emissions permit trading, although use of such policies is hardly foreordained. Ignoring this factor means that the costs reported in the economic models probably understate the costs societies will actually incur in GHG control. By the same reason, studies of international policies that assume ideal conditions of implementation and compliance are overoptimistic.

A subtle but important influence on the cost of GHG control is whether emission-reducing policies also raise revenues (such as a carbon tax) and what is done with those revenues. When the revenue generated by a carbon tax or other policy is used to reduce other taxes, this revenue recycling offsets some of the negative effect on incomes and labor force participation of the increased cost of energy. However, it may be more effective at stimulating employment and economic activity in countries with chronically high unemployment than in the United States. The issue of revenue recycling applies also to policies that would reduce CO_2 through carbon permits or caps. If CO_2 permits are auctioned, then the revenues can be recycled through cuts in existing taxes; freely offered CO_2 permits do not allow the possibility of revenue recycling. The difference in net social costs of GHG control in the two cases can be dramatic. The analysis by Parry, Williams, and Goulder (1999) finds that reducing CO_2 emissions with auctioned permits and revenue recycling can have net costs less than the benefits of GHG control indicated by the IA models. In contrast, with a system of freely provided CO_2 permits, *any* level of emissions reduction yields environmental benefits (according to the IA models) that fall short of society's costs of abatement.

Most cost analyses presume that the relevant energy and technology markets work reasonably efficiently (other than the commonly recognized failure of private markets to provide for all the basic R&D that society wants, because this is a kind of public good). This assumption is more or less reasonable for most developed industrial economies. Even in these countries, one can identify problems such as direct and indirect energy subsidies that encourage excessive GHG emissions. Problems of market inefficiency are far more commonplace in the developing countries and in countries in transition toward market systems; accordingly, one expects incremental CO_2 control costs to be lower (even negative) in those countries (Jepma and others 1996; Lopez 1999). However, the institutional barriers to accomplishing GHG control in these economic systems may negate the potential efficiency gains.

Thus far, we have concentrated on CO_2 control. Because CO_2 is only one of several GHGs, and because CO_2 emissions can be sequestered or even eliminated by using certain technologies, emissions targets related to climate change can be met in several ways. Some recent analyses suggest that the costs of other options alternatives compare very favorably with the costs of CO_2 reduction. For example, counting the results of forest-based sequestration and the reduction of non-CO_2 gases toward total GHG reduction goals could lower the cost to the United States of meeting its Kyoto Protocol emissions target by roughly 60% (Reilly and others 1999). But care is needed in interpreting some of the cost estimates. In particular, low estimates for the cost of carbon sequestration may not adequately capture all the opportunity cost of different land uses (see Sedjo and others 1997; Newell and Stavins 2000).

Uncertainty, Learning, and the Value of New Information

Another key factor in choosing the timing and intensity of climate change mitigation is the opportunity to learn more about both the risks of climate change and the costs of mitigation. Several studies show that the value of more and better information about climate risks is substantial (see Manne and Richels 1992; Peck and Teisberg 1993; Chao 1995; Kolstad 1996). This value arises because one would like to avoid putting lots of resources into mitigation in the short term, only to find out later that the problems related to climate change are not serious. However, one also would like to minimize the risk of doing too little mitigation in the short term, only to find out later that very serious consequences of climate change will cost much more to avert because of the delay. Manne and Richels (1992) showed that it generally pays to do a little bit of abatement in the short run under these conditions—

to hedge against the downside without making too rapid a commitment. One virtue of some delay in emissions control is that it allows us to learn more about the severity of the risk of climate change and the options for responding to it. If the risk turns out to be worse than expected, mitigation can be accelerated to make up for lost time. To be sure, the strength of this argument depends on how costly it is to accelerate mitigation and on the degree of irreversibility of climate change. Analysts will continue to debate these points for some time to come.

Summary

In this section, we have explained that benefits and costs matter, for both efficiency and equity reasons, and that benefits and costs must and can be considered in the context of the uncertainties that surround climate change. Economic analyses provide several rationales for pursuing only gradual abatement of GHG emissions. Because damages accrue gradually, catastrophes are uncertain and off in the future, and unit mitigation costs are likely to fall over time (especially with well-designed climate policies), it makes sense to proceed gradually. To the extent that innovation is slower than desired with this approach, government programs targeted at basic R&D can help. The IA models indicate that rapid abatement does not maximize the present value of all society's resources.

We have not argued that current benefit–cost analyses are the last word on the subject. Opportunities certainly exist to improve the measurement of benefits and costs and to track the incidence of costs and risks across groups and over time. In practice, policy decisions will turn on a broader set of considerations than a single expected benefit–cost ratio. However, the arguments in favor of purposeful but gradual reduction in GHGs seem strong.

Whatever climate change policy goals are agreed upon, it makes sense to adopt cost-effective policies. Doing so means not committing excessive resources to meeting the climate policy goals, but preserving greater resources for other worthy goals, such as education and health. We discuss this topic in more detail in the next section.

Designing Climate Policy Instruments

Good climate change policies reflect the inherent trade-off between the stringency of a target (however defined) and the flexibility to meet this goal. Different policy tools can inflate or attenuate the costs of hitting any given target. Inflexible, inefficient policies will inflate costs without additional reductions in climate risk. Well-designed policies will lower the cost of achieving any particular targets and thereby make more stringent targets affordable. In this section, we emphasize economic policy tools that work cost-effectively by creating incentives for GHG mitigation while maintaining flexibility in the means used.

Creating Economic Incentives: Taxing and GHG Trading

Economic tools help cut the costs of achieving a GHG emissions target because they generate a market price for GHG emissions, which otherwise are treated as a free good. This price creates tangible financial reasons to reduce carbon emissions while providing flexible means to do so at low cost. Emissions taxes and GHG permit trading are economists' favorite incentives. Consumers respond to the price signals that these policies represent by switching to less-carbon-intensive fuels (for example, natural gas for coal); increasing energy efficiency per unit of output by using less-energy-intensive technologies; adopting technologies to reduce the emissions of other GHGs (assuming they are covered in the tax program); reducing the production of what become high-cost, carbon-intensive goods; increasing the sequestration of carbon through reforestation; and developing and refining new technologies (for example, renewable energy resources) for avoiding GHG emissions.

Carbon can be taxed indirectly by taxing fossil fuels. Taxing fossil fuels works because their carbon content is easily ascertained, and no viable option for end-of-pipe carbon abatement (for example, scrubbing) currently exists. A fossil fuel tax could be collected in several ways: as a severance tax on domestic fossil fuel output, plus an equal tax on imports; as a tax on primary energy inputs levied on refineries, gas transportation systems, and coal shippers; or as a tax downstream, on consumers of fossil fuels. However, the farther upstream the tax is levied (that is, closer to the producers of fossil fuels), the less carbon leaks out through uncovered activities such as oil field processing. Implementing such a tax would be relatively straightforward in the United States and most other developed countries, given existing tax collection systems, but more challenging in developing countries that have less effective institutions for levying taxes and monitoring behavior.

Carbon trading is somewhat more complicated than a carbon tax. One has to decide where to assign property rights for carbon: downstream, upstream, or some combination of the two. In principle, a downstream approach encompasses all emissions. In practice, however, all people in the United States who heat their homes with fossil fuel and/or drive a car would be required to buy and sell carbon permits. Operating and overseeing such a market would be an administrative nightmare. In contrast, an upstream system would be easier to administer because the number of market actors is smaller. Comprehensive policy would have to account for imported refined products as well as domestic fossil energy supplies and to address noncombustion uses of fossil fuels (for example, chemical feedstocks). One possibility is a system in which emissions of large sources are regulated directly and small sources are regulated through limits on their fossil fuel supplies. Or, a carbon tax could be levied on the energy used by smaller sources.

Questions about how to distribute permits also complicate carbon trading. A government could sell permits to the highest bidders in an auction-style system, hand them out gratis according to some formula such as grandfathering (that is, the government assigns permits to existing emitters relative to a historical base year), or combine the two approaches somehow. The choice forces policymakers to address trade-offs among goals of economic efficiency, distributional equity, and political feasibility. Efficiency increases with greater auctioning because the revenues can be used to offset existing distortionary taxes. Gratis permit allocation can target the distribution of a valued commodity toward the people most adversely affected by the policy (for example, low-income households and coal miners) or to those wielding the greatest political influence over the distribution of trading profits and losses. This option no doubt could increase the political feasibility of a trading policy. Bovenberg and Goulder (forthcoming) provide some simulation analyses that suggest that the cost of compensating fossil fuel–producing companies and their shareholders for losses resulting from reduced sales under a carbon-trading system is not very large. The cost increases, of course, if policy also seeks to compensate fossil fuel–intensive industries and the affected workers.

Which GHGs to address beyond CO_2 is another issue that both trading and taxation policies must address. For instance, the appropriate tax on natural gas entering the pipeline system could account for leakage and the greater relative potency of methane. Levies also could be placed on methane releases from coal mines and landfills and on human-manufactured gases on the basis of their expected venting to the atmosphere through sources such as automobile air conditioners. Some gases will be more difficult than others to control. A prime example is how to capture decentralized sources of agricultural methane that would be costly to measure.

Tax or trading systems also could be extended to carbon sequestration activities such as reforestation programs, which could earn tax credits or garner additional emissions permits. The challenge is to define a credible baseline to measure the amount of carbon sequestered by the forest. For example, one does not want a system that rewards carbon sequestration that would have occurred anyway as part of forest rotation practices, or a system that encourages deforestation so that landowners could then claim credit for replanting trees. For additional discussion about carbon sequestration, see Sedjo and others 1997.

Rules for banking and borrowing carbon permits are another key component of a trading system. Banking lowers costs by allowing traders to hedge against risks in emissions patterns (for example, a colder than average winter), and to smooth out fluctuations in abatement costs over time. With borrowing, traders have more flexibility to respond to unexpected short-term increases in abatement costs, thereby spreading the economic risk of compliance across time. The Kyoto Protocol provides a very limited amount of such flexibility by allowing Annex I countries to average their emissions over a five-year commitment period (2008–12).

Banking and borrowing raise the more fundamental issue of how to set credible long-term targets while facilitating short-term adjustments. In principle, policy could set a long-term GHG concentration target and let private actors reach the target most cost-effectively by adjusting their abatement strategies to minimize costs over time (see Kosobud and others 1994; Peck and Teisberg 1998). But critics doubt the credibility of such long-term targets. They also argue that a firm's natural tendency to delay emissions control to the future could impose unacceptable future climate change costs and make targets unenforceable (see Leiby and Rubin 2000).

This issue may be overwhelmed by the larger question: When should GHG reduction take place—now or later? As indicated earlier, many analyses suggest substantial cost savings from a more gradual path of emissions control than is envisaged in the Kyoto Protocol (however, some critics question how compelling these findings are). Even critics of intertemporal GHG trading need to address the possibility of policy targets with more flexibility over time and more gradual controls.

International Implementation of GHG Trading

GHG trading can be extended around the globe. Theory says that global trading can generate mutual gains by allowing low-cost abaters to profit from selling permits to grateful high-cost abaters. The Kyoto Protocol allows for both formal GHG trading among the Annex I developed countries and bilateral trading through the Clean Development Mechanism (CDM). Under the CDM, emissions reduction activities in noncapped, non-Annex I nations can generate emission reduction credits for Annex I nations. Annex I trading could involve tying together domestic emissions trading programs or a project-level approach in which participants can generate emission credits from emission-reducing actions in other Annex I countries (so-called joint implementation).[15] These various endeavors could be organized and financed by Annex I investors, the developing countries themselves, and international third parties.

The CDM could generate both low-cost emissions reductions for developed countries and tangible benefits to the host country through the transfer of efficient, low-carbon technology. However, many obstacles remain. The key immediate question is how to design a credible monitoring and enforcement system that does not impose such high transaction costs that it chokes off CDM trades. People will not start a project if the time, effort, and financial outlays needed to search out, negotiate, and obtain governmental approvals are too onerous. For additional discussion, see Goldemberg 1998; Grubb and others 1999; Jepma and van der Gaast 1999; Haites and Yamin 2000.

The United States has been a strong advocate for international trading in international negotiations. Other countries, notably in Western Europe and the developing world, have been cooler toward decentralized private-sector emission trading. Some nations like trading, but only if strict rules are imposed, which in a sense may ultimately be self-defeating. European negotiators have advocated trading limits, which restrict the degree to which Kyoto Protocol targets could be met through international flexibility mechanisms. Such "supplementarity" constraints (as they are termed in the debates) have been stoutly resisted by the United States for fear that they would unduly restrict opportunities for cost-effective emissions control and delay the evolution of effective GHG permit markets.

Trade-offs between Taxing and Trading GHGs

Choosing between taxing and trading GHGs requires consideration of several key trade-offs. First, a tax generates revenue for the government, which need not be the case with permit trading. Second, taxes fix the price and allow the emissions levels to vary, putting the risk on the environment, because the firms know the cost of emission reduction, whereas permits fix the emission target and allow the price to vary, putting the risk on the regulated firms, because the firm no longer knows the cost of a permit with certainty. As such, a permit system fits more naturally into the Kyoto Protocol (UNFCCC 1999b), which focuses on fixed emissions targets and timetables.

A downside to GHG trading is that society does not know what the actual abatement cost will be for a fixed quantity of emissions. When costs are uncertain and potentially severe, society may be better off with a tax-based approach that caps the cost of emissions control but does not ensure hitting a specific emissions target (Weitzman 1974; Pizer 1997; Newell and Pizer 1998). The exception would be cases in which a strong reason exists to limit GHG concentrations below a certain limit because of the risk of catastrophic damages. But no solid evidence exists at this time on which to base such a judgment. It is also possible to adopt a hybrid policy based on emissions trading but with a safety valve in case costs go too high. In practice, this policy would involve the government issuing additional permits if the price went beyond some predetermined level (which could change over time).[16]

Technology and Market Reform Policies

Incentive-based policies such as taxing and GHG trading work to encourage the diffusion of existing low-carbon technology and the development of new technology. They beg the question of whether additional nonprice policies are necessary to promote climate-friendly technology advances and investment. Proponents of such policies argue that economic incentives are inadequate to change behavior to a degree sufficient to reduce climate risk. They advocate public education and demonstration programs; institutional reforms, such as changes in building codes and utility regulations; and technology mandates, such as fuel economy standards for automobiles and the use of renewable energy sources for power generation.

No one doubts that such approaches eventually might reduce GHG emissions. At issue is the cost-effectiveness of such programs. Advocates of technology mandates often argue that the subsequent costs are negligible because the realized energy cost savings more than offset the initial investment costs. But as we noted earlier, this view does not address several factors that impinge on technology choices, and it implies a widespread lack of rational decisionmaking by energy users.

The economic perspective emphasizes searching for real inefficiencies that impede low-cost choices as opposed to barriers that reflect unavoidable direct or hidden costs, such as the capacity of technology to predictably meet the needs of its users. Most economic analysis recognizes that energy use suffers from inefficiencies but remains skeptical that such large no-regret gains actually exist. Economic analyses also acknowledge a role for government when consumers have inadequate access to information or if existing regulatory institutions are poorly designed. This role can include subsidies to basic R&D to compensate for an imperfect patent system, reform of energy sector regulation and reduction of subsidies that encourage uneconomic energy use, and provision of information about new technological opportunities. For additional discussion and competing perspectives on these issues, see Geller and Nadel 1994; Jaffe and Stavins 1994; Metcalf 1994; Levine and others 1995; Jaffe and others 1999.

Finally, in developing countries, barriers in the energy sector stall the diffusion of cost-effective technology. These barriers often are compounded by other economy-wide policy and infrastructure problems. When barriers to technology diffusion exist, the most effective solution typically is not found in regulatory mandates or ill-focused rules for technology adoption. Rather, solutions are found in institutional or broader market reforms, such as greater availability of information, expansion of financing opportunities, and reforms in energy sector pricing and other areas. For additional discussion, see Blackman 1997; Lopez 1999.

Coherent International Architecture Matters

Good domestic policy will be only as effective as the stability of an international agreement that defines a common purpose across nations. As such, the third factor that matters for climate change policy is a coherent international architecture. Speaking about the Rio negotiations, Prime Minister Gro Bruntland said, "We knew the basic principles on which we needed to build: cost-effectiveness, equity, joint implementation, and comprehensiveness. But not how to make them operational" (as quoted in Schmalensee 1996).

Because the source of the risk is widespread, responsibility for resolving the problem ultimately must be shared. But the more widespread the responsibility, the greater the challenge of maintaining a stable agreement, because nations have more incentive to free ride on the actions of other nations. This challenge is compounded by national differences related to income, vulnerability to climate change, and capacities to respond. Two related elements of cooperative and noncooperative economic behavior underlie the numerous intricacies of international diplomacy aiming to design and implement a climate agreement: the paradox of international agreements and the engagement of developing nations.

The Paradox of International Agreements

The problem of achieving effective and lasting agreements can be stated simply: A self-enforcing deal is easiest to close when the stakes are small, or when no other option exists (a clear and present risk). Nations have a common interest in responding to the risk of climate change, yet many are reluctant to reduce GHG emissions voluntarily. They hesitate because climate change is a global public good—no nation can be prevented from enjoying climate protection, regardless of whether it participates in a treaty. Each nation's incentive to reduce emissions is thus limited because it cannot be prevented from enjoying the fruits of other nations' efforts. This incentive to free ride reflects the divergence between national actions and global interests.

No global police organization exists to enforce an international climate agreement. As such, an agreement must be voluntary and self-enforcing—all sovereign parties must have no incentive to deviate unilaterally from the terms of the agreement. But a self-enforcing agreement is hardest to achieve in the gray area between low and infinite stakes. By free riding, some nations can be better off refusing an agreement. The greater the global net benefits of cooperation, the stronger the incentive to free ride; therefore, a self-enforcing agreement is harder to maintain. A self-enforcing agreement is most easily maintained when the global net benefits are not much bigger than no agreement—hence, the paradox. For more discussion, see Hoel 1992; Carraro and Siniscalco 1993; Barrett 1994; and Bac 1996.

If self-enforcement is insufficient, signatories who have ongoing relationships can try to alleviate free riding on climate change policy by retaliating with threats such as trade sanctions (see Chen 1997). But the force of linkage and deterrence is blunted in several respects. A nation's incentive not to participate in reducing GHG emissions depends on the balance between short-term gains from abstaining relative to the long-term cost related to punishment. Participating nations must see a gain in actually applying punishment, otherwise their threats of retaliation will not be credible. Credibility problems arise when, for example, retaliation through trade sanctions damages both the enforcer and the free rider. Moreover, because many forms of sanctions exist, nations would need to select a mutually agreeable set of approaches—probably another involved negotiation process. For an illustration, see Dockner and Van Long 1993.

Even if a self-enforcing agreement involved only two or three big emitting markets (for example, the United States and the European Union) and many small nations refused to agree, total emissions probably would remain higher than global targets. For their part, many decisionmakers in industrialized countries worry about the consequences to their economies of reducing emissions while developing countries face no limits. This situation could adversely affect comparative advantages in the industrialized world, whereas leakage of emissions from controlled to uncontrolled countries would limit the environmental effectiveness of a partial agreement. Estimates of this carbon leakage vary from a few percent to more than one-third of the Annex B reductions, depending on model assumptions regarding substitutability of different countries' outputs and other factors (Weyant and Hill 1999).

Designing Climate Agreements to Draw In Developing Nations

Developing nations have many pressing needs, such as potable water and stable food supplies, and less financial and technical capacity than rich countries to mitigate or adapt to climate change. These nations have less incentive to agree to a policy that they see as imposing unacceptable costs. The international policy objective is obvious, but elusive: finding incentives to motivate nations with strong and diverse self-interests to move voluntarily toward a collective goal of reduced GHG emissions.

Equity is a central element of this issue, because differences in perceptions about what constitutes equitable distributions of effort complicate any agreement. No standard exists for establishing the equity of any particular allocation of GHG control responsibility. Simple rules of thumb, such as allocating responsibility based on equal per capita rights to emit GHGs (advantageous to developing countries) and allocations that are positively correlated to past and current emissions (advantageous to developed countries) are unlikely to command broad political support internationally. The same problem arises with dynamic graduation formulas, which seek to gradually increase the control burden of developing countries as they progress economically. However, these dynamic approaches do offer more negotiating flexibility (see Burtraw and Toman 1992; Rose and Stevens 1993; Manne and Richels 1995; Schelling 1995; Rose and others 1998; Yang 1999).

Direct side payments through financial or low-cost technical assistance can increase the incentive to join the agreement. Incentive-based climate policies can help by reducing the cost of action for all countries. In particular, both buyers and sellers benefit from trade in emissions permits. Emissions trading also allows side payments through the international distribution of national emissions targets. More reluctant countries can be enticed to join with less stringent targets while other countries meet more stringent targets to achieve the same overall result. These points often are lost when critics argue that emissions trading will weaken international agreement because a seller country can fail to meet its domestic target and export "phony" emissions permits.

Side payments through emissions trading result when countries are given national quotas in excess of their expected emissions, an allocation sometimes called "headroom." Such an allocation was provided to Russia and Ukraine in the Kyoto Protocol and came to be called "hot air" by critics, who feared it would slow international progress by giving advanced industrial countries such as the United States a cheap way out of cutting their own emissions. But had this cost-reducing option not been part of the package, it is unclear whether the United States and other countries would have agreed to the protocol or could achieve its goals in practice (see Wiener 1999). Nevertheless, international reallocations of wealth in permit trading give rise to broader domestic political debates. Imagine, for instance, the domestic debate if the United States administration decided to transfer many billions of dollars annually to Russia, or perhaps China in a subsequent agreement, for emissions permits (see Victor and others 1998).

Critics claim that tradable permits have a Catch-22 that threatens the future of the Kyoto Protocol and longer-term agreements. Without trading, mitigation costs are too high to be politically acceptable; with trading, the distribution of these costs is too unfair to be politically acceptable. So, some observers promote individually administered national carbon taxes as the only reasonable option (Cooper 1998). However, this approach is not a panacea for distributional concerns, in that the initial allocation of rights and responsibilities is implicit in *any* international control agreement, including taxes. Moreover, the argument for taxes rests on the willingness of the developing world to implement substantially higher energy taxes than exists today. Although developing countries theoretically would reap some advantages of increasing energy taxes (for example, more reliable revenue than from income taxes), it is unclear whether the advantages are so compelling in practice. (Wiener 1999 offers several efficiency and political economy arguments in favor of a quantity-based over a tax-based approach.) Without such participation, the tax approach becomes an inefficient partial agreement like the Kyoto Protocol.

Concluding Remarks

Climate change poses risks to society. We have reviewed what researchers know about this risk and have discussed the benefits and costs of different protection strategies. Several lessons emerge that underscore the similarities and differences between climate change and other environmental issues. As with other issues—though perhaps to a greater degree in some cases—efficiency and complex equity issues must be addressed. Economic incentives are necessary for cost-effective and credible policies. Policymakers need to better understand the political and economic trade-offs between flexibility and stringency in the design of climate policies, and people need to recognize and account for the serious uncertainties that exist. In addition, international participation is necessary to effectively address climate issues, and significant challenges exist to establishing agreements that are substantial in their aims and credible in their implementation.

We also have identified several gaps between what the economics of climate change would tell us about policy and the actual direction of U.S. and international policy debates. In terms of the three themes we have followed through this chapter, economic analysis would suggest the following:

- Benefits and costs matter, as does uncertainty.
  - —There needs to be some balance of concern between the irreversible consequences of climate change and the costs of misplaced mitigation investment.
  - —A gradual approach to the implementation of GHG control targets to take advantage of cost savings and opportunities for learning is desirable.
- Well-designed, cost-effective climate policies are essential.
  - —Incentive-based mechanisms warrant a warm embrace, both domestically and internationally.
  - —A greater emphasis is needed on price-based approaches over strict quantity targets in the short to medium term to manage the risk of uncertain response costs.
  - —Targeted efforts to compensate the greatest losers with the least waste for political expediency should be undertaken.
  - —Climate policies should be coupled to broader economic reform opportunities to maximize win–win opportunities.
- Coherent international architecture is key to success.
  - —Serious discussion is needed of common ground for common but differentiated participation of developed and developing countries based on shared burdens and mutual benefit.

In practice, the policy debate has tended to emphasize climatological and socioeconomic damage risks over risks related to economic response costs. It has focused on strict and ambitious quantitative emissions targets for a subset of countries, without a clear path for implementing or broadening the agreement, and has been somewhat dismissive of those who see their own self-interest as not aligned with the emissions control targets being advocated. The debate has downplayed the importance of response costs by emphasizing potential "free lunch" opportunities in the technological arena, has sought to limit the operation of cost-reducing incentive-based systems (especially outside the United States), and has played up the tensions between developed and developing countries.

Some of these contrasts might be rationalized on the basis of factors such as strong aversion to risk of climate change, concerns about credible implementation of policies in practice, and honest disagreements in assessing the costs of GHG control. But a great deal of the difference, as we see it, reflects the politics of the issue. It remains to be seen how these political issues will play out, particularly whether the Kyoto Protocol or a successor agreement will be successfully ratified and implemented.

Finally, many uncertainties remain that affect climate policy design. To reduce the related uncertainties and improve the feasibility of future policy, policymakers need to better understand several points:

- the risks of climate change from a socioeconomic as well as a scientific perspective;
- the nature of public concern about climate change risks;
- the trade-offs between adaptation and emissions control, and the importance of different forms of infrastructure (especially in developing countries) for enhancing the capacity for adaptation;
- the costs of GHG control in an international context, accounting for trade and financial flows under different patterns of participation in international abatement efforts;
- how large the "energy efficiency gap" is in practice, and the consequences for assessing the cost of GHG abatement;
- the incentives for technical progress created by different climate policies, and the opportunity costs of inducing innovation toward GHG control versus other applications;
- the processes of international negotiation and coalition formation as they apply to climate agreements, in theory and practice; and
- the distributional impacts of different policy regimes.

Most of these questions will persist well beyond a third edition of this book. Starting to address them now can only increase the economic soundness and ultimately the reliability of climate change policy into the future.

Notes

1. Human-created sources of methane release include natural gas supply leaks, some coal mines, decomposition in landfills, and agricultural sources (for example, rice paddies and domestic animals).

2. Particularly vexing is the inability of models to better capture several factors: how climate change operates on a less than continental scale, in order to assess regional changes; how conventional pollutants such as very fine "aerosol" particles offset climate change by reflecting back sunlight; and how human activity on land can create "carbon sinks" to sequester greenhouse gases in biomass (for example, reforestation).

3. For additional discussion of these issues, see Morrisette and others 1991; Kempton and others 1995; Toman and others 1999; and Krosnick and others forthcoming. Climate change has been a great concern in western Europe; for a summary of positions taken, see Grubb and others 1999. In most developing countries, climate change must compete with more immediate pressing environmental and poverty-related concerns.

4. Annex I nations were named after a list in an appendix to the Framework Convention. The agreed-to targets varied across countries; the United States agreed to a 7% reduction, whereas western Europe undertook an overall cut of 8% (divided unequally among E.U. members in subsequent negotiations) and Japan accepted a less steep reduction of 6%. Special provisions were made in defining the obligations of the industrialized countries of central and eastern Europe and the former Soviet Union, the emissions of which already are below 1990 levels. Annex I countries are Australia, Austria, Belarus, Belgium, Bulgaria, Canada, Croatia, Czech Republic, Denmark, European Union, Estonia, Finland, France, Germany, Greece, Hungary, Iceland, Ireland, Italy, Japan, Latvia, Liechtenstein, Lithuania, Luxembourg, Monaco, Netherlands, New Zealand, Norway, Poland, Portugal, Romania, Russian Federation, Slovakia, Slovenia, Spain, Sweden, Switzerland, United Kingdom of Great Britain and Northern Ireland, and United States of America.

5. However, variations in long-term heat-trapping capacity do not immediately translate into variations in potential damage. For example, methane has a high heat-trapping potential but a shorter residence time in the atmosphere than CO_2; thus, if damages from climate change are growing over time because of greenhouse gas (GHG) accumulation generally, near-term methane releases will be less consequential than near-term CO_2 releases, whereas the opposite would be true if emissions occurred near the time of peak climate change and impacts (Reilly and Richards 1993; Schmalensee 1993; Hammitt and others 1996; Smith and Wigley 2000a, 2000b). Ideally, for policy purposes, different GHGs should be traded off against each other on the basis of their relative contribution to socioeconomic impacts, not only their chemical properties; however, there is no agreement on what damage-based equivalence factors should be.

6. We appreciate the assistance of Jonathan Wiener in identifying these measures. The broad-ranging benefits included, in the judgment of some proponents, decreased foreign oil imports and increased use of U.S. energy sources and technologies as well as environmental benefits. It is also useful to keep in mind that the United Nations Framework Convention was ratified in 1992 (UNFCCC 1999a), during the Bush Administration.

7. The pre–Kyoto Protocol results from the President's Interagency Analysis Team (IAT) are within this range as well (see Yellen 1998). One exception is that the IAT estimates that would reduce emissions to 1990 levels by 2010 would cost Americans 900,000 jobs by 2005 and 400,000 jobs by 2010.

8. For additional material and some contrasting views related to this discussion, see Heal and Chichilnisky 1993; Arrow and others 1996; Howarth 1996; Munasinghe and others 1996; Lind and Schuler 1998; and Portney 1998.

9. For example, as people become more informed about climate change, it is safe to presume that the importance they attach to the issue will change. Critics of the economic methodology argue that this process reflects in part a change in preferences through various social processes, not only a change in information. Moreover, in conditions of great uncertainty, the legitimacy of a policy decision may depend even more than normally on whether the processes used to determine it are deemed inclusive and fair, as well as on the substantive evidence for the decision.

10. For additional discussion of this modeling approach, see Nordhaus 1993; 1994a; Tol 1995; Peck and Teisberg 1996; Weyant and others 1996; Kolstad 1998.

11. One ongoing question concerns the cost of adjusting to a changing climate versus the long-term cost of a changed climate. Another is whether the effects of climate change (for example, in encouraging the spread of human illness through a greater incidence of tropical diseases, reducing river flows that concentrate pollutants, and increasing the incidence of heat stress) are being underestimated.

12. The percentages of GDP are not reported by Weyant and Hill (1999) but are inferred from graphs presented there.

13. Goulder and Schneider (1999) note that opportunity costs may be associated with inducing more technical innovation in greenhouse gas mitigation: To the extent that fewer research and development resources are made available in the economy as a whole for other innovation activities, productivity growth in the economy as a whole would be lower than otherwise.

14. More flexible approaches are more cost-effective when abatement costs in the economy are heterogeneous because such approaches allow more abatement to be carried out by those actors with the lowest costs, using the most effective technologies. Thus, for example, an abatement policy that relies in part on automobile fuel efficiency standards may impose excessive costs if the incremental cost of greenhouse gas abatement is lower in the power sector.

15. Hahn and Stavins (1999) describe the practical difficulties of operating a transaction-specific, credit-based joint implementation program internationally with heterogeneous domestic greenhouse gas measures. They point out the trade-off between international cost-effectiveness and domestic policy sovereignty that it engenders.

16. A version of this idea is sketched in Pizer (1997), and a policy for U.S. implementation is suggested in Kopp and others (1999). If permits are internationally traded, regulations would have to prevent entities in the United States from selling off all their "base" permits to trigger the safety valve.

References

Arrow, Kenneth J., and others. 1996. Intertemporal Equity, Discounting, and Economic Efficiency. In *Climate Change 1995: Economic and Social Dimensions of Climate Change*, edited by James P. Bruce, Horsang Lee, and Erik F. Haites. Contribution of Working Group III to the Second Assessment Report of the Intergovernmental Panel on Climate Change. New York: University of Cambridge Press, 125–44.

Bac, Mehnet. 1996. Incomplete Information and Incentives to Free Ride on International Environmental Resources. *Journal of Environmental Economics and Management* 30(3):301–15.

Barrett, Scott. 1994. Self-Enforcing International Environmental Agreements. *Oxford Economic Papers* 46:878–94.

Blackman, Allen. 1997. The Economics of Technology Diffusion: Implications for Greenhouse Gas Mitigation in Developing Countries. RFF Climate Issues Brief No. 5, October. Washington, DC: Resources for the Future.

Bovenberg, A. Lans, and Lawrence H. Goulder. Forthcoming. Neutralizing the Adverse Industry Impacts of CO_2 Abatement Policies: What Does It Cost? In *Behavioral and Distributional Effects of Environmental Policies: Evidence and Controversies*, edited by Carlo Carraro and Gilbert Metcalf. Chicago, IL: University of Chicago Press.

Burtraw, Dallas, and Michael A. Toman. 1992. Equity and International Agreements for CO_2 Containment. *Journal of Energy Engineering* 118(2):122–35.

Burtraw, Dallas, Alan Krupnick, Karen Palmer, Anthony Paul, Michael Toman, and Cary Bloyd. 1999. Ancillary Benefits of Reduced Air Pollution in the U.S. from Moderate Greenhouse Gas Mitigation Policies in the Electricity Sector. RFF Discussion Paper 99-51. September. Washington, DC: Resources for the Future.

Camerer, Colin F., and Howard Kunreuther. 1989. Decision Processes for Low Probability Events: Policy Implications. *Journal of Policy Analysis and Management* 8(4):565–92.

Carraro, Carlo, and Dominico Siniscalco. 1993. Strategies for the International Protection of the Environment. *Journal of Public Economics* 52(3):309–28.

Chao, Hung-po. 1995. Managing the Risk of Global Climate Catastrophe: An Uncertainty Analysis. *Risk Analysis* 15(1):69–78.

Chen, Zhiqi 1997. Negotiating an Agreement on Global Warming: A Theoretical Analysis, *Journal of Environmental Economics and Management* 32(2):170–88.

Cline, William R. 1992. *The Economics of Global Warming.* Washington, DC: Institute for International Economics.

Clinton, William J., and Al Gore. 1993. *Climate Change Action Plan.* Washington, DC: Executive Office of the President.

Cooper, Richard. 1998. Toward a Real Global Warming Treaty. *Foreign Affairs* 77(2, March/April):66–79.

Council of Economic Advisers. 1998. *The Kyoto Protocol and the President's Policies to Address Climate Change: Administration Economic Analysis.* July. Washington, DC: Executive Office of the President.

Crocker, Thomas D., and Jason F. Shogren. 1997. Endogenous Risk and Environmental Program Evaluation. In *Environmental Program Evaluation: A Primer*, edited by Gerrit Knaap and Tschangho J. Kim. Urbana, IL: University of Illinois Press, 255–69.

Darmstadter, Joel, and Michael A. Toman, eds. 1993. *Assessing Surprises and Nonlinearities in Greenhouse Warming: Proceedings of an Interdisciplinary Workshop.* Washington, DC: Resources for the Future.

Dockner, Englebert, and Ngo Van Long. 1993. International Pollution Control: Cooperative versus Noncooperative Strategies. *Journal of Environmental Economics and Management* 24:13–29.

EIA (Energy Information Administration). 1998. *Impacts of the Kyoto Protocol on U.S. Energy Markets and Economic Activity.* Washington, DC: EIA.

Frankhauser, Samuel, Richard S. J. Tol, and David W. Pearce. 1998. Extensions and Alternatives to Climate Change Impact Valuation: On the Critique of IPCC Working Group III's Impact Estimates. *Environment and Development Economics* 3(Part 1):59–81.

Geller, Howard, and Steven Nadel. 1994. Market Transformation Strategies to Promote End-Use Efficiency. *Annual Review of Energy and the Environment.* 19:301–46.

Gjerde, Jon, Sverre Greppud, and Snorre Kuendakk. 1999. Optimal Climate Policy under the Possibility of a Catastrophe. *Resource and Energy Economics* 21(3–4):289–317.

Goldemberg, Jose. 1998. *Issues and Options: The Clean Development Mechanism.* New York: United Nations Development Programme.

Goulder, Lawrence H., and Koshi Mathai. 2000. Optimal CO_2 Abatement in the Presence of Induced Technological Change. *Journal of Environmental Economics and Management* 39(1):1–38.

Goulder, Lawrence H., and Stephen H. Schneider. 1999. Induced Technological Change and the Attractiveness of CO_2 Abatement Policies. *Resource and Energy Economics* 21(3–4):211–53.

Grubb, Michael J. 1997. Technologies, Energy Systems and the Timing of CO_2 Emissions Abatement: An Overview of Economic Issues. *Energy Policy* 25(2):159–72.

Grubb, Michael J., T. Chapuis, and H. D. Minh. 1995. The Economics of Changing Course: Implications of Adaptability and Inertia for Optimal Climate Policy. *Energy Policy* 23(4/5):417–32.

Grubb, Michael J., Christiaan Vrolijk, and Duncan Brack. 1999. *The Kyoto Protocol: A Guide and Assessment.* London: Royal Institute of International Affairs.

Ha-Duong, M., Michael J. Grubb, and Jean-Charles Hourcade. 1997. Influence of Socioeconomic Inertia and Uncertainty on Optimal CO_2 Emission Abatement. *Nature* 390:270–73.

Hahn, Robert W., and Robert N. Stavins. 1999. What Has Kyoto Wrought? The Real Architecture of International Tradeable Permit Markets. March. RFF Discussion Paper 99-30. Washington, DC: Resources for the Future.

Haites, Erik, and F. Yamin. 2000. The Clean Development Mechanism: Proposals for Its Operation and Governance. *Global Environmental Change* forthcoming.

Hammitt, James K., Atul K. Jain, and John L. Adams. 1996. A Welfare-Based Index for Assessing Environmental Effects of Greenhouse Gas Emissions. *Nature* 381:301–3.

Heal, Geoffrey, and Graciela Chichilnisky. 1993. Global Environmental Risks. *Journal of Economic Perspectives* 7(4):65–86.

Hoel, Michael. 1992. International Environmental Conventions: The Case of Uniform Reductions of Emissions. *Environmental and Resource Economics* 2(2):141–59.

Hourcade, Jean-Charles, and others. 1996a. A Review of Mitigation Cost Studies. In *Climate Change 1995: Economic and Social Dimensions of Climate Change,* edited by James P. Bruce, Horsang Lee, and Erik F. Haites. Contribution of Working Group III to the Second Assessment Report of the Intergovernmental Panel on Climate Change. New York: University of Cambridge Press, 263–96.

Hourcade, Jean-Charles, and others. 1996b. A Review of Mitigation Cost Studies. In *Climate Change 1995: Economic and Social Dimensions of Climate Change,* edited by James P. Bruce, Horsang Lee, and Erik F. Haites. Contribution of Working Group III to the Second Assessment Report of the Intergovernmental Panel on Climate Change. New York: University of Cambridge Press, 297–306.

Howarth, Richard H. 1996. Climate Change and Overlapping Generations. *Contemporary Economic Policy* 14:100–111.

——. 1998. An Overlapping Generation Model of Climate–Economy Interactions. *Scandinavian Journal of Economics* 100(3):575–91.

IPCC (Intergovernmental Panel on Climate Change). 1996a. *Climate Change 1995: The Science of Climate Change.* Contribution of Working Group I to the Second Assessment Report of the Intergovernmental Panel on Climate Change. New York: Cambridge University Press.

——. 1996b. *Climate Change 1995: Impacts, Adaptations, and Mitigation of Climate Change: Scientific-Technical Analysis.* Contribution of Working Group II to the Second Assessment Report of the Intergovernmental Panel on Climate Change. New York: Cambridge University Press.

——. 1996c. *Climate Change 1995: Economic and Social Dimensions of Climate Change.* Contribution of Working Group III to the Second Assessment Report of the Intergovernmental Panel on Climate Change. New York: Cambridge University Press.

——. 1998. *The Regional Impacts of Climate Change: An Assessment of Vulnerability.* New York: Cambridge University Press.

Interlaboratory Working Group (IWG). 1997. *Scenarios of U.S. Carbon Reductions: Potential Impacts of Energy Technologies by 2010 and Beyond.* September. Report LBNL-40533 and ORNL-444. Berkeley, CA, and Oak Ridge, TN: Lawrence Berkeley National Laboratory and Oak Ridge National Laboratory.

Jacoby, Henry, Ronald Prinn, and Richard Schmalensee. 1998. Kyoto's Unfinished Business. *Foreign Affairs* 77(4, July/August):54–66.

Jaffe, Adam B., and Robert N. Stavins. 1994. The Energy-Efficiency Gap: What Does It Mean? *Energy Policy* 22(1):804–10.

Jaffe, Adam B., Richard G. Newell, and Robert N. Stavins. 1999. Energy-Efficient Technologies and Climate Change Policies: Issues and Evidence. December. RFF Climate Issue Brief No. 19. Washington, DC: Resources for the Future.

Jepma, Catrinus J., and Wytze van der Gaast, eds. 1999. *On the Compatibility of Flexibility Instrument.* Dordrecht, The Netherlands: Kluwer Academic Publishers.

Jepma, Cristinus J., and others. 1996. A Generic Assessment of Response Options. In *Climate Change 1995: Economic and Social Dimensions of Climate Change,* edited by James P. Bruce, Horsang Lee, and Erik F. Haites. Contribution of Working Group III to the Second Assessment Report of the Intergovernmental Panel on Climate Change. New York: University of Cambridge Press, 225–62.

Kane, Sally, and Jason Shogren. 2000. Adaptation and Mitigation in Climate Change Policy. *Climatic Change* forthcoming.

Kempton, Willett, James S. Boster, and Jennifer A. Hartley. 1995. *Environmental Values in American Culture.* Cambridge, MA: MIT Press.

Kolstad, Charles D. 1996. Learning and Stock Effects in Environmental Regulation: The Case of Greenhouse Gas Emissions. *Journal of Environmental Economics and Management* 31:1–18.

——. 1998. Integrated Assessment Modeling of Climate Change. In *Economics and Policy Issues in Climate Change,* edited by W. Nordhaus. Washington, DC: Resources for the Future, 263–86.

Kopp, Raymond J., Richard Morgenstern, William Pizer, and Michael Toman. 1999. A Proposal for Credible Early Action in U.S. Climate Policy (Feature). *Weathervane* (http://www.weathervane.rff.org/features/feature060.html; accessed March 18, 2000).

Kosobud, Richard, Thom Daly, David South, and Kevin Quinn. 1994. Tradable Cumulative CO_2 Permits and Global Warming Control. *Energy Journal* 15(2):213–32.

Krosnick, John A., Holbrook, A. L., and P. S. Visser. Forthcoming. The Impact of the Fall 1997 Debate about Global Warming on American Public Opinion. *Public Understanding of Science.*

Leiby, Paul., and Jonathan. Rubin. 2000. Bankable Permits for the Control of Stock Pollutants: The Greenhouse Gas Case. February 2. Draft paper. Orono, Maine: University of Maine.

Levine, Mark D., Jonathan G. Koomey, James E. McMahon, Alan H. Sanstad, and Eric Hirst 1995. Energy Efficiency Policy and Market Failures. In *Annual Review of Energy and the Environment* 19:535–55.

Lichtenstein, Sarah., and others. 1978. The Judged Frequency of Lethal Events. *Journal of Experimental Psychology* 4(6):551–78.

Lind, Robert C., and Richard E. Schuler. 1998. Equity and Discounting in Climate-Change Decisions. In *Economics and Policy Issues in Climate Change*, edited by William D. Nordhaus. Washington, DC: Resources for the Future, 59–96.

Lopez, Ramon. 1999. Incorporating Developing Countries into Global Efforts for Greenhouse Gas Reduction. January. RFF Climate Issue Brief 316. Washington, DC: Resources for the Future.

Lutter, Randall, and Jason Shogren. 1999. Reductions in Local Air Pollution Restrict the Gain from Global Carbon Emissions Trading. November 22. AEI Working Paper. Washington, DC: American Enterprise Institute.

Manne, Alan S. 1996. Hedging Strategies for Global Carbon Dioxide Abatement: A Summary of the Poll Results EMF 14 Subgroup—Analysis for Decisions under Uncertainty. In *Climate Change: Integrating Science, Economics, and Policy*, edited by Nebojsa Nakicenovic and others. Laxenburg, Austria: International Institute for Applied Systems Analysis.

Manne, Alan S., and Richard Richels. 1992. *Buying Greenhouse Insurance: The Economic Costs of* CO_2 *Emission Limits.* Cambridge, MA: The MIT Press.

——. 1995. The Greenhouse Debate: Economic Efficiency, Burden Sharing, and Hedging Strategies. *The Energy Journal* 16(4):1–37.

——. 1997. On Stabilizing CO_2 Concentrations—Cost-Effective Emission Reduction Strategies. *Environmental Modeling and Assessment* 2(4):251–265.

Mendelsohn, Robert. 1999. *The Greening of Global Warming.* Washington, DC: AEI Press.

Mendelsohn, Robert, and James E. Neumann, eds. 1999. *The Impact of Climate Change on the United States Economy.* Cambridge, U.K.: Cambridge University Press.

Metcalf, Gilbert E. 1994. Economics and Rational Conservation Policy. *Energy Policy* 22(10):819–25.

Morrisette, Peter M., Joel Darmstadter, Andrew J. Plantinga, and Michael A. Toman. 1991. Prospects for a Global Greenhouse Gas Accord. *Global Environmental Change* 1(3):209–23.

Munasinghe, Mohan, and others. 1996. Applicability of Techniques of Cost–Benefit Analysis to Climate Change. In *Climate Change 1995: Economic and Social Dimensions of Climate Change*, edited by James P. Bruce, Horsang Lee, and Erik

F. Haites. Contribution of Working Group III to the Second Assessment Report of the Intergovernmental Panel on Climate Change. New York: University of Cambridge Press, 145–77.

Narain, Urvashi, and Anthony Fisher. 1999. Irreversibility, Uncertainty, and Catastrophic Global Warming. Gianni Foundation Working Paper 843. Berkeley, CA: University of California, Department of Agricultural and Resource Economics.

NAS (National Academy of Sciences). 1991. *Policy Implications of Greenhouse Warming.* Washington, DC: National Academy Press.

Newell, Richard G., and William A. Pizer. 1998. Regulating Stock Externalities under Uncertainty. RFF Discussion Paper 99-10. Washington, DC: Resources for the Future.

Newell, Richard G., and Robert N. Stavins. 2000. Climate Change and Forest Sinks: Factors Affecting the Costs of Carbon Sequestration. *Journal of Environmental Economics and Management* forthcoming.

Nordhaus, William D. 1993. Rolling the "DICE": An Optimal Transition Path for Controlling Greenhouse Gases. *Resource and Energy Economics* 15(1):27–50.

——. 1994a. *Managing the Global Commons: The Economics of Climate Change.* Cambridge, MA: MIT Press.

——. 1994b. Expert Opinion on Climatic Change. *American Scientist* 82:45–51.

——. 1998. *Roll the DICE Again: The Economics of Global Warming.* December 18. New Haven, CT: Yale University.

Nordhaus, William D., and Zili Yang. 1996. A Regional Dynamic General-Equilibrium Model of Alternative Climate-Change Strategies. *American Economic Review* 86(4):741–65.

NRC (National Research Council). 2000. *Reconciling Observations of Global Temperature Change.* Washington, DC: National Academy Press.

OTA (Office of Technology Assessment). 1991. *Changing by Degrees: Steps to Reduce Greenhouse Gases.* OTA-0-482. Washington, DC: U.S. Government Printing Office.

Parry, Ian W. H., Roberton C. Williams III, and Lawrence H. Goulder 1999. When Can Carbon Abatement Policies Increase Welfare? The Fundamental Role of Distorted Factor Markets. *Journal of Environmental Economics and Management* 37:52–84.

Pearce, David, and others. 1996. The Social Cost of Climate Change: Greenhouse Damage and the Benefits of Control. In *Climate Change 1995: Economic and Social Dimensions of Climate Change,* edited by James P. Bruce, Horsang Lee, and Erik F. Haites. Cambridge, U.K.: Cambridge University Press.

Peck, Stephen C., and Thomas J. Teisberg. 1993. Global Warming Uncertainties and the Value of Information: An Analysis Using CETA. *Resource and Energy Economics* 15(1):71–97.

——. 1996. Uncertainty and the Value of Information with Stochastic Losses from Global Warming. *Risk Analysis* 16:227–35.

——. 1998. A Property Rights Approach to Climate Change Mitigation. Unpublished manuscript, June 16. Palo Alto, CA: Electric Power Research Institute.

Pielke, Roger A. Jr. 1998. Rethinking the Role of Adaptation in Climate Policy. *Global Environmental Change* 8(2):159–70.

Pizer, William A. 1997. Prices vs. Quantities Revisited: The Case of Climate Change. RFF Discussion Paper 98-02. Washington, DC: Resources for the Future.

——. 1999. The Optimal Choice of Climate Change Policy in the Presence of Uncertainty. *Resource and Energy Economics* 21(3–4):255–87.

Portney, Paul. 1998. Applicability of Cost–Benefit Analysis to Climate Change. In *Economics and Policy Issues in Climate Change*, edited by William D. Nordhaus. Washington, DC: Resources for the Future, 111–27.

Portney, Paul R., and John P. Weyant, eds. 1999. *Discounting and Intergenerational Equity.* Washington, DC: Resources for the Future.

Reilly, John M., and Kenneth R. Richards. 1993. Climate Change Damage and the Trace Gas Index Issue. *Environmental and Resource Economics* 3(1):41–61.

Reilly, John M., Ronald Prinn, J. Harrisch, J. Fitzmaurice, H. Jacoby, D. Kicklighter, J. Melillo, P. Stone, A. Sokolov, and C. Wang. 1999. Multi-Gas Assessment of the Kyoto Protocol. *Nature* 401:549–55.

Rose, Adam, and Brandt Stevens. 1993. The Efficiency and Equity of Marketable Permits for CO_2 Emissions. *Resource and Energy Economics* 15(1):117–46.

Rose, Adam, Brandt Stevens, Jae Edmonds, and Marshall Wise. 1998. International Equity and Differentiation in Global Warming Policy: An Application to Tradeable Emission Permits. *Environmental and Resource Economics* 12:25–51.

Rosenberg, Norman J. 1993. *Towards an Integrated Assessment of Climate Change: The MINK Study.* Boston, MA: Kluwer Academic Publishers.

Roughgarden, Tim, and Schneider, Stephen H. 1999. Climate Change Policy: Quantifying Uncertainties for Damages and Optimal Carbon Taxes. *Energy Policy* 27:415–29.

Schelling, Thomas C. 1995. Intergenerational Discounting. *Energy Policy* 23:395–401.

Schmalensee, Richard. 1993. Comparing Greenhouse Gases for Policy Purposes. *Energy Journal* 14(1):245–55.

——. 1996. Greenhouse Policy Architectures and Institutions. November 13. Report. Cambridge, MA: MIT Joint Program on the Science and Policy of Global Change.

Sedjo, Roger A., R. Neil Sampson, and Joe Wisniewski, eds. 1997. *Economics of Carbon Sequestration in Forestry.* New York: CRC Press.

Shogren, Jason. 1999. *The Benefits and Costs of Kyoto.* Washington, DC: American Enterprise Institute.

Smith, J. B., N. Bhatti, G. V. Menzhulin, R. Benioff, M. I. Budyko, M. Campos, B. Jallow, and F. Rijsberman, eds. 1996. *Adapting to Climate Change: Assessments and Issues.* New York: Springer-Verlag.

Smith, Steven. J., and Thomas M. L. Wigley. 2000a. Global Warming Potentials: 1. Climatic Implications of Emissions Reductions. *Climatic Change* 44:445–57.

——. 2000b. Global Warming Potentials: 2. Accuracy. *Climatic Change* 44:459–69.

Sohngen, Brent, and Robert Mendelsohn. 1999. The U.S. Timber Market Impacts of Climate Change. In *The Impacts of Climate Change on the UnitedStates Economy*, edited by Robert Mendelsohn and James E. Neumann. Cambridge, U.K.: Cambridge University Press, Chapter 5.

Tol, Richard S. J. 1995. The Damage Costs of Climate Change Toward More Comprehensive Calculations. *Environment and Resource Economics* 5:353–74.

——. 1999. The Marginal Costs of Greenhouse Gas Emissions. *Energy Journal* 20(1):61–81.

Toman, Michael A. 1998. Sustainable Decision-Making: The State of the Art from an Economics Perspective. In *Valuation and the Environment: Theory, Method and Practice*, edited by Martin O'Connor and Clive Spash. Northampton, MA: Edward Elgar, 59–72.

Toman, Michael A., Richard D. Morgenstern, and John Anderson. 1999. The Economics of "When" Flexibility in the Design of Greenhouse Gas Abatement Policies. *Annual Review of Energy and the Environment* 24:431–60.

UNFCCC (United Nations Framework Convention on Climate Change). 1999a. *Convention on Climate Change*. UNEP/IUC/99/2. Geneva, Switzerland: Published for the Climate Change Secretariat by the UNEP's Information Unit for Conventions (IUC). http://www.unfccc.de (accessed March 14, 2000).

——. 1999b. *The Kyoto Protocol to the Convention on Climate Change*. UNEP/IUC/99/10. France: Published by the Climate Change Secretariat with the Support of UNEP's Information Unit for Conventions (IUC). http://www.unfccc.de (accessed March 14, 2000).

——. 1999c. Guide to the Climate Change Negotiation Process. http://www.unfcc.de/resource.

U.S. DOE (Department of Energy). 1998. *Annual Energy Review 1997*. July. Washington, DC: U.S. DOE, Energy Information Administration.

——. 1999a. *Annual Energy Review 1998*. July. Washington, DC: U.S. DOE, Energy Information Administration.

——. 1999b. *Annual Energy Outlook 2000*. December. Washington, DC: U.S. Doe, Energy Information Administration.

U.S. EPA (Environmental Protection Agency). 1999. *Inventory of U.S. Greenhouse Gas Emissions and Sinks 1990–1997*. March. Washington, DC: U.S. EPA.

Victor, David G., Nebojsa Nakicenovic, and Nadejda Victor. 1998. The Kyoto Protocol Carbon Bubble: Implications for Russia, Ukraine, and Emissions Trading. Interim Report IR-98-094. Laxenburg, Austria: International Institute for Allied Systems Analysis.

Viscusi, W. Kip. 1992. *Fatal Tradeoffs*. Cambridge: Oxford University Press.

Weitzman, Martin L. 1974. Prices vs. Quantities. *Review of Economic Studies* 41(4):477–91.

——. 1998. Why the Far-Distant Future Should Be Discounted at Its Lowest Possible Rate. *Journal of Environmental Economics and Management* 36(3):201–8.

——. 1999. Just Keep Discounting, but…. In *Discounting and Intergenerational Equity*, edited by Paul R. Portney and John P. Weyant. Washington, DC: Resources for the Future.

Weyant, John P., and J. Hill. 1999. Introduction and Overview. *The Energy Journal* Special Issue: vii–xiiv.

Weyant, John P., and others. 1996. Integrated Assessment of Climate Change: An Overview and Comparison of Approaches and Results. In *Climate Change 1995: Economic and Social Dimensions of Climate Change*, edited by James P. Bruce, Horsang Lee, and Erik F. Haites. Contribution of Working Group III to the Second Assessment Report of the Intergovernmental Panel on Climate Change. New York: University of Cambridge Press, 366–439.

Wiener, Jonathan B. 1999. Global Environmental Regulation: Instrument Choice in Legal Context. *Yale Law Journal* 108(4):677–800.

Wigley, Thomas M. L., Richard Richels, and James A. Edmonds. 1996. Economic and Environmental Choices in the Stabilization of Atmospheric CO_2 Concentrations. *Nature* 379(6562):240–43.

Yang, Zili. 1999. Should the North Make Unilateral Technology Transfers to the South? North–South Cooperation and Conflicts in Responses to Global Climate Change. *Resource and Energy Economics* 21(1):67–87.

Yellen, Janet. 1998. The Economics of the Kyoto Protocol. March 5. Statement before the Committee on Agriculture, Nutrition, and Forestry, U.S. Senate.

Yohe, Gary W. 1993. Sorting Out Facts and Uncertainties in Economic Response to the Physical Effects of Global Climate Change. In *Assessing Surprises and Nonlinearities in Greenhouse Warming: Proceedings of an Interdisciplinary Workshop*, edited by Joel Darmstadter and Michael Toman. Washington, DC: Resources for the Future, 109–32.

six

Water Pollution Policy

A. Myrick Freeman III

The water in our oceans, lakes, rivers, and streams supports a wide range of uses. Water can be withdrawn for drinking and other domestic purposes, for industrial processes, or for irrigation. It can support fish populations for commercial exploitation and recreational fishing. It can be used for boating and swimming. And it can be used to flush away the wastes from factories and municipal sewers. To varying degrees, most of these uses depend on the quality of the water. Yet the use of a body of water as a waste receptor can seriously degrade water quality and impair—even preclude—other uses.

In 1969, the Cuyahoga River in Cleveland, Ohio, burst into flames, dramatizing the deplorable conditions that had come to characterize many of our nation's bodies of water. A subsequent Ralph Nader Task Force Report, *Water Wasteland*, published in 1971 (see Zwick and Benstock 1971), focused the nation's attention on these problems and may have helped to spur Congress to enact major revisions in federal water pollution law in 1972. These revisions are known as the Federal Water Pollution Control Act Amendments of 1972 (FWPCA-72). This law established goals with target dates for discharge reductions and improvements in water quality. It also called for the elimination of all discharges of pollutants into navigable waters by 1985. Congress adopted revisions to the law in 1977 and 1987.[1] The amended act is now known as the Clean Water Act (CWA).

For at least two reasons, now is a good time to take a new look at our federal water pollution control policies. First, we now have more than a decade of experience with the Clean Water Act since the last revisions of 1987. And second, the target dates in the act for achieving the water quality goals and the elimination of pollution discharges have now long since passed; thus, it is important to know what progress has been made toward these goals and whether the goals themselves provide a useful or meaningful basis for federal water pollution–control policy.

In this chapter, I first provide a brief review of the history and evolution of federal water pollution–control policy, highlighting some of the changes and concerns that have guided policy in the direction of greater federal responsibility for goal-setting, implementation, and financing. Second, I discuss the key features of FWPCA-72 that define the present approach to water pollution–control policy. I also discuss the revisions to this framework that have been adopted through amendments in 1977, 1981, and 1987. Third, I review, using several measures of performance, what has been accomplished in controlling discharges and improving the quality of our nation's waters, to the extent that these can be known.

Finally, I turn to an economic assessment of water pollution–control policy in which I compare benefits and costs and consider the problem of cost-effectiveness. Some of these economic principles are then applied to evaluate the past performance and possible modifications of the federal program for subsidizing construction of municipal sewage treatment plants and the evolving federal approach to the control of nonpoint-source pollution.

History and Evolution of Water Pollution Policy

The history of the federal government's involvement in controlling water pollution begins with the Refuse Act of 1899. This act, the purpose of which was to prevent impediments to navigation, stated that "it shall not be lawful to throw, discharge, or deposit … any refuse matter of any kind … into any navigable water of the United States" (quoted in National Commission on Water Quality 1976, 285–6). Summaries of the key features of this and subsequent federal laws dealing with water pollution–control policy are provided in Table 6-1. The Refuse Act is a useful reminder that at one time, the sheer physical volume of waste discharged into waterways—for example, sludge, fiber, and sawdust from paper mills and sawmills—threatened to block some rivers and channels.[2]

The first federal legislation to deal explicitly with conventional forms of water pollution was the Water Pollution Control Act of 1948.[3] This act authorized the federal government to engage in research, investigation, and surveys dealing with water pollution problems—activities not unlike those called for in the first federal legislation on air pollution control (see Chapter 4, Air Pollution Policy). The act also authorized the federal government to make loans to municipalities for the construction of municipal sewage treatment facilities. But there was no federal authority to establish water quality standards, limit discharges, or engage in any form of enforcement.

The Water Pollution Control Act Amendments of 1956 established a federal program of direct grants to municipalities to share in the costs of constructing sewage treatment facilities. This act also established the "enforcement conference" as a mechanism for imposing cleanup requirements on individual dischargers. If a serious water pollution problem were recognized in interstate waters, the Public Health Service could convene a meeting of state and local officials, major polluters, and other interested parties to make recommendations as to who should clean up and by how much. However, the combination of discretionary action on the part of the federal agency and reliance on consensus and volunteerism on the part of participants in the conferences meant that this act made little contribution to the establishment and enforcement of meaningful pollution control requirements on individual dischargers.

The Water Quality Act of 1965 was the first federal law to mandate state actions with respect to water pollution–control policy. Under the standard-setting provisions of the act, each state was required to draw up minimum water quality standards for the portions of interstate waters within its borders. Although the law was not specific on this point, the standard-setting requirement gave states the opportunity to weigh the benefits and costs of attaining different levels of water quality. After setting water quality standards, states were to determine the maximum discharges of various pollutants consistent with meeting these standards. The total allowable discharges then would be divided among the major sources of discharges on some basis. Dischargers would receive permits specifying the quantity of wastes they could legally discharge. Detecting and punishing violations of permits were also the responsibilities of the states.

For three reasons, this system of standard-setting and enforcement by states came to be viewed by many as ineffective and unworkable (see Zwick and Benstock 1971). The first was the difficulty in determining how much each individual source must cut back its discharges for the predetermined water quality standards to be met along an entire river basin. In principle, a water quality model could have been developed for each river basin and used to determine the set of discharge reductions that would meet the water quality standards at minimum aggregate pollution control cost (see Kneese and Bower 1968). But in practice, the development and use of such models is costly and time-consuming and was beyond the analytical capabilities of

TABLE 6-1. Federal Water Pollution Control Laws.

Title and year of enactment	*Key provisions*
The Refuse Act, 1899	*Goals:* Protection of navigation *Means:* Barred discharge or deposit of refuse matter in navigable waters without permit *Federal vs. state responsibility:* Federal permits and enforcement *Financing of municipal sewage treatment:* None
Water Pollution Control Act, 1948	*Goals:* Encouragement of water pollution control *Means:* Authority for federal research and investigation *Federal vs. state responsibility:* Left to state and local governments *Financing of municipal sewage treatment:* Authorized federal loans for construction, but no funds were appropriated
Water Pollution Control Act Amendments, 1956	*Goals:* Authorized states to establish water quality criteria *Means:* Federally sponsored enforcement conferences to negotiate cleanup plans *Federal vs. state responsibility:* Federal discretionary responsibility to initiate enforcement conferences for interstate waters *Financing of municipal sewage treatment:* Authorized federal grants to cover up to 55% of construction costs
Water Quality Act, 1965	*Goals:* Attainment of ambient water quality standards required to be established by states *Means:* State-established implementation plans placing limits on discharges from individual sources *Federal vs. state responsibility:* State responsibility for setting standards, developing implementation plans, and enforcement; federal oversight through approval and strengthened enforcement conference procedures *Financing of municipal sewage treatment:* No significant change
Federal Water Pollution Control Act, 1972	*Goals:* Fishable and swimmable waters *Means:* Enforcement of technology-based effluent standards on individual dischargers *Federal vs. state responsibility:* Federal responsibility for establishing effluent limits for categories of sources, and for issuing and enforcing terms of permits to individual dischargers; state option to take over responsibility for permits and enforcement *Financing of municipal sewage treatment:* Federal share increased to 75% and total authorization substantially increased ($18 billion over 3 years)

continued on next page

TABLE 6-1. (continued).

Title and year of enactment	*Key provisions*
Clean Water Act, 1977	*Goals:* Postponed some deadlines established in the 1972 act; increased control of toxic pollutants *Means:* No significant changes *Federal vs. state responsibility:* No significant changes *Financing of municipal sewage treatment:* No significant changes (authorizations for an additional $25.5 billion in federal grants over 6 years)
Municipal Wastewater Treatment Construction Grant Amendments, 1981	*Financing of municipal sewage treatment:* Reduced federal share to 55%, changed allocation priorities, and lowered authorizations to $2.4 billion per year for 4 years
Water Quality Act, 1987	*Goals:* Further postponement of deadlines for technology-based effluent standards *Financing of municipal sewage treatment:* Transition from federal grants to contributions to state revolving loan funds

Sources: Adapted from Kneese and Schultze 1975 (30, table 3.1) and Tietenberg 2000, with additional information provided by the author.

many state pollution control agencies. Hence, many state agencies applied simple rules of thumb such as requiring secondary treatment or its equivalent for pollutant discharges from all sources (see Tietenberg 2000).

A second reason involved the enforcement of individual discharge standards when water quality violations were detected. The problem was establishing a legally satisfactory way by which one or more dischargers could be held responsible for causing a violation of the standards. In other words, some means had to be implemented to determine who was at fault when water quality fell below the minimum acceptable level.

The third reason for viewing the system as unworkable was that the responsibility for enforcement was left primarily with the individual states. States varied enormously in their commitment to pollution control objectives; the talent of their personnel; the resources they could make available for implementation, monitoring, and enforcement; and their willingness to resist the temptation to compete for new industry by offering "friendly" regulatory environments.

Responding to these concerns, Congress passed a major revision to federal water pollution–control policy in 1972. FWPCA-72 established new federal goals and standards for water quality, set deadlines for cleanup actions, and created new procedures and mechanisms for regulation and enforcement. These revisions represented a major departure from the earlier approach in three main respects: goals, methods, and federal responsibility.

FWPCA-72 established a national goal for water pollution policy: the attainment of fishable and swimmable waters by July 1, 1983, and the elimination of all discharges of pollutants into navigable waters by 1985. The means selected for achieving this goal was a system of technology-based effluent standards to be established by the U.S. Environmental Protection Agency (EPA). These standards would define the maximum quantities of pollutants that each source would be allowed to discharge. By basing effluent standards strictly on technological factors (such as what kind of pollution abatement equipment was available) rather than water quality objectives, the 1972 amendments did away with the need for regulators to estimate the assimilative capacity of water bodies and the relationship between individual discharges and water quality. FWPCA-72 called for the same effluent standards to be applied to all dischargers within classes and categories of industries, rather than a plant-by-plant determination of acceptable discharges on the basis of water quality considerations.

The third major departure of FWPCA-72 from past policy was that, at least initially, the major responsibility for issuing permits to dischargers was shifted to a federal agency: EPA. All dischargers would have to hold permits and comply with their terms, which were to embody the technology-based effluent standards to be set by EPA. The law called for EPA to turn over the responsibility for issuing permits to individual states when these states met certain conditions.

While making major changes in the objectives and approach to enforcement, FWPCA-72 continued and strengthened the program of federal subsidies to cities and towns for treating municipal wastes. The 1972 amendments raised the federal share of treatment plant construction costs to 75% and substantially increased the amounts authorized to be spent.

The Clean Water Act of 1977 represented a midcourse correction to the path outlined in the 1972 act. Two major changes were embodied in the 1977 act. The first was postponing several deadlines established in the 1972 act for compliance with effluent standards by individual dischargers. The second was making a clearer distinction between the so-called conventional pollutants (organic matter and suspended solids, for example) that were the focus of FWPCA-72 and the so-called toxic water pollutants, about which concern had been mounting. The 1977 act also established new procedures and deadlines for determining effluent limitations for toxic pollutants and ensuring compliance with these limitations by individual dischargers.

The most recent changes in the law were made as part of the reauthorization of the Clean Water Act in early 1987. The Water Quality Act of 1987 was passed over President Ronald Reagan's veto as the first major item of business of the 100th Congress. The president's principal objection was to the continued high level of federal aid for the construction of municipal sewage treatment facilities. Congress authorized $9.6 billion in construction grants through fiscal year 1990, at which time the grant program ended. An additional $8.4 billion in federal contributions to state revolving loan funds was authorized through 1994. The funds were to support continued construction of sewage treatment plants after the end of the federal grant program. The 1987 act also further postponed the deadlines for compliance with effluent standards and established new requirements for states to develop and implement programs for controlling so-called nonpoint sources of pollution. Such sources include runoff from cultivated agricultural land, silvicultural activities, and urban areas.

Three major trends in the evolution of a national policy toward water pollution control are evident in this brief review of federal laws. The first is the shift in the focus of responsibility for setting goals, implementing policy, and enforcing pollution control requirements from the state and local levels to the federal government. Before 1948, there was no federal role in water pollution–control policy except for controlling impediments to navigation. But by 1972, the federal government had assumed the responsibility for setting pollution control objectives and for overseeing implementation and enforcement.

The second trend is the shift in financial responsibility for constructing municipal sewage treatment plants from state and local governments to a significant federal cost-sharing program. Municipalities have been constructing sewage treatment works throughout this century, with no help from the federal government. In 1956, the first federal construction grant program was authorized. And by 1972, the federal share of the cost had been increased to 75% of capital costs; authorizations were generous enough so that every municipal project that met minimum requirements could expect to receive its federal grant sooner or later. This trend was reversed somewhat by amendments in 1981 and 1987, but a significant federal role was ensured at least through 1994.

The third major trend is the change in the nature of the objectives of water pollution–control policy. In the early 1900s, the major justification for the construction of municipal sewage treatment facilities was to reduce the flow of disease-carrying human wastes into rivers and streams, where they could pose threats to human health through contamination of drinking water, shellfish beds, and other resources. But beginning at least with the 1965 act, federal policy reflected an increased concern with the protection of so-called in-stream uses of water bodies (such as swimming and boating) as well as ecological values. The water quality standards that were to be adopted under the 1965 Water Quality Act were meant to ensure water of sufficient quality to support various existing or desired uses of bodies of water, including boating, fishing, and swimming. The 1972 amendments made fishable and swimmable water a national policy objective. And the Clean Water Act of 1977 gave increasing emphasis to "the protection and propagation of a balanced population of shellfish, fish, and wildlife in the establishment of effluent limitations."[4]

FWPCA-72 and Its Amendments

Because of the importance of the changes brought about by the Federal Water Pollution Control Act of 1972, and because subsequent legislation has not significantly altered the basic goals and means of federal policy, I examine here in more detail the key features of the act and its amendments. Because FWPCA-72 takes quite different approaches to point-source pollution and nonpoint-source pollution, I conclude this section with a discussion of those provisions of the act that relate to nonpoint sources.

Goals

The first sentence of FWPCA-72 states that the goal of the act is to "restore and maintain the chemical, physical, and biological integrity of the nation's waters." However, none of the major implementation provisions of the act was designed specifically to achieve this ambitious goal. Two statements of operational objectives follow: the elimination of all discharge of pollutants into navigable waters by 1985 and the achievement of "fishable and swimmable" waters by 1983. The act makes no provision for considering the costs of attaining either of these objectives.

The act also retains the system of state-established standards for in-stream water quality that was enacted in 1965. But under FWPCA-72, the states are now required to review these standards at least every three years and submit any revisions to EPA for approval. The standards are to specify the uses of each water body that are to be protected and to establish maximum allowable concentrations of pollutants consistent with the designated uses. To be consistent with the objectives of the act, the minimum water quality standards must be sufficient to allow swimming and some types of fishing, but stricter standards are allowed—for example, standards that would protect pristine water quality in undeveloped rivers. Nothing in the act prohibits states from considering benefits and costs when they establish water quality standards.

In addition to these features, FWPCA-72 states "that the discharge of toxic pollutants in toxic amounts be prohibited" nationwide, that federal financial assistance be provided for the construction of municipal treatment plants, and that area-wide planning and technological innovation be encouraged.

Means

The principal means for attaining water quality objectives under FWPCA-72 are the establishment and enforcement of technology-based effluent standards. These standards are quantitative limits imposed on all dischargers where the quantities are determined by reference to present technology. To put it simply, standards are set on the basis of what can be done with available technology, rather than what needs to be done to achieve ambient water quality standards or to balance benefits and costs. Because production processes, quantities and composition of waste loads, and treatment technologies vary substantially across industries, separate discharge standards must be developed for the different categories of industry. In FWPCA-72, these standards are referred to as effluent limitations.

Technology-based standards in general must spell out the degree of technological sophistication to be embodied in a standard. The limitations could be based on present standard operating practice, best current practice, best available or demonstrated technology, and so forth. If the legislation governing the standards called only for technological considerations, it would be a purely technology-based system. But often, as in the case of FWPCA-72, technological requirements are tempered by economic considerations and qualifications such as "at reasonable cost." These two aspects, the technological and the economic, present a major difficulty in the technological approach to defining effluent standards. Although technology-based effluent standards may appear to be definitive and objective, in practice both the definition of the technology to be used and the economic qualifications that are usually attached are imprecise and ambiguous. "Best practical," "best available," and "reasonable cost" are not objective, scientific terms. Such language grants tremendous discretion to the officials charged with the development of technology-based standards and places a large responsibility on them to interpret a bewildering assortment of engineering, scientific, and economic information.[5]

Technology-based standards were to be achieved in two stages. By 1977, industrial dischargers were required to meet effluent limitations on the basis of the best practicable control technology (BPT). In determining what level was practicable, EPA was to consider the costs of the technologies relative to the benefits they were to achieve. By the same year, publicly owned treatment works were to provide secondary treatment (which typically removes 80–90% of the loadings of organic materials) to their effluents. The combination of these two 1977 requirements is often abbreviated BPT/ST. By 1983, effluent limitations for industrial dischargers were to be based on the best available technology economically achievable (BAT).

The drafters of this legislation recognized the possibility that pollution loads in some rivers might be so severe that fishable and swimmable water quality could not be attained even if dischargers met the mandated effluent limitations. For that reason, the act called for EPA to impose more stringent effluent limitations on point sources whenever necessary to achieve the desired water quality standards. In other words, the act requires better than best in such cases.

The Clean Water Act of 1977 maintained the basic structure of technology-based standards and effluent limitations. But this act provided for somewhat different criteria than FWPCA-72 for establishing effluent limitations for conventional pollutants (organic material, suspended solids, bacteria, and pH) as well as for toxic pollutants. These new effluent limitations were to be based on best conventional pollution technology (BCT). The Clean Water Act of 1977 also modified some of the deadlines for achieving stated effluent limitations.

Effluent limitations become the basis for discharge permits to be held by all dischargers. These permits limit the allowable discharges of individual polluters to the quantities that are consistent with the relevant technology-based effluent limitation. Enforcement of permit terms is based on two kinds of compliance monitoring: the evaluation of self-reported discharge data, and on-site inspections by government enforcement personnel.[6]

Nonpoint Sources of Pollution

Most of the major provisions of FWPCA-72 deal with the control of discharges from point sources—that is, from factories and municipal sewage systems. Yet urban and rural sources such as storm water runoff, cropland erosion, and runoff from construction sites, pastures, feed lots, and woodlands are also major sources of pollution in many water bodies.[7] Estimates of total discharges of five pollutants in 1972, before the implementation of FWPCA-72, are shown in Table 6-2. In aggregate, nonpoint sources

TABLE 6-2. Estimates of National Discharges from Point and Nonpoint Sources, 1972, before FWPCA (millions of pounds per year).

	Five-day biochemical oxygen demand	*Total suspended solids*	*Total dissolved solids*	*Total phosphorus*	*Total nitrogen*
Point sources					
Industrial	8,252	50,355	290,184	353	559
Municipal	5,800	6,000	31,847	101	1,111
Total	14,052	56,355	322,031	454	1,670
Nonpoint sources	18,901	3,422,321	1,536,458	2,986	12,480
National total	32,953	3,478,676	1,858,489	3,440	14,150
Nonpoint sources as a percentage of total discharges	57%	98%	83%	87%	88%

Source: Gianessi and Peskin 1981 (804).

accounted for between 57 and 98% of total national discharges of these substances. Of course, to the extent that effluent limits established in accordance with the law have reduced point-source discharges of these substances, the relative importance of nonpoint sources will have significantly increased since then.

The principal tool for dealing with nonpoint sources in FWPCA-72 is Section 208, which calls for the development and implementation of area-wide waste treatment management plans. Section 208 makes explicit reference to nonpoint sources and also authorizes federal grants to share in the cost of developing area-wide management plans. In 1977, the Clean Water Act Amendments authorized a program of grants, to be administered through the Soil Conservation Service of the Department of Agriculture, that would cover up to 50% of the costs to rural land owners of implementing and maintaining "best management practices" to control nonpoint-source pollution. The Water Quality Act of 1987 gave the states the responsibility for developing plans and specifying best management practices for controlling nonpoint-source pollution.

Accomplishments of the Programs since 1972

It is of great interest to know what has been accomplished as a result of all this legislative and regulatory activity and how well the machinery created by

FWPCA-72 and subsequent amendments has worked in controlling pollution and improving water quality. For two reasons, this question turns out to be a difficult one to answer with any degree of confidence.

The first reason has to do with the choice among the various measures of performance and accomplishment and with the availability of relevant data. For example, FWPCA-72 established a number of deadlines for the promulgation of effluent limitations, the issuance of permits, and other procedures. Thus, one indicator of accomplishment would be agency performance—that is, how well EPA has performed in meeting the deadlines. Another set of measures concerns the performance of polluters in complying with pollution control requirements. (Have the required treatment systems been installed on time and according to the terms of permits? And have actual discharges been within the limits established by permits?) However, neither agency nor polluter performance measures are directly related to the real objectives of the act: the reduction of pollution and the improvement of water quality. To evaluate them, it is necessary to examine what is known about changes in water quality.

The second reason concerns the difficulty of distinguishing between improvements that have resulted from implementation of the acts and improvements that have occurred for other reasons. The relevant question is not how much some measure of water quality has changed over time at some location; this is a before-versus-after question. The right question is a with-versus-without question: How much better was actual water quality in, say, 1997 than it would have been in that year without the cleanup requirements imposed by the 1972 act, but with the same economic conditions, weather, rainfall, and so forth? To answer it, we would have to be able to predict what the discharges of pollutants would have been in the absence of the 1972 act, but with the provisions of the 1965 Water Pollution Control Act still in place.

Two kinds of forces are at work that would produce different answers to the with-versus-without and before-versus-after questions. On one hand, economic growth and increases in industrial production likely would have led to increases in pollution discharges since 1972, in the absence of regulation. Thus, in the absence of the act, water quality might have declined since 1972. If so, the answer to a before-versus-after question could underestimate the accomplishments of the 1972 act. On the other hand, to the extent that state and federal agencies might have become more effective in using the machinery created in 1965, reductions in pollution and improvements in water quality might have occurred after 1972 even in the absence of new legislation. If so, attributing all of the observed improvements (or prevention of degradation) to the 1972 act might overstate its true accomplishments. These problems should be kept in mind as we interpret the data presented below.

Agency Performance

FWPCA-72 required that all sources of water pollution reduce their discharges so as to be in compliance with effluent limitations by specific dates: BPT effluent limitations by 1977, BAT limitations by 1983. Before sources could comply with these requirements, however, EPA had two major tasks to perform. The first was to write the effluent limitations for a variety of industrial and municipal sources.[8] The second was to translate the terms of these effluent limitation guidelines into specific requirements to be imposed on each source and to be spelled out in the permits issued to the sources. One measure of the accomplishments under the 1972 act, then, is how well the agency performed in writing the effluent limitation regulations and issuing permits to individual sources.

The 1972 act required that EPA promulgate effluent limitations as regulations within one year of the date of enactment. This requirement turned out to be unreasonable, and EPA had not issued a single regulation by the deadline. In fact, the agency still had not promulgated all of the BPT regulations by 1977, the deadline for sources to be in compliance with the regulations, and more than 250 court cases were challenging various provisions of those regulations that had been issued by then. At least partly in response to these problems, in 1977 Congress pushed back the deadline for complying with BAT requirements for a specified list of toxic pollutants to July 1, 1984, and allowed up to three additional years for compliance with BAT requirements for other toxic chemicals.

In regard to EPA's second task, EPA reported that as of March 31, 1976, it had issued permits to only 67% of all industrial dischargers. The percentage was higher for major industrial sources—more than 90% (Council on Environmental Quality 1976), but many of these permits were written before the applicable effluent limitations were promulgated! Some of the issued permits were inconsistent with the effluent limitations eventually adopted, which meant that some dischargers were in nominal compliance with the terms of their permits but not in compliance with the applicable technology-based effluent standards.

Dischargers' Performance

The second measure of accomplishment is the extent to which individual dischargers have complied with the terms of their relevant effluent limitations and discharge permits. It has been estimated that as of the 1977 deadline, about 80% of industrial dischargers were in compliance with the relevant BPT effluent limitations and that by 1981, 96% of industrial sources were in compliance (Council on Environmental Quality 1982, 86). The compliance rate with the secondary treatment requirement from municipal dischargers was substantially lower in both 1977 and 1981. It also has been estimated that full compliance with the BPT regulations by industry would result in approximate reductions of 65% in industrial discharges of oxygen-demanding organic material, 80% in industrial discharges of suspended solids, 21% in oil and grease discharges, and 52% in discharges of dissolved solids (Council on Environmental Quality 1982, 85–86). There are no comparable estimates of compliance rates or percentages of pollutant removed for the BCT and BAT effluent limitations.

The term "compliance" as used by EPA generally means the installation of treatment equipment capable of meeting the BPT effluent limitations when properly operated. These data on compliance do not say anything about actual discharges. To determine the degree of effective compliance, it is necessary to examine the discharges of polluters and to compare them with the terms of their permits and relevant effluent limitations.

The U.S. General Accounting Office (GAO) attempted to do this over an eighteen-month period in 1981–82. It had to rely on discharge data that had been supplied by the dischargers rather than on independently determined or verified discharge data. Nevertheless, the GAO study found a significant noncompliance problem. Discharge data for about one-third of all industrial and municipal dischargers in six states were examined, and results are summarized in Table 6-3. Eighty-two percent of the sources had at least one month of noncompliance during the eighteen-month period. Moreover, about 24% of the sample was in "significant noncompliance" for at least four consecutive months, during which discharges exceeded permitted levels by at least 50%. The performance of municipal sources was poorer than that of

TABLE 6-3. Self-reported Noncompliance with Discharge Permits in Six States, October 1, 1981–March 31, 1982.

Type of discharger	*Number of dischargers sampled*	*Percentage of total number of sources (%)*	*Noncompliance (% of sample)*[a]	*Significant noncompliance (% of sample)*[b]
Municipal	274	34	85	32
Industrial	257	36	79	16
Total	531	35	82	24

Note: The six states were Iowa, Louisiana, Missouri, New Jersey, New York, and Texas.

[a] At least one reported monthly average reading exceeded the permit limit during the eighteen-month period.

[b] At least four consecutive monthly average readings exceeded the permit limit by 50% or more.

Source: U.S. GAO 1983 (7–10).

industrial sources, especially in cases of significant noncompliance (U.S. GAO 1983, 13–14). I must reemphasize that the GAO findings were based on self-reported discharges. Because sources may have incentives to understate actual discharges, the extent of noncompliance could have been even greater than that found by the study.

The results of the GAO study demonstrate the need for more vigorous monitoring and enforcement efforts. Several studies have shown that unscheduled inspections of facilities by enforcement officials and an increase in the enforcement actions undertaken by EPA yield significant decreases in the rates of discharge of pollutants and incidence of noncompliance (Magat and Viscusi 1990; Strock 1990; Laplante and Rilstone 1996). However, as recently as 1994, GAO found that about one of every six major industrial and municipal sources inspected was in significant violation of the terms of their discharge permits (U.S. GAO 1996). And in 1998, the agency's own inspector general found widespread failures in the permitting and enforcement processes at both the federal and state levels (*New York Times* 1998).

Changes in Water Quality

It is important to try to determine whether the 1972 and 1977 laws have resulted in levels of water quality across the country that are better than they would have been, other things being equal, without these laws—and if so, how much better. None of the available data can answer these questions conclusively, but we can draw some inferences from several sets of data. The data are of three types:

- *observations* of actual changes in water quality from monitoring stations and other sources;
- *predictions* of changes in water quality made in response to changes in pollutant discharges that are based on water quality models that hold other variables (such as the level of economic activity) constant; and
- *assessments* of the proportions of the nation's bodies of water that are of sufficiently high quality to support uses such as fishing and swimming.

Water Quality Measures. The only nationwide system for gathering data regarding various measures of water quality is the National Stream Quality Accounting Network operated by the U.S. Geological Survey (USGS). These data show modest effects of FWPCA-72. Some of these data are listed in Table 6-4. Overall, the percentage of readings in violation of a dissolved oxygen standard of 5.0 milligrams per liter is low, varying from less than 1% to 5% per year from 1975 to 1994. About one-third of all readings of fecal

TABLE 6-4. Trends in Measured Water Quality.

	Readings in violation of standard[a] *(%)*				
Water quality measure	*1975*	*1980*	*1985*	*1990*	*1994*
Fecal coliform bacteria	36	31	28	26	29
Dissolved oxygen	5	5	3	2	<1
Total phosphorus	5	4	3	3	4

Note: This table is based on data from the U.S. Geological Survey's National Stream Quality Accounting Network (NASQAN).

[a] For fecal coliform bacteria, a violation is >200 cells/100 milliliters. For dissolved oxygen, a violation is <5.0 milligrams/liter. For total phosphorus, a violation is <1.0 milligrams/liter.

Source: Council on Environmental Quality 1997, 455, table 40.

coliform bacteria have been in violation of water quality standards, and this percentage changed little over the same period.

Two comprehensive analyses of trends in a large number of water quality measures based on these data provide similar results for the period 1974–87 (Smith and others 1987; Lettenmaier and others 1991). Monitoring stations where significant improvements in bacteria, dissolved oxygen, and phosphorus were found outnumbered stations that found deterioration, but fewer than 20% of the stations found improvements in these measures of water quality. For suspended solids, the numbers of stations that reported improvements and declines were approximately equal. Monitoring stations that found deteriorating levels of nitrates outnumbered those that found improvements by at least a factor of 4.

Water Quality Modeling. Researchers at Resources for the Future (RFF) modeled the effects of the 1972 amendments on several measures of water quality (Gianessi and Peskin 1981; Gianessi and others 1981). The RFF water quality network model is based on inventories of waste generated at point sources and estimates of actual removal rates as of 1972. The model also incorporates estimates of nonpoint-source discharges. The inventories of wastes generated and discharged are combined with a model of how pollution moves through water to predict values for four water quality parameters at more than 1,000 locations in the continental United States. Estimates of increased treatment levels as a consequence of BPT and secondary treatment effluent limitations can be used to predict changes in discharges and, hence, the effects on measures of water quality across the country.

The researchers analyzed two scenarios. The first was based on the estimated 1972 discharges of polluting substances and predicts the percentage of locations achieving assumed water quality standards. In the second scenario, the model predicts water quality at each location, assuming all point sources of pollution to be in compliance with the relevant BPT industrial effluent limitation or secondary treatment standard for municipal treatment plants.

As Table 6-5 shows, in the first scenario reflecting conditions before the passage of the act, 83% of all locations were predicted to have met the standard for dissolved oxygen, whereas 68% were predicted to have met the standard for biochemical oxygen demand. Also, relatively few locations were predicted to have attained the assumed standard for the nutrients phosphorus and nitrogen. The model predicts that BPT/ST will result in increases in the number of locations meeting the standards for each of the four water quality parameters, but the absolute and percentage increases in locations meeting the standards are surprisingly small. The model predicts only a 6% increase in the number of locations satisfying the dissolved oxygen standard, in large part because of the high percentage of locations already meeting the standard before application of BPT/ST. On the other hand, for the two parameters where there is greatest room for improvement—phosphorus and nitrogen—the application of BPT/ST has a relatively small effect on the number of locations in violation, because the point sources affected by BPT/ST are relatively unimportant sources of these pollutants (see Table 6-2). In summary, to the extent that the RFF model accurately predicts water quality, it appears that the first, or BPT/ST, phase of FWPCA-72 had relatively little effect on these measures of water quality in many areas.

Measures of water quality such as dissolved oxygen or total phosphorus may not have much meaning to most people. What matters most to them is

TABLE 6-5. Effect of BPT/ST Controls on Four Water Quality Parameters (1972 baseline).

	Locations meeting assumed water quality standard			
	Dissolved oxygen[a]	*Biochemical oxygen demand*[b]	*Total phosphorus*[c]	*Total Kjeldahl nitrogen*[d]
Estimated 1972 baseline discharges	83	68	27	30
Predicted, with BPT/ST controls	88	75	32	32
Change in number of locations (%)	6	10	19	7

[a] The standard is >5.0 milligrams/liter. [b] The standard is 9.0 milligrams/liter.
[c] The standard is 0.180 milligrams/liter. [d] The standard is 0.90 milligrams/liter.
Source: Gianessi and Peskin 1981, 805–7.

how changes in these measures affect various uses of a water body. One such use of rivers and lakes is recreational fishing. To the extent that reduced pollution and improved water quality result in more recreational opportunities and higher quality recreation, recreational fishermen are better off. Another team of researchers at RFF developed a method for classifying water bodies by the quality of fishing opportunities they present and for translating changes in measures of water quality (as predicted by the RFF water quality network model) into changes in the number of acres of surface waters available for various categories of fishing (Vaughan and Russell 1982, 38–48). According to this approach, a body of water can be placed in one of four categories, depending on water temperature, pH, and the concentrations of dissolved oxygen and total suspended solids. The four categories are

- not fishable;
- fishable for warm-water rough fish, such as carp and catfish;
- fishable for warm-water game fish or panfish, such as perch and bass; and
- fishable for cold-water game fish, such as trout.

Using the estimates of actual discharges for 1972, this model predicted that only 4.2% of the waters included in the model fell into the unfishable category in 1972. The prediction was consistent with the results of an RFF survey of state recreation and fishery officials, which showed that only 7.4% of the nation's waters were judged to be unfishable as of 1972 (Vaughan and Russell 1982, table 2–5).

Estimates of the total U.S. fishable water area in 1972 and of the allocation of the total fishable area to the three categories of fishing activity (with and without pollution controls) are listed in Table 6-6. The implementation of BPT and secondary treatment was predicted to increase the total fishable area by only 0.35%. The major benefit of the BPT/ST regulations comes from improving the quality of fishing in already fishable areas. The area of water suitable only for warm-water rough fish is predicted to decline by 66.8% as more than 2 million acres are improved to support warm-water game fish. There also is a modest increase in the waters that support cold-water game fish.

More recently, another team took a similar approach and reached similar conclusions (Bingham and others 1998). The researchers classified water bodies according to whether they were of sufficient quality to support swimming (highest quality), fishing, only boating, or no recreation activity. They used a detailed water quality model that covered more than 600,000 miles of rivers in the United States to predict levels of the three water quality parameters that determined the classification of each mile of river: biochemical oxygen demand, total suspended solids, and fecal coliform bacteria.

TABLE 6-6. Estimates of Changes in Fishable Waters due to BPT/ST Pollution Control (in thousands of acres).

	Total fishable acres	*Warm-water rough fish acres*	*Warm-water game fish acres*	*Cold-water game fish acres*
Estimated 1972 baseline discharges	30,615	3,429	20,941	6,245
Predicted, with BPT/ST controls	30,721	1,137	22,611	6,974
Change in acres (%)	+0.35	–66.8	+8.0	+11.7

Source: Vaughan and Russell 1982 (54, table 2-9).

They predicted water quality levels under two scenarios. The first was based on actual levels of discharges from point sources in the mid-1990s. The second was the discharges that would have been expected with the same level of economic activity but with rates of treatment of wastes set at the 1972 levels, that is, the levels to be expected in the absence of the FWPCA-72. Some results are shown in Table 6-7. The percentage of miles predicted to be unfishable without the act is substantially higher than in the previous study (36% versus 4.2%). But this difference can be attributed at least in part to the higher levels of economic activity and therefore untreated waste discharges in 1990 than in 1972. As in the other studies, the estimated impact of the act on number of river miles meeting various quality standards is relatively small. With the act, the number of miles meeting swimmable, fishable, and boatable standards increased by only 6.3%, 4.2%, and 2.8%, respectively.

Assessments of Waters that Support Beneficial Uses. EPA requires the states, Native American tribes, and some other jurisdictions to conduct a water quality inventory every two years. Each reporting entity reports whether the assessed water bodies are sufficiently free of pollution to support the designated uses for that body, for example, drinking water supply, fishing, contact recreation, and so forth. The first such inventory was conducted in 1973. It indicated significant improvements in most major waterways over the preceding decade, at least in regard to organic wastes and bacteria (see Council on Environmental Quality 1974; 1975). At least some states had significant regulatory and construction programs under way before 1972, and these early efforts apparently were somewhat successful.

Some results from the most recent inventory (1996) are presented in Table 6-8. The percentages of waters with quality too low to support desig-

TABLE 6-7. Estimates of Changes in Swimmable, Fishable, Boatable, and Other Waters Attributed to the FWPCA-72 (in miles).

	Swimmable	*Fishable*	*Boatable*	*None*
Without the act in the mid-1990s	223,121	406,154	460,916	171,636
With the act in the mid-1990s	237,067	423,117	473,998	158,564
Difference (%)	+6.3	+4.2	+2.8	–7.6

Source: Bingham and others 1998.

TABLE 6-8. Waters that Support Designated Uses on the 1996 National Water Quality Inventory (in percent of area or length assessed).

	River and streams (miles)	*Lakes, ponds, and reservoirs (acres)*	*Estuaries (square miles)*
Fully supporting	56%	51%	58%
Threatened	8%	10%	4%
Good but impaired	36%	39%	38%
Percentage of total waters assessed	19%	40%	72%

Source: U.S. EPA 1998.

nated uses are low and comparable to the proportion found to be too low for boating in the most recent water quality modeling effort described earlier. The proportions that fully support the designated uses are in the range of 51–58%. The most frequently reported source of pollution that is impairing water quality in rivers, streams, lakes, and ponds was agriculture (50–60% of impaired miles or acres), and municipal point sources were a distant second. For impaired estuarine waters, urban runoff and storm sewers headed the list of sources (see U.S. EPA 1998).

EPA recently announced a shift in emphasis of its enforcement program away from compliance with individual discharge permits and toward attainment of established water quality standards. In those bodies of water that do not currently support fishing and swimming, states would be required to establish plans for attaining these standards (*New York Times* 1999). This move is an important step toward strengthening the link between goals and means. But it leaves open the question of the appropriateness of the goals in economic terms (benefits versus costs).

Summary of Accomplishments since 1972

This review of the available evidence on accomplishments since 1972 suggests the following three conclusions:

- As measured by ability to support designated uses or to meet certain water quality standards, average water quality was not so bad in the United States in 1972. Large areas of water could support designated uses or meet water quality standards. However, some localities—particularly large, industrialized metropolitan areas—were experiencing serious water pollution problems.
- Water quality has improved since 1972. In terms of aggregate measures or national averages, the change has not been dramatic, but local success stories report substantial cleanup in what had been seriously polluted water bodies.
- Despite passage of FWPCA-72, we are losing ground to pollution in some areas. Some water bodies show trends of declining water quality.

Two qualifications to these conclusions must be kept in mind. The first is that when we have looked at trends in water quality and the ability of waters to meet designated uses, we have been answering a before-versus-after question. We cannot tell from this evidence whether pollution might have gotten much worse without FWPCA-72 as a result of economic growth and increases in the production and discharge of polluting materials. At least some deterioration in water quality over time would have been inevitable.

Second, all of the above conclusions regarding water quality refer to only the conventional pollutants and related measures of water quality, such as bacteria, nutrients, and dissolved oxygen. The picture could be different if it included toxic pollutants, but almost no data are available from which trends in toxic pollution could be established. This deficiency must be remedied.

Economic Issues and Problems

In examining federal water pollution–control policy from an economic perspective, we look at the costs of controlling water pollution as well as what we are getting for our money—that is, the benefits. Pollution control is costly. Devoting more of society's scarce resources of labor, capital, and administrative and technical skills to pollution control necessarily means that fewer resources are available for other things also valued by society—either collectively or as individuals. Because pollution control is costly, it is in society's interest to be "economical" in its selection of pollution control objectives.

Society must take an economical approach in two senses. First, whatever pollution control objectives are chosen, the means of achieving them should be selected so as to minimize cost. Using more resources than are necessary to achieve pollution control objectives is wasteful; yet as I will show, current water pollution control policies are wasteful in several aspects. Second, society should choose its environmental objectives wisely. If we are to make the most of our endowment of scarce resources, we should compare what we receive from devoting resources to pollution control with what we give up by taking resources from other uses. We should undertake pollution control activities only if the results are worth more in some sense than the values we forgo by diverting resources from other uses.

Government intervention to control pollution is justified on grounds of economic efficiency if the beneficial effects (broadly defined) to society as a whole from such action outweigh the costs. Examination of costs and beneficial effects should become an integral part of the process of establishing pollution control objectives.

In this section, I review available information on the benefits and costs of federal water pollution–control policy. I then search for ways in which the present set of policies might be modified to improve the relationship between benefits and costs. This topic leads us to consider modifying pollution control objectives and requirements where costs at the margin exceed benefits and to search for ways to reduce the costs of achieving given pollution control targets—that is, to improve cost-effectiveness. I apply some of the principles and insights developed here to two specific aspects of current policy: the municipal construction grant program and the program for controlling nonpoint sources of pollution.

Comparisons of Benefits and Costs

To arrive at a conceptually sound estimate of the benefits of a pollution control regulation, it is necessary to have information about four separate relationships. The first is between the specific details of the regulation and changes in the quantities of polluting substances discharged. The second is between changes in the quantities discharged and changes in the relevant measures of water quality. The third is between changes in water quality and changes in the uses people make of that water (for example, an increase in recreational activity or a reduction in treatment costs for municipal and industrial wastes might change water quality). The fourth is between changes in uses and the monetary values that people place on these changes. There may be nonuse values, as well. Although imprecise, monetary values can be determined by applying various economic models to estimate the maximum sums of money individuals would be willing to pay to attain the improved water quality and the increased uses that would be made possible.[9] These relationships are illustrated in Figure 6-1.

Unfortunately, no studies of the aggregate national benefits of FWPCA-72 have dealt with all four sets of relationships in a fully satisfactory manner. Some studies deal carefully with one or two of the relationships and/or one particular type of use, or examine the benefits of controls at one particular water body. Lacking fully satisfactory national aggregate benefit estimates, the analyst who wishes to make a benefit–cost comparison for the overall federal water pollution–control policy must do so through some kind of synthesis and extrapolation from the most soundly based of existing studies.

I prepared one such estimate for the Council on Environmental Quality in 1979 and revised it in 1982 (Freeman 1979; 1982). In the two reports, I presented my estimates of the national aggregate benefits of attaining the effluent guidelines and municipal wastewater treatment goals of the act as of 1985. This year was chosen because several of the studies on which I based my estimate used it as the target date. Thus, the figures I reported could be interpreted as my estimate, as of 1982, of where I thought we were going to be in 1985. My estimate was based on a synthesis of more than half a dozen empirical studies of damages associated with water pollution and/or the benefits of water pollution control that were conducted during the 1970s and early 1980s. These studies were of varying quality. Many of these studies did not formally model changes in water quality. And most of them did not formally model the "without the act" scenario; that is, they took the 1972 levels of water quality as the baseline without taking account of the fact that water quality might worsen after 1972 without the act.

The results of my review, synthesis, and judgment are presented in Table 6-9. These figures can be interpreted as the likely annual benefits of achieving BPT/BAT effluent standards. In reviewing these results, remember that a substantial range of uncertainty surrounds my best judgement or most likely point estimates for each of the categories. The upper and lower bounds differ from the most likely estimate by a factor of about 2. Also note the importance of recreation benefits. At $11.1 billion per year, they account for about half of the total $22.6 billion.

Looking at these figures with the benefit of hindsight, I would make several changes. First, I would substantially reduce the estimates of the benefits associated with improved marine sports fishing. This activity attracts a substantial number of participants and generates significant economic value. However, there is relatively little evidence to indicate that this value has been significantly affected by conventional water pollutants (Freeman 1995). I also suspect that the benefits associated with boating are too high. It is my conjec-

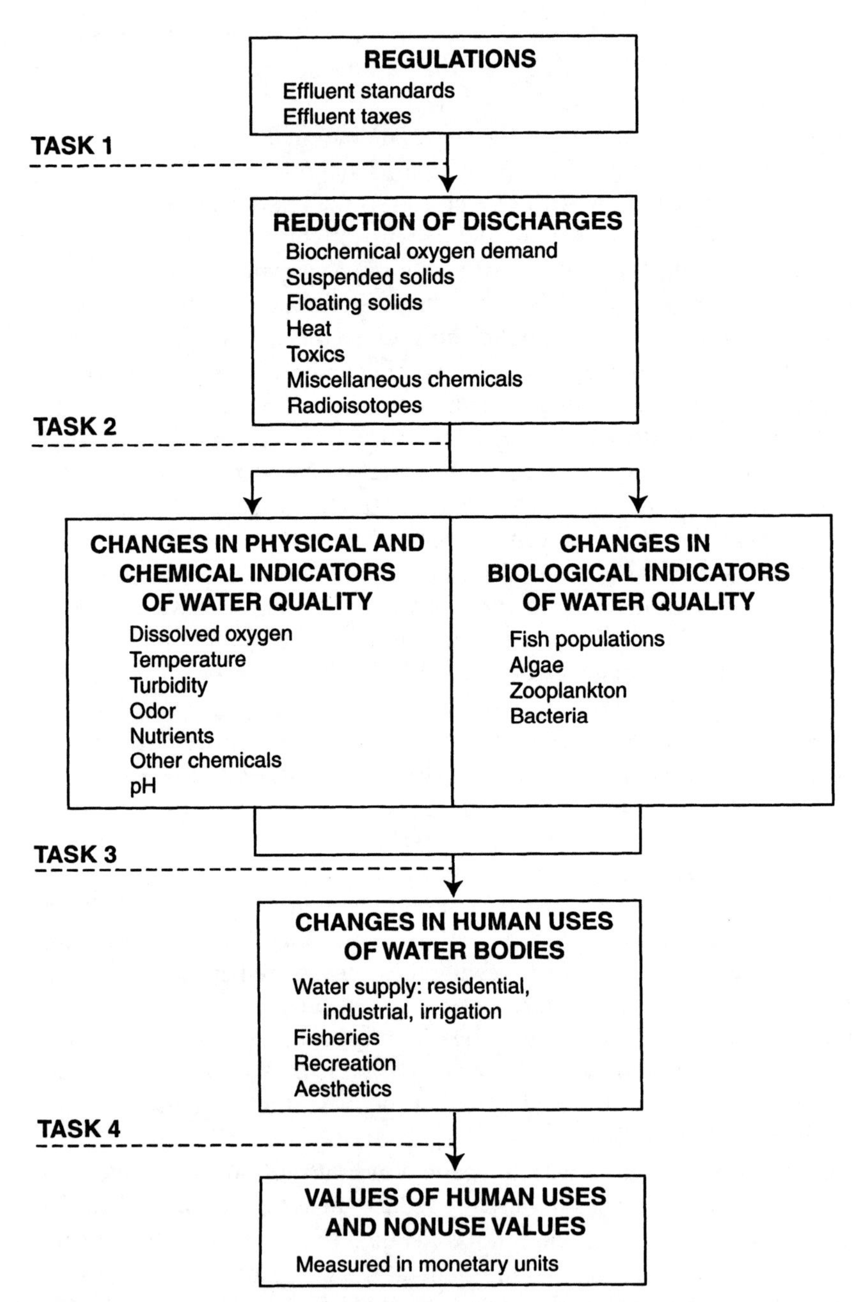

FIGURE 6-1. Tasks Involved in Estimating the Benefits of Water Pollution-Control Policies.

TABLE 6-9. Annual Benefits in 1985 of Achieving BPT and BAT/BCT Pollution Control (in billions of 1996 dollars).

Category	*Low*	*High*	*Best*
Recreation			
Freshwater fishing	1.2	3.4	2.4
Marine sports fishing	0.2	7.2	2.4
Boating	2.4	4.8	3.6
Swimming	0.5	4.8	2.4
Waterfowl hunting	0.0	0.7	0.2
Subtotal	4.3	20.9	11.1
Nonuser benefits: aesthetics, ecology, and property value	1.2	9.6	2.9
Commercial fisheries	1.0	2.9	1.9
Diversionary uses			
Drinking water/health	0.0	4.8	2.4
Municipal treatment costs	1.4	2.9	2.2
Households	0.2	1.2	0.7
Industrial supplies	1.0	1.9	1.4
Subtotal	2.7	10.8	6.7
Total	9.1	44.3	22.6

Note: 1996 dollars have been adjusted using the consumer price index. Subtotals and totals may not sum exactly due to rounding.
Source: Freeman 1982 (161, table 8-3; 170, table 9-1).

ture that where boating activities have been impaired by diminished water quality, the source of the problem probably has been nutrients rather than discharges of conventional pollutants.

Finally, I suspect that the aesthetic benefits associated with the FWPCA-72 are underestimated. I have no hard evidence to support this conjecture, but I suspect that improvements in water quality along urban waterfronts has created opportunities for commercial development and for increases in what might be called incidental recreational uses. Although the "per use" values may be small, the aggregate benefits might be significant.

An estimate of the annual costs of achieving these standards is presented in EPA's "Cost of Clean" report (U.S. EPA 1990). The estimate is based on reported investments in pollution abatement facilities, equipment, and operating expenses as reported by business and government. Market mechanisms such as changes in prices and outputs are likely to both shift the burden of firms' expenditures and change the magnitude of the burden. At least for nonmarginal changes in cost, firms will raise prices and will experience decreases in quantities demanded. These price changes result in losses of consumers' surplus and can lead to changes in producers' surpluses and factor incomes as well. Thus, true costs may not be accurately measured by summing the expenditures of firms and governments (see Portney 1981; Hazilla and Kopp 1990).

Annual costs for 1985 are estimated to be $42.4 billion in 1996 dollars (U.S. EPA 1990). Comparing this amount with the estimate of benefits shows that costs likely substantially outweighed the benefits as of 1985. But because of the reported uncertainty in the benefit estimate and the unknown degree of uncertainty in the cost estimate, the possibility that benefits exceeded costs cannot be ruled out.

Two other estimates of the benefits attributable to recreation and nonuse values associated with attaining the goals of the act have been published. The first is based on a contingent valuation survey administered in 1983 to a national sample (Carson and Mitchell 1993). The authors asked individuals (a random sample of the U.S. population) how much they were willing to pay

- to ensure that water quality would not fall below the level that would support boating on all water bodies;
- to improve all waters from boatable to fishable; and
- to improve all waters from fishable to swimmable.

The sum of the responses to these three questions was interpreted as the willingness to pay to move from a no-control baseline to a state in which all waters were of swimmable quality.

The median and mean responses per household were approximately $190 and $430 per year (1996 dollars). The authors applied the household mean willingness to pay to the 1990 population of the United States to estimate aggregate willingness to pay of about $35 billion with an uncertainty range of $29 billion to $54 billion. This figure is substantially higher than my estimate for water-based recreation. But two differences in our methods can account for at least part of the difference. First, my estimate was implicitly based on a smaller population than that of the United States in 1990. And second, the baseline water quality used in my estimate was substantially higher than the no-control baseline assumed by Carson and Mitchell.

The second study combines EPA's Water Quality Model (described earlier) with the results of Carson and Mitchell's contingent valuation survey (Bingham and others 1998). After predicting the number of miles of rivers that would attain boatable, fishable, and swimmable quality in the mid-1990s, the authors calculated the number of people living "proximate" to each stretch of these rivers. They used Carson and Mitchell's estimates of mean willingness to pay for changes in water quality to calculate the benefits of attaining that level of water quality for each river stretch. For their benefit calculation, they took as a baseline the no-treatment scenario rather than actual water quality levels in 1972. The total willingness to pay for the U.S. urban population was about $9 billion per year (1996 dollars). As with the Carson and Mitchell estimate, this figure is not directly comparable with my estimate of recreation benefits because of the difference in the assumed baseline. But the comparison with Carson and Mitchell's own estimate of $35 billion per year shows the importance of looking at what water quality levels were expected to be attained rather than assuming attainment of swimmable quality in all U.S. waters. It also is important to note that this $9 billion figure counts only benefits of in-stream uses and existence value for the control of conventional pollutants discharged to rivers and streams. It does not include benefits to lakes, ponds, estuaries, or marine waters; benefits from control of toxic compounds; or benefits associated with diversionary uses of water.

Annual costs in 1990 were estimated to be $54.4 billion (U.S. EPA 1990). This estimate is not directly comparable to either of the two benefit estimates for two reasons. First, neither of the benefit estimates includes potential benefits to commercial fisheries and diversionary uses, which could range from $5 billion to $10 billion per year. Second, both of the benefit figures take no treatment as the baseline, whereas the cost estimate is the increase in cost associated with going beyond 1972 levels of treatment. However, the rough order of magnitude of these estimates tends to support the conclusion that the act does not appear to have achieved benefits commensurate with its costs. Therefore, it is important to seek ways to modify present policies so as to improve the benefit–cost relationship.

Broadly speaking, two such avenues are to be investigated. The first is adjusting targets or pollution control requirements where, at the margin, the costs of current controls are substantially different from the benefits. The second seeks ways to reduce the costs of achieving the existing goals.

Adjusting the Targets and Standards

Present water pollution–control policy establishes a national interim target of fishable and swimmable water quality and an ultimate stated objective of zero pollutant discharge. It also imposes pollution reduction requirements on individual dischargers based on effluent guidelines that are uniform within industrial categories. If we were to adopt as an alternative the principle that pollution control policies should be designed to maximize the net benefits from pollution control activities, then pollution control requirements for individual dischargers would emerge as the result of a two-part analytic process. The first part would be to establish a set of water quality standards for each water body, such that in each case, the marginal benefits of raising water quality to that point would just equal the marginal cost of doing so. In cases where marginal pollution control costs were high, the resulting water quality standard might be lower than the fishable/swimmable national target. But in other cases, this economic benefit–cost approach might lead to very high standards for water quality.

The second part of the analytic process would be to determine the individual effluent reductions necessary to meet the water quality standards for each water body and each segment of river. These reductions might vary across dischargers not only because of differences in industrial processes and control technologies but also because of differences in costs and impacts on water quality. This approach to policymaking could save resources by imposing less stringent effluent limitations where the marginal costs of achieving fishable/swimmable water quality were greater than the marginal benefits of doing so. It should be noted that economically based water quality standards would have to be reviewed periodically and revised to reflect changes in the economic and other factors that determine benefits and costs.

The Water Quality Act of 1965 provided a framework through which such an economic approach could have been implemented. But Congress chose to replace that framework with the present one in 1972. The choice of a criterion for policymaking is essentially a political question. The adoption of an explicitly economic framework for policymaking would be a major change in philosophy. At least for now, Congress has settled the question in favor of the present technology-based approach.

Even if the present framework for policymaking is left essentially unchanged, several opportunities remain for making the kinds of adjustments suggested by the National Commission on Water Quality. One possibility is to authorize more formal and substantive analysis of economic factors in establishing BAT and BCT effluent limitations. Congress has given some latitude to EPA in setting these limitations. For example, in describing the factors to be taken into account in establishing BCT standards, Congress called for "consideration of the reasonableness of the relationship between the costs of obtaining the reduction in effluents and the effluent reduction benefits derived…."[10] This language appears to allow the balancing of benefits and costs in setting standards. However, the language that defines BAT requirements includes a reference to the cost of achieving effluent reductions but does not explicitly authorize a consideration of benefits.

A second way that economic considerations could influence pollution control standards would be to allow officials who write individual discharge permits greater latitude to deviate from the effluent limitation guidelines after case-by-case comparisons of water quality benefits and costs. For example, where fishable/swimmable water quality has already been achieved, little may be gained by imposing additional BCT or BAT requirements on sources that discharge into that water body. Yet present law does not provide any authority for waiving BAT requirements where fishable/swimmable water quality targets are already being met.

The 1977 amendments to FWPCA-72 did create one opportunity for relaxing pollution requirements on publicly owned treatment works. One section of the 1977 act permits a waiver of secondary treatment requirements for municipal facilities that discharge into ocean waters where fishable/swimmable requirements are satisfied and where it can be shown that the assimilative capacity of the ocean waters will allow for the absorption and dispersion of these wastes without impairing other uses of the water.

There are also likely to be bodies of water where BAT and BCT limitations will not be sufficient to achieve fishable/swimmable waters; the costs of BAT and BCT would be incurred without realizing the benefits of fishable/swimmable water. One reason for such nonattainment is that even with BAT and BCT in place, the total volume of industrial and municipal discharges might still be too great to permit achievement of the fishable/swimmable standard. The 1972 act anticipates this kind of situation by requiring even stricter effluent limitations for these cases. However, the EPA administrator is required to examine the relationship between costs and benefits and may choose not to require the additional effluent limitation if he or she finds no reasonable relationship between the economic and social costs and the benefits to be obtained.

A more sophisticated approach to dealing with these cases would be to take advantage of seasonal variations in designated water uses or seasonal variations in water quality. For example, where cold weather precludes water-based recreation, water quality standards and pollution control requirements might be relaxed during winter to save on operating costs. Or where existing water quality standards are violated only during some seasons (as when there is low water flow or high water temperature), the most stringent control requirements could be imposed only during the problem seasons.[11] Such intermittent controls may be difficult to administer.

A second possible reason for not attaining fishable/swimmable water quality with BAT and BCT limitations could be pollution from nonpoint sources, because they are not presently covered by any effluent limitations, technology-based or otherwise. The larger the nonpoint sources of pollution relative to point sources in any water body, the less impact control of point sources will have on ambient water quality. Thus, significant control of nonpoint sources may be necessary to achieve fishable/swimmable water in these cases. And it may be possible to substantially reduce the costs of achieving fishable/swimmable water by imposing stricter requirements on nonpoint sources while relaxing the pollution control requirements on point sources.

Improving Cost-Effectiveness

By cost-effectiveness, economists mean the degree to which an activity, such as pollution control, is carried out to achieve the stated targets at the lowest possible total cost. The importance of achieving cost-effective pollution control policies should be self-evident. Any cost savings frees resources to produce other goods and services of value to people. If some change in the allocation of cleanup requirements among dischargers lowers the total cost of controlling pollution without degrading water quality, then society is clearly better off.

A pollution control policy is cost-effective only if it allocates the responsibility for cleanup among sources so that the incremental or marginal cost of achieving a one-unit improvement in water quality at any location is the same for all sources. Differences in the marginal cost of improving water quality can arise from both variations in the marginal cost of treatment or waste reduction across sources and variations among sources in the effects of lower discharges on water quality.

To illustrate the first point, suppose that two adjacent factories discharge the same substance into a lake. In this case, a one-unit decrease in discharges gives the same incremental benefit to water quality whether it is achieved by Factory A or Factory B. Now, suppose that to achieve fishable/swimmable water quality in the lake, discharges of the polluting substance must be reduced by 50 tons per day. One way to achieve the target is to require each factory to clean up 25 tons per day. But suppose that with this allocation of cleanup responsibility, Factory A's marginal cost of cleanup is $10 per ton per day, whereas Factory B's marginal cost is only $5 per ton per day? Allowing Factory A to reduce its cleanup by one ton per day saves it $10. If Factory B is required to clean up an extra ton per day, total cleanup is the same and the water quality standard is met, and the total cost of pollution control is reduced by $5 per day. Additional savings are possible by continuing to shift cleanup responsibility to Factory B and away from Factory A (thus reducing Factory A's marginal cost). This shift should continue until Factory B's rising marginal cost of control is made equal to Factory A's now lower marginal cost. Discharges of a substance that have different impacts on water quality depending on the location of the source also must be taken into account in finding the least-cost or cost-minimizing pattern of discharge reductions.

From an economic standpoint, a major criticism of technology-based standards is that they are almost certain to result in total cost that is higher than necessary for any particular level of water quality. Nothing in the logic or the procedures for setting technology-based limits ensures that the conditions for cost minimization will be satisfied. One analysis of the marginal cost of removing oxygen-demanding organic material to meet the BPT standards found a thirtyfold range of marginal costs within the six industries examined (Magat and others 1986, table 6–1). In this instance, spending an extra dollar for treatment in an industry with low marginal costs would buy thirty times more pollution removal than spending the same dollar in an industry with high marginal costs. One set of studies prepared for the National Commission on Water Quality showed potential for substantial cost savings (30–35% for the nation as whole) through the selection of cost-minimizing effluent reductions to achieve the same water quality improvement (Luken and others 1976).[12] On the basis of my earlier calculation of the annual costs of water pollution control, such reductions would mean a potential savings of $10 billion to $12 billion per year.

Bubbles. A first step toward obtaining the economic advantages of transferable discharge permits is the application of the "bubble" concept to water pollution control.[13] In a major industrial facility such as an integrated steel mill, several separate activities or processes may be going on, each subject to a different BCT or BAT requirement. Many of these processes discharge the same substances, yet the incremental costs of pollution control may be quite different across activities (see Green 1979). As a result, the total cost of controlling the aggregate discharge from the plant is often higher than necessary. In such cases, plant managers should be allowed to adjust treatment levels at different activities if they can lower total treatment costs, as long as the total amount of a pollutant discharged from the plant does not exceed the aggregate of the effluent limitations for individual processes. EPA now allows such bubble trade-offs at integrated steel mills, provided that the trade-offs result in a net reduction of the total amount of pollutants discharged (see Water Quality Committee 1983). Perhaps present law should be modified to facilitate similar intraplant trades in all industrial categories.

Effluent Taxes. A more comprehensive step would be to place a tax or charge on each unit of each pollutant discharged and to allow each discharger to choose the degree of cleanup that minimizes its total cost (that is, cleanup cost plus tax).[14] The effluent charge strategy has long been attractive to economists because it creates an incentive for firms by making pollution itself a cost of production. It also provides an incentive for innovation and technological change in pollution control. A system of effluent charges also can contribute significantly to cost-effectiveness. Where several sources are discharging into the same water body, effluent charges would encourage the dischargers to minimize the total cost of reducing pollution, because each discharger would control discharges up to the point where its marginal cost of control is equated with the given charge. If all dischargers face the same charge, they will have equated their marginal cost of pollution control. This is the condition for cost minimization in reducing charges.

It is also possible to use water quality models to design charge systems that are differentiated by location to take into account the differing effects of discharges on water quality across sources. With the appropriate set of differentiated charges, the total cost of meeting any set of water quality standards within a river basin or region can be minimized. Some studies have shown that costs might be reduced by as much as 20–50% by appropriately designed systems of effluent charges (see Kneese and Bower 1968, 158–64; Herzog 1976). To calculate systems of optimal differentiated effluent charges for a particular river requires lots of data and sophisticated economic and water quality modeling. It probably is not practical to recommend that such calculations be carried out for all sources of pollution on all rivers and streams in the United States. But a simpler effluent charge system, such as the one adopted in the Federal Republic of Germany in 1976, may be feasible and desirable (see Brown and Johnson 1984).[15]

Tradable Pollution Permits. Another approach to water pollution control that has essentially the same incentive and cost-minimizing effects is a system of tradable or marketable discharge permits (see Chapter 3, Market-Based Environmental Policies).[16] Under this system, the pollution control authority issues a limited number of pollution permits or tickets. Each ticket entitles its owner to discharge one unit of pollution during a specified time period. The authority either distributes the tickets free of charge to polluters or other parties on some basis or auctions them off to the highest bidders. Dischargers might also buy and sell permits among themselves. A system of marketable discharge permits allocates pollution rights that would meet the ambient water quality standards while minimizing the total costs of pollution control. When dischargers are located at different points along a river, the terms at which permits are exchanged would have to reflect the different locations of sources and the varying impacts of their discharges on water quality.

One advantage of a system of transferable discharge permits over effluent charges is that the discharge permits are a less radical departure from the existing system. Because all sources are presently required to obtain permits that specify maximum allowable discharges, it would be a relatively straightforward matter to rewrite permits in a divisible format (ten one-ton permits instead of one ten-ton permit) and allow exchanges to be made. A more modest step would be to allow two (or more) sources to propose reallocating cleanup requirements between them if they found it to their mutual advantage and if water quality would not decrease. A source that has low incremental costs of control should be willing to increase its cleanup provided that another firm compensated it for its increase in pollution control costs. And a source that has high pollution control costs would find it cheaper to pay for another source's cleanup than to clean up itself. EPA recently issued a policy statement that encourages states to adopt effluent trading systems within watersheds. The agency has estimated that potential combined cost savings from this new policy could amount to $1.3 billion to $15 billion (reported in Council on Environmental Quality 1997, 40; see also U.S. EPA 1996).

The Municipal Construction Grant Program

From 1956 until 1991, the federal government subsidized the construction of wastewater treatment facilities by cities and towns. Although this subsidy program has ended, it is useful to consider what was accomplished. One indicator is the substantial progress in increasing the degree of treatment and the proportion of municipal wastewater receiving treatment. Some data are shown in Table 6-10. The population living in homes connected to municipal wastewater systems (reflecting the trend of urbanization) increased by almost 68 million. The percentage of the population whose wastes received no treatment fell from 63% to about 1% between 1960 and 1988, and the proportion of the population whose wastes received at least secondary treatment (a standard imposed by FWPCA-72) rose from only 4% to 84% during the same period.

TABLE 6-10. Population Served by Municipal Wastewater Systems, with Level of Treatment.

Population	*1960*	*1970*	*1980*	*1988*
Total U.S. population (in millions)	180	203	224	247.6
Population served by wastewater system (in millions)	110	145	159	177.7
Percentage of total population	61%	71%	71%	72%
Percentage of served population with:				
No treatment	63%	41%	1%	1%
Primary treatment	33%	NA	26%	15%
At least secondary treatment	4%	NA	73%	84%

Note: NA = not available.

Source: Council on Environmental Quality 1982, 295, table A-61; 1990, 454, table 27.

Although the progress in increasing the treatment of municipal wastes was associated with increased federal financial aid going back to 1956, it is not possible to conclude that federal aid was fully responsible for the improvement. Available evidence suggests that to some extent, municipalities used federal dollars to reduce their own financial burden for constructing facilities that would have been built in any event. Some evidence of this displacement of municipal spending by federal cost sharing is indicated in Figure 6-2. In 1982 constant dollars, state and local spending peaked in 1972 and had declined by more than 50% by 1982. One econometric study of federal and municipal spending estimated that each additional dollar of federal spending reduced municipal spending by $0.67 (Jondrow and Levy 1984).

Some people have argued that the federal subsidy was necessary to induce municipalities to comply with federal regulations that required secondary treatment. However, the evidence on displacement does not support this justification. It appears that the "stick" of federal regulation was relatively more important than the "carrot" of federal aid in influencing decisions about municipal pollution control. If this was the case, then the ques-

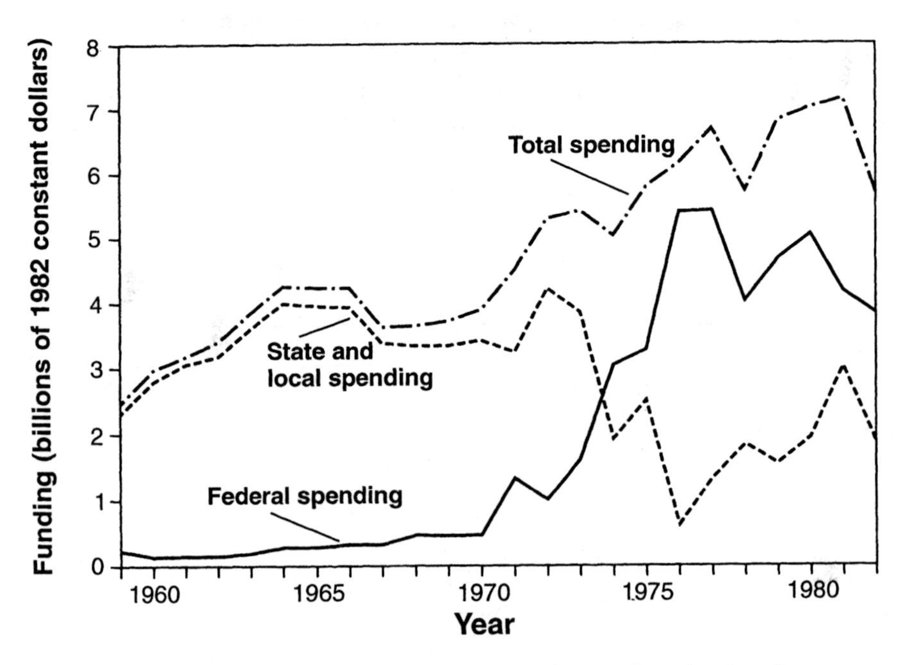

FIGURE 6-2. Substitution of State and Local Funding with Federal Dollars.

Source: U.S. EPA 1984b, 3-2.

tion of the level and form of federal financial assistance is essentially a political one with substantial equity implications.

Nonpoint Sources of Pollution

Several features of nonpoint-source pollution must be considered in any discussion and evaluation of policy options. Among them is the wide variety of sources of such pollution, which include urban storm sewers that discharge directly to water bodies, urban runoff into streams and drainage ditches, nonurban sources such as agricultural and silvicultural runoff from erosion, and surface and subsurface mining activities.

The various sources, in turn, are associated with various pollutants, including oxygen-demanding organic material, fecal bacteria, suspended solids and sediments, heavy metals, toxic organic compounds, and two plant nutrients—phosphorus and nitrogen—that can degrade water quality when present in excessive concentrations. The variety of sources and pollutants also means that conventional end-of-pipe technologies often will not be possible and that several alternative control technologies will be required. These approaches may include changes in agricultural, forestry, and mining practices to prevent pollutants from being carried away by normal runoff.

Another feature of nonpoint-source pollution is its episodic character: It occurs primarily in connection with periods of rainfall or the melting of accumulated snow. Point-source pollution, in contrast, is a more-or-less steady flow of pollutants from municipal sewer systems and many industrial processes.

A water quality modeling study discussed earlier (Gianessi and Peskin 1981) demonstrated what has been known for a long time: that some degree of control of nonpoint-source pollution will be essential if water quality standards are to be achieved throughout the country. The study analyzed the effects of three different control programs on four water quality parameters: phosphorus, nitrogen, biochemical oxygen demand, and dissolved oxygen. The three programs were BPT/ST at point sources, elimination of nonpoint-source discharges from all nonirrigated cropland, and a combination of the two programs. As indicated in Table 6-11, for each of the pollutants, control of sediment from nonpoint sources improves water quality (as measured by the number of stations meeting standards) more than BPT/ST alone. The difference is substantial in the case of the two nutrients.

Even in water bodies where significant control of nonpoint-source pollution is not required for attaining water quality standards, it may be desirable on grounds of cost-effectiveness to control it anyway. Wherever the incremental costs of nonpoint-source control are less than the costs of controlling

TABLE 6-11. Effects of BPT/ST, Agricultural Sediment Controls, and Combined Controls on Water Quality.

	Locations meeting assumed water quality standards (%)			
	1972 baseline	*BPT/ST*	*Agricultural sediment control*	*Both controls*
Total phosphorus[a]	27	32	39	48
Total Kjeldahl nitrogen[b]	30	32	43	48
Biochemical oxygen demand[c]	68	75	76	83
Dissolved oxygen[d]	83	88	89	93

[a] The standard is 0.180 milligrams/liter. [b] The standard is 0.90 milligrams/liter.
[c] The standard is 9.0 milligrams/liter. [d] The standard is >5.0 milligrams/liter.
Source: Gianessi and Peskin 1981, 805–7.

point sources to meet the BAT/BCP standards, the total cost of meeting water quality standards could be reduced by relaxing the control requirements on point sources and substituting the relatively less costly control of nonpoint sources of the same pollutants.

Turning to the regulatory framework, the present federal policy for nonpoint-source pollution control implies a belief that the problem is best dealt with by state and local agencies. Section 208 of FWPCA-72 requires state planning agencies to include consideration of nonpoint sources in developing area-wide waste management plans, and the Water Quality Act of 1987 gives the responsibility for controlling nonpoint-source pollution to the states. But neither provision establishes authority for federal regulation.

In one respect, the decision to leave nonpoint-source control planning to the state and local agencies may be a virtue. If Congress had followed the same approaches as it did with point sources—that is, reliance on technology-based effluent standards—we might now be saddled with an inflexible and highly costly set of national uniform design requirements for the control of nonpoint sources. An EPA report stated,

> The basic approach taken by the Clean Water Act for managing point sources—that is, the application of uniform technological controls to classes of dischargers—is not appropriate for the management of nonpoint sources. Flexible, site-specific, and source-specific decision making is the key to effective control of nonpoint sources. Site-specific decisions must consider the nature of the watershed, the nature of the water body, the nature of the nonpoint source(s), the use impairment caused by the nonpoint source(s), and the range of management practices available to control non-

mine whether the costs per life saved for standards for different contaminants are comparable within the list of substances covered under the Safe Drinking Water Act. Second, we must compare the cost per life saved for the Safe Drinking Water Act programs with the costs per life saved in other environmental and safety regulations. Although I have not seen a comprehensive analysis of these questions, a study of the results of the Safe Drinking Water Act by researchers at RCG/Hagler, Bailly, Inc., sheds some light on the subject (Raucher and others 1993). Their analysis was limited to MCLs for contaminants that pose a cancer risk. The researchers first listed costs and cancer deaths avoided for the program as a whole. They used these data to arrive at a cost per cancer death avoided of about $4 million. This calculation compares favorably with the value of statistical life used by EPA in several recent assessments ($4.8 million) and is at the low end of the range of such costs found in some other studies (Van Houtven and Cropper 1996; Viscusi 1996). The RCG/Hagler, Bailly, Inc., researchers then listed costs and deaths avoided for a subset of the most cost-effective MCLs. The cost per death avoided for these contaminants was $2.5 million. From these data, it is possible to estimate the cost per life saved associated with the remaining MCLs not included in this subset. This value is very high: $108 million per death avoided. The results of these studies demonstrate that attaining MCLs for some substances can be very costly.

value of benefits due to improved water quality is estimated to be in the range of $1.9 to $5.5 billion (Ribaudo 1989).

Air Emissions and Water Quality

Some water quality problems originate as emissions into the atmosphere. One example is the mercury emitted in trace quantities from coal-burning power plants and trash-to-energy incinerators. These emissions travel great distances before settling back to earth. The mercury then migrates to bodies of water, where it is picked up by microorganisms. It accumulates in the organisms and becomes concentrated through the food chain. This process has resulted in mercury levels in fish that exceed safe limits for human consumption.

Another example is the nitrates that can lead to excessive algae growth and death of submerged aquatic vegetation in estuaries. In the Chesapeake

Bay, a major source of nitrates is the deposition of nitrate particles that are formed by the nitrogen oxides emitted from coal-burning power plants and from automobiles. Reducing nitrogen oxide air emissions may be part of a cost-effective strategy for reducing nutrient levels in the Chesapeake Bay. The benefits of the resulting improvements in water quality should be counted as part of the benefits of reducing nitrogen oxide emissions in any benefit–cost assessment of the air pollution–control regulations.

Conclusions

Three major themes traverse this discussion of water pollution–control policy: the importance of comparing benefits and costs, the value of seeking cost-effective control programs, and the potential role of economic incentives such as charges or marketable discharge permits.

We have seen that, in aggregate, the costs of the present policy appear to substantially outweigh the benefits. Yet if the goal of fishable and swimmable water quality is to be met everywhere, even more costs will have to be incurred. If it is accepted that the resources presently devoted to water pollution control are scarce, involve opportunity costs, and may have more valuable uses in other activities, then some water pollution goals and standards might have to be a reconsidered on a case-by-case basis. This approach may mean accepting less than fishable/swimmable water quality where the costs of obtaining it are inordinately high. And it could mean requiring less than the best available technology or the best conventional control technology where additional control would bring very few additional benefits.

I argue that one way to improve the benefit–cost relationship in the existing policy is to seek more cost-effective means of achieving given standards. The emphasis on equal treatment of dischargers or uniformity of cleanup requirements has made the cost of reaching present water quality objectives substantially higher than necessary. Thus, fewer of society's resources are available for other valuable uses. More emphasis should be given to the development of cost-effective means of achieving targets.

I discussed the potential role of charges or marketable discharge permits in moving toward a more cost-effective pollution control policy. Yet even if charge strategies are not adopted, substantial savings could be realized through more selective rather than uniform application of discharge standards such as best available technology, and by considering trade-offs between controls on point and nonpoint sources that discharge to the same water body.

Finally, we have seen that progress toward attaining water pollution control objectives has been slow. Timetables have not been kept, and deadlines have been reached and passed without full compliance with the legislated objectives. These shortfalls in implementation are due in substantial part to the complexity of the task. But a major share of the responsibility for the slow pace of progress must be assigned to the inappropriate incentive structures created by the regulatory approach to pollution control. The various kinds of charges for pollution present opportunities for restructuring pollution control incentives. Marketable discharge permits or charges on all pollution dischargers, for example, could strengthen the incentives for proper operation of treatment systems and for innovation and technological change in pollution reduction methods.

Notes

1. The 1977 amendments are discussed in Freeman 1978.

2. For accounts of a short-lived effort to use the Refuse Act as the basis for a national water pollution–control program in the early 1970s, see National Commission on Water Quality 1976, 285–301, and Davies and Davies 1975.

3. Much of the following discussion of the history of federal policy is drawn from Kneese and Schultze 1975, 30–45.

4. See 33 U.S.C. 466, Sections 301 (g)(1)(C) and 301 (h)(2).

5. For further discussion of technology-based standards, see Freeman 1980.

6. For further discussion of monitoring and enforcement problems, see Russell 1990.

7. For information about the sources and consequences of nonpoint-source pollution, see Carpenter and others 1998.

8. For a description of the procedures for issuing effluent limitations regulations and the process used by the EPA in developing these regulations, see Magat and others 1986, chapter 3.

9. For a detailed exposition of the methods of estimating pollution control benefits, see Freeman 1993.

10. See 33 U.S.C. 466, Section 304(b)(4)(A).

11. For a discussion of this kind of policy innovation, with some estimates of potential cost savings, see Downing and Sessions 1985.

12. Earlier studies that compared the costs of uniform treatment policies with cost-minimizing alternatives found the former to be sometimes two to three times larger than the latter (see Kneese and Bower 1968, 158–164, 224–235).

13. The bubble concept, first applied to multiple-stack sources of air pollution, was given this name because it treats a collection of stacks or sources as if they were encased in a bubble. Pollution control requirements are applied to aggregate emissions leaving the bubble, rather than each individual source.

14. Allen Kneese and Charles Schultze are perhaps the best-known advocates of greater reliance on charges and other economic incentives (see Kneese and Schultz

1975, 30–45; Anderson and others 1977). For a more critical perspective, see Rose-Ackerman 1973; 1977; and Russell 1979.

15. For discussion of other European policies that incorporate effluent charges in some form, see Johnson and Brown 1976 and Bower and others 1981.

16. A system of tradable permits for emissions of sulfur dioxide is the cornerstone of the program to control acid deposition adopted as part of the Clean Air Act Amendments of 1990.

17. In February 1998, President Bill Clinton announced a "Clean Water Initiative," which included increased attention to nonpoint-source pollution and establishment of water quality standards for nitrogen and phosphorus.

18. For a useful discussion of policy options in the context of agricultural nonpoint-source pollution, see Harrington and others 1985.

References

Anderson, Frederick R., Allen V. Kneese, Phillip D. Reed, Serge Taylor, and Russell B. Stevenson. 1977. *Environmental Improvement through Economic Incentives.* Baltimore, MD: The Johns Hopkins University Press for Resources for the Future.

Bingham, Tayler H., Timothy R. Bondelid, Brooks M. Depro, Ruth C. Figueroa, A. Brett Hauber, Suzanne J. Unger, and George L. Van Houtven. 1998. *A Benefits Assessment of Water Pollution Control Programs since 1972.* Revised draft report to the U.S. Environmental Protection Agency. Research Triangle Park, NC: Research Triangle Institute.

Bower, Blair T., Rémi Barré, Jochen Kühner, and Clifford S. Russell with Anne J. Price. 1981. *Incentives in Water Quality Management: France and the Ruhr Area.* Washington, DC: Resources for the Future.

Brown, Gardner M. Jr., and Ralph W. Johnson. 1984. Pollution Control by Effluent Charges: It Works in the Federal Republic of Germany, Why Not in the United States? *Natural Resources Journal* 24(4):929–66, October.

Carpenter, Stephen, Nina F. Caraco, David L. Correll, Robert W. Howarth, Andrew Sharpley, and Val H. Smith. 1998. Nonpoint Source Pollution of Surface Waters with Phosphorus and Nitrogen. *Issues in Ecology* 3; http://esa.sdsc.edu (last accessed March 3, 2000).

Carson, Richard T., and Robert Cameron Mitchell. 1993. The Value of Clean Water: The Public's Willingness to Pay for Boatable, Fishable, and Swimmable Quality Water. *Water Resources Research* 29(7, July):2445–54.

Council on Environmental Quality. 1974. *Environmental Quality.* Washington, DC: Council on Environmental Quality, 280–9.

——. 1975. *Environmental Quality.* Washington, DC: Council on Environmental Quality, 350–5.

——. 1976. *Environmental Quality—1976.* Washington, DC: Council on Environmental Quality, 15.

——. 1982. *Environmental Quality—1982.* Washington, DC: Council on Environmental Quality.

——. 1990. *Environmental Quality—1989.* Washington, DC: Council on Environmental Quality.

——. 1997. *Environmental Quality—25th Anniversary Report.* Washington, DC: Council on Environmental Quality.

Davies III, J. Clarence, and Barbara S. Davies. 1975. *The Politics of Pollution,* 2nd ed. New York: Bobbs–Merrill, 208–10.

Downing, Donna, and Stuart Sessions. 1985. Innovative Water Quality–Based Permitting: A Policy Perspective. *Journal of the Water Pollution Control Federation* 57(5):358–65, May.

Freeman, A. Myrick III. 1978. Air and Water Pollution Policy. In *Current Issues in U.S. Environmental Policy,* edited by Paul R. Portney. Baltimore, MD: The Johns Hopkins University Press for Resources for the Future, 45–65.

——. 1979. The Benefits of Air and Water Pollution Control: A Review and Synthesis. Report to the Council on Environmental Quality, Executive Office of the President. December. Brunswick, ME: Bowdoin College.

——. 1980. Technology–Based Standards: The U.S. Case. *Water Resources Research* 16(1, February.):21–27

——. 1982. *Air and Water Pollution Control: A Benefit–Cost Assessment.* New York: John Wiley.

——. 1993. *The Measurement of Environmental and Resource Values: Theory and Methods.* Washington, DC: Resources for the Future.

——. 1995. The Benefits of Water Quality Improvements for Marine Recreation: A Review of the Empirical Evidence. *Marine Resource Economics* 10(4, Winter): 385–406.

Gianessi, Leonard P., and Henry M. Peskin. 1981. Analysis of National Water Pollution Control Policies: 2. Agricultural Sediment Control. *Water Resources Research* 17(4, August):803–21.

Gianessi, Leonard P., Henry M. Peskin, and G. D. Young. 1981. Analysis of Water Pollution Control Policies: 1. A National Network Model. *Water Resources Research* 17(4, August):796–802.

Green, Robert C. 1979. Water Pollution Control for the Iron and Steel Industry. In *Benefit–Cost Analysis of Social Regulations,* edited by James C. Miller III and Bruce Candle. Washington, DC: American Enterprise Institute.

Harrington, Winston, Alan J. Krupnick, and Henry M. Peskin. 1985. Policies for Nonpoint Source Water Pollution Control. *Journal of Soil and Water Conservation* 40(1, January–February):27–32.

Hazilla, Michael, and Raymond J. Kopp. 1990. The Social Cost of Environmental Quality Regulations: A General Equilibrium Analysis. *Journal of Political Economy* 98(4, August):853–73.

Herzog, Henry W. Jr. 1976. Economic Efficiency and Equity in Water Quality Control: Effluent Taxes and Information Requirements. *Journal of Environmental Economics and Management* 2(3, February):170–84.

Johnson, Ralph W., and Gardner M. Brown Jr. 1976. *Cleaning Up Europe's Waters: Economics, Management, and Policies.* New York, Praeger.

Jondrow, James, and Robert A. Levy. 1984. The Displacement of Local Spending for Pollution Control by Federal Construction Grants. *American Economic Review* 74(2, May):174–8.

Kneese, Allen V., and Blair T. Bower. 1968. *Managing Water Quality: Economics, Technology, Institutions.* Baltimore, MD: The Johns Hopkins University Press for Resources for the Future, Chap. 11.

Kneese, Allen V., and Charles L. Schultze. 1975. *Pollution, Prices, and Public Policy.* Washington, DC: Brookings Institution.

Laplante, Benoit, and Paul Rilstone. 1996. Environmental Inspections and Emissions of the Pulp and Paper Industry in Canada. *Journal of Environmental Economics and Management* 31(1, July):19–36.

Lettenmaier, Dennis P., Eric R. Hooper, Colin Wagoner, and Kathleen Faris. 1991. Trends in Stream Quality in the Continental United States, 1978–1987. *Water Resources Research* 27(2, March):327–39.

Luken, Ralph A., Daniel J. Basta, and Edward H. Pechan. 1976. *The National Residuals Discharge Inventory.* Washington, DC: National Research Council, Chap. 9.

Magat, Wesley A., Alan J. Krupnick, and Winston Harrington. 1986. *Rules in the Making: A Statistical Analysis of Regulatory Agency Behavior.* Washington, DC: Resources for the Future.

Magat, Wesley A., and W. Kip Viscusi. 1990. Effectiveness of the EPA's Regulatory Enforcement: The Case of Industrial Effluent Standards. *Journal of Law and Economics* 33(2):331–60.

National Commission on Water Quality. 1976. *Report to the Congress.* Washington, DC: National Commission on Water Quality, March.

New York Times. 1998. EPA and States Found to be Lax on Pollution Law. June 7, p. 1.

New York Times. 1999. EPA Wants Water Quality as New Gauge for Cleanup. August 15, p. 13.

Portney, Paul R. 1981. The Macroeconomic Impacts of Federal Regulation. In *Environmental Regulations and the U.S. Economy,* edited by Henry M. Peskin, Paul R. Portney, and Allen V. Kneese. Baltimore, MD: The Johns Hopkins University Press for Resources for the Future.

Raucher, Robert S., and others. 1993. *An Evaluation of the Federal Drinking Water Program under the Safe Drinking Water Act as Amended in 1986.* Prepared for the American Water Works Association. Boulder, CO: RCG/Hagler, Bailly, Inc.

Ribaudo, Marc O. 1989. *Water Quality Benefits from the Conservation Reserve Program.* Agricultural Economic Report No. 606. Washington, DC: U.S. Department of Agriculture, Economic Research Service.

Rose-Ackerman, Susan. 1973. Effluent Charges: A Critique. *Canadian Journal of Economics* 6:512–28.

——. 1977. Market Models for Pollution Control: Their Strengths and Weaknesses. *Public Policy* 25(3):383–406.

Russell, Clifford S. 1979. What Can We Get from Effluent Charges? *Policy Analysis* 5(2, Spring):155–80.

——. 1990. Monitoring and Enforcement. In *Public Policies for Protecting the Environment,* edited by Paul R. Portney. Washington, DC: Resources for the Future.

Smith, Richard A., Richard B. Alexander, and M. Gordon Wolman. 1987. Water-Quality Trends in the Nation's Rivers. *Science* 235(March 22):1607–15.

Strock, James M. 1990. Final Summary Report: Enforcement Effectiveness Case Studies. Memorandum from James M. Strock, Assistant Administrator, Office of Enforcement, U.S. EPA, to the Steering Committee on the State/Federal Enforcement Relationship, Deputy Regional Administrators, Associate Enforcement Counsels, Office of Enforcement Directors, Office of Enforcement. Washington, DC: U.S. EPA.

Tietenberg, T. H. 2000. *Environmental and Natural Resource Economics*, 5th ed. Reading, MA: Addison-Wesley, 446.

U.S. EPA (Environmental Protection Agency). 1984a. *Non-Point Source Pollution in the U.S.* Report to Congress. Washington, DC: U.S. EPA, Office of Policy Analysis, xiii–xiv.

——. 1984b. *Study of the Future Federal Role in Municipal Wastewater Treatment—Report to the Administrator.* Washington, DC: U.S. EPA.

——. 1990. *Environmental Investments: The Cost of a Clean Environment.* Washington, DC: U.S. EPA.

——. 1996. Effluent Trading in Watershed Policy Statement. *Federal Register* 61(28, February 9):4994–6.

——. 1998. *National Water Quality Inventory—1996 Report to Congress.* Washington, DC: U.S. EPA.

U.S. GAO (General Accounting Office). 1983. *Waste Water Dischargers Are Not Complying with EPA Pollution Control Permits.* Washington, DC: U.S. GAO.

——. 1996. *Water Pollution: Many Violations Have Not Received Appropriate Enforcement Attention.* Washington, DC: U.S. GAO.

Van Houtven, George, and Maureen L. Cropper. 1996. When is a Life Too Costly to Save? The Evidence from U.S. Environmental Regulations. *Journal of Environmental Economics and Management* 30(3):344–68.

Vaughan, William J., and Clifford S. Russell. 1982. *Freshwater Recreational Fishing: The National Benefits of Water Pollution Control.* Washington, DC: Resources for the Future.

Viscusi, W. Kip. 1996. The Dangers of Unbounded Commitments to Regulate Risk. In *Risks, Costs, and Lives Saved: Getting Better Results from Regulation*, edited by Robert W. Hahn. New York: Oxford University Press.

Water Quality Committee. 1983. Annual Review of Significant Activities—1983. *Natural Resources Lawyer* 17(2):277–8.

Zwick, David, and Marcy Benstock. 1971. *Water Wasteland: Ralph Nader's Study Group Report on Water Pollution.* New York: Grossman Publishers, 264–84.

seven

Hazardous Waste and Toxic Substance Policies

Hilary Sigman

Concerns about toxic chemicals helped spawn modern environmentalism and with it the growth of environmental regulation in the United States. In *Silent Spring*, Rachel Carson wrote movingly about the ecological effects of the pesticide DDT. Published in 1962, the book raised the specter of widespread environmental destruction as a result of synthetic chemicals and helped shape modern views about our relationship with the environment.

Nonetheless, federal policies for hazardous wastes and toxic substances lagged behind policies for air and water pollution. The outlines of current policies were not in place until the 1980s, almost a decade later than modern air and water programs. However, several major environmental statutes now address these issues.

In this chapter, I discuss two related sets of federal public policies.[1] The first set addresses hazardous wastes. The Resource Conservation and Recovery Act (RCRA) regulates the current management of hazardous wastes. Although it addresses contamination from current industrial processes, RCRA does not fully address the legacy of contamination from the past. Cleanup of the latter occurs under the Comprehensive Environmental Response, Liability, and Compensation Act (CERCLA), better known as Superfund.

I also discuss policies that address toxic substances. These policies have the general goal of controlling exposure to toxic substances at any stage, unlike RCRA and Superfund, which focus on chemicals at the end of their useful life. Although the scope sounds broad, toxic substance policies have had more limited application in practice than other environmental policies.

Three important toxic substance programs are in place in the United States. The Toxic Substances Control Act (TSCA) gives the U.S. Environmental Protection Agency (EPA) authority to test and regulate all chemicals in commerce in the United States. The Federal Insecticide, Fungicide, and Rodenticide Act (FIFRA) has similar goals but applies specifically to pesticides. The Emergency Planning and Community Right-to-Know Act (EPCRA) addresses industrial releases of toxic chemicals into many different environmental media. Unlike TSCA and FIFRA, which impose restrictions on polluters, EPCRA makes information about their discharges available to the public.

Public Policy for Hazardous Waste Management

It is tempting to think of hazardous waste as another form of pollution, like air or water pollution. However, it really is a precursor of pollution: The effects of hazardous waste on the environment depend on the way it is managed, whether through recycling, treatment, or disposal. Hazardous waste management can affect several different environmental media.

Management of Hazardous Wastes

Waste management includes three general categories of activities: disposal, treatment, and recycling. Facilities can dispose of waste on land, typically by placing it in landfills or injecting it into deep underground wells. They also can treat waste by using chemical processes that vary with the kind of waste. For example, treatment can render waste inert through vitrification, a process that converts waste to a glasslike substance. Few treatments avoid the need to dispose of some residuals later, but they may result in a waste product that is less toxic or less mobile than the pretreated substance. Finally, facilities can avoid disposal by reusing organic waste as fuel or by recycling some hazardous waste, especially waste that contains metals and spent solvents.

Table 7-1 is a summary of the major hazardous waste management practices used in 1997. Underground injection accounts for the largest share of waste management by weight. However, this statistic probably overstates the relative environmental importance of underground injection because this method manages wastewaters. Wastewaters often have vast quantities of water mixed with only trace amounts of hazardous chemicals—however, all of this mixture is categorized as hazardous waste. Other management meth-

TABLE 7-1. Summary of Hazardous Waste Management Methods, with 1997 Quantities.

Management method	*Description*	*Quantity (million tons)*	*Percentage of total*
Disposal			
Landfills	Solid and contained-liquid waste placed into lined disposal areas and covered with soil	1.5	3.98%
Underground injection	Wastewaters injected into deep wells or salt caverns	26.2	69.50%
Surface impoundments	Liquids and sludges placed in pits, ponds, or lagoons for treatment, storage, or disposal	1.0	2.65%
Land treatment	Wastes applied to soils for degradation or immobilization by soil microbes	0.01	0.03%
Other disposal		0.3	0.80%
Treatment			
Incineration	Wastes burned at high temperatures	1.7	4.51%
Stabilization	Wastes solidified or immobilized	1.4	3.71%
Sludge treatment		0.4	1.06%
Recycling and reuse			
Recycling	Recovery, especially of solvents and metals	2.1	5.57%
Energy recovery	Waste reused as fuel or mixed with fuels	3.2	8.49%

Notes: Residuals from treatment not included in disposal quantities. Recycling and recovery underreport true quantities.

Source: Quantities from U.S. EPA 1999b, 2–10.

ods typically involve wastes that are more concentrated. Excluding underground injection, recycling and reuse as fuel are the leading management approaches, followed by landfilling and incineration.[2]

Hazardous waste management affects the environment in several ways. Land disposal of hazardous waste may contaminate groundwater. Liquids that move through landfilled wastes, such as rainwater, can carry hazardous substances into underground aquifers. Scientists are still learning about the conditions that allow substances to migrate into the groundwater and the behavior of contaminated water once there. Groundwater contamination can affect human health when people extract groundwater for drinking water or irrigation. Also, some groundwater flows into surface waters, such as streams and wetlands, and thus may cause ecological damage.

Although groundwater protection is the primary motivation for hazardous waste regulation, hazardous waste management may have other environmental costs. Land disposal can affect soils and surface water (such as lakes and streams) if hazardous substances migrate into these environmental media. Landfills also can release air pollutants, causing air quality problems.

Treatment can have harmful consequences, too. Incineration nearly destroys the waste (leaving behind only a small amount of ash) and thus often provides an alternative to land disposal. However, like land disposal, it can give rise to harmful air emissions. For example, EPA estimates that air emissions of dioxins from hazardous waste facilities may cause 49–140 cases of cancer annually (U.S. EPA 1990).

Regulatory Strategy under RCRA

RCRA (often pronounced "rick-ra") regulates current hazardous waste management. It emerged in 1976 as amendments to an older law, the 1965 Solid Waste Disposal Act. RCRA was the first legislation to give the federal government authority to regulate hazardous wastes and nonhazardous solid wastes (see Chapter 8, Solid Waste Policy, for a discussion of the latter provisions).

When Congress initially passed RCRA, it gave EPA great flexibility in designing a program for hazardous waste regulation. The agency struggled with the task. It did not issue the first set of regulations until 1980 and had not gotten very far in issuing permits to facilities when Congress significantly amended RCRA in 1984.[3]

EPA's speed and approach dissatisfied Congress, which decided to take matters into its own hands. In the 1984 Hazardous and Solid Waste Amendments, Congress specified extraordinarily detailed regulatory requirements and fixed timetables for EPA to issue regulations.[4] These changes greatly increased the stringency of the regulations but maintained some of the earlier framework.

Characterizing Hazardous Waste. When Congress gave EPA the authority to regulate hazardous wastes in 1976, EPA first had to determine which wastes were hazardous. This question does not have a simple answer. Congress defined hazardous wastes very generally, as wastes that "pose a substantial present or potential hazard to human health or the environment."[5] Potentially hazardous wastes may come from many sources, including industrial facilities, households, and commercial businesses. Wastes vary greatly in the threats they pose. Not only do wastes contain a dizzying variety of chemicals, but the composition of the waste, the concentration of the chemicals, and the way the waste is managed also affect the degree of risk. In large enough quantities, many common substances (table salt, for example) can pose threats to human health or the environment.

Initially, EPA planned to require all generators to test their waste. Waste would qualify as hazardous if tests proved positive for one of many characteristics, including potential to accumulate in living organisms, toxicity to plants, and tendency to cause genetic mutations. This approach is desirable because it regulates waste based on the risks it poses. Facilities thus would have incentives to change the properties of their waste to eliminate its hazardous characteristics.

However, EPA backed away from this strategy because of its complexity for regulated firms and lack of consensus on testing procedures for many of the characteristics. Instead, it established a mixed system for determining whether wastes are hazardous. EPA considers wastes hazardous if they test positive for one of four characteristics—corrosivity, ignitability, reactivity, or toxicity—*or* if EPA explicitly lists them as hazardous waste. EPA often classifies "listed" wastes by their use rather than their chemical composition; this approach makes the regulations simpler for firms but may lead to the regulation of relatively benign wastes. Table 7-2 lists the categories that waste generators reported for their wastes in 1997.[6] As the data indicate, much of regulated waste tests positive for toxicity.

Many wastes that might appear to meet RCRA's definition for hazardous waste are nonetheless exempt from these regulations. For example, wastewaters regulated by the Clean Water Act are not subject to RCRA regulation. In addition, Congress chose to exclude from coverage mining wastes and wastes from the production and development of fossil fuels. The logic for excluding these wastes is probably more political than environmental. RCRA also excludes some wastes if they are recycled, rather than treated or disposed, in an attempt to encourage recycling. This exclusion creates some thorny issues in defining what constitutes recycling. Finally, it does not regulate hazardous

TABLE 7-2. Types of Waste Regulated by RCRA, 1997.

Type of waste		*Quantity (million tons)*	*Percentage of reported waste*
"Characteristic" waste	Toxic	27.28	67%
	Corrosive	10.00	25%
	Reactive	4.87	12%
	Ignitable	5.97	15%
"Listed" waste		18.86	46%

Note: Total percentages do not equal 100 because wastes may fall into more than one category.
Source: U.S. EPA 1999b.

waste generated by households, even though paints, herbicides, and batteries can be as dangerous as industrial wastes.

Facilities that generate small quantities of waste also received special exemptions before the 1984 amendments. In its initial regulations, EPA opted not to include facilities that generate less than 1000 kilograms of regulated wastes per month. However, in 1984, Congress extended most hazardous waste regulations to facilities that generate between 100 and 1,000 kilograms of hazardous waste per month, known as small-quantity generators. In 1984, there were about 113,000 small-quantity generators, mostly automobile repair shops (Abt Associates 1985). By comparison, there were only about 14,000 large-quantity generators, so this policy change greatly increased the number of facilities that had to comply with RCRA.

Federal RCRA rules are not the only basis for deciding which wastes are hazardous. Several states have expanded their definitions beyond the federal requirements.[7] Some states have decided to bring additional wastes under regulation, most commonly, PCBs (polychlorinated biphenyls) and waste oil. A few states, including California, also specify additional tests for characterizing wastes as hazardous. Waste generators and waste management facilities must follow RCRA requirements for wastes that their states designate as hazardous.

Standards for Treatment and Disposal Facilities. Of course, the definition of hazardous waste matters only because of the restrictions that follow if waste meets these criteria. (Although hazardous waste is subject to the most stringent requirements, generators of other waste are not entirely off the hook: Industrial wastes are regulated, but under much less strict solid waste regulations.) At the outset, the primary goal of RCRA was to set standards for facilities that manage hazardous wastes. Standards for treatment, storage, and disposal facilities began with the interim regulations issued in 1980 and were replaced by final regulations in 1982. Although the requirements have changed over time, they continue to have the same basic structure.

For most treatment facilities, the regulations set standards for the removal of contaminants from the waste and for emissions. For example, incinerators—facilities that burn waste—must destroy 99.99% of the principal organic constituents of most wastes (and a larger percentage of constituents of dioxin-bearing wastes). They also must achieve certain air emissions standards, which the 1990 Clean Air Act Amendments have strengthened. Finally, incinerators must manage their ash as a hazardous waste.

For land disposal facilities, RCRA attempts to lower the likelihood that wastes contaminate groundwater. According to detailed specifications by Congress in the 1984 amendments, land disposal facilities are required to have thick liners (barriers to keep contaminated liquids, called leachate, from leaving the landfill), leachate collection systems, and groundwater monitoring to ensure that there is no leakage.

In addition, all treatment, storage, and disposal facilities face financial responsibility requirements. They must either carry insurance against sudden and accidental pollution incidents or have adequate assets to cover the cleanup of such incidents. Disposal facilities have additional financial requirements. They must have liability insurance for gradual releases as well as sudden and accidental releases. They also must guarantee that they can finance care of the facility after it closes. (Such care may include monitoring the groundwater for contamination and maintaining the structures that contain the waste.)

Congress hoped that these financial requirements would keep the government from bearing cleanup costs, as it has at many Superfund sites. The financial requirements also may strengthen facilities' incentives to reduce the risk of environmental contamination (Boyd 1997). Without these requirements, firms could declare bankruptcy to limit their losses in the event of damage; as a result, those firms might take few steps toward preventing environmental contamination. The financial responsibility requirements force firms to either have a larger stake in the risk of damages or obtain insurance. In the latter case, insurance companies have incentives to require precautions by firms that they insure.

Land Disposal Restrictions. EPA had only begun to implement the first standards for treatment, storage, and disposal facilities when Congress decided to restrict waste management more severely. In the late 1970s, widespread concern developed about the health effects of abandoned hazardous waste sites. This concern, which also prompted the Superfund program, led Congress to tighten RCRA's regulation of land disposal. Trying to avoid repeating past mistakes, it ordered EPA to eliminate nearly all land disposal of untreated wastes.

The land disposal restrictions prohibit the disposal of any hazardous waste on land—options which include landfills, land treatment, surface impoundments, and underground injection—unless those wastes are first treated. If possible, treatment must destroy hazardous constituents, for example, by incineration. When the hazardous constituents cannot be destroyed, their mobility must be reduced. The treatment standards require use of the "best demonstrated available technology"; for some wastes, any treatment that reduces contaminants to the same extent that this technology does is acceptable.

Frustrated with the slow initial progress on RCRA regulations, Congress established strict deadlines for EPA to establish treatment standards on different groups of wastes. If EPA did not keep up with Congress's schedule, "hammer" provisions would ban all land disposal of hazardous wastes, even treated wastes. Congress specified that the first regulations should address solvents and dioxin-bearing wastes by November 1986 and a group of wastes already banned from land disposal in California (known as the "California list" wastes) by June 1987. The remaining regulated wastes were divided into three groups (referred to as "thirds") and restricted sequentially in 1988 through 1990. Since then, EPA has continued to issue regulations pertaining to wastes newly classified as hazardous.

Cleanup Programs. The 1984 RCRA amendments also added two programs directed at cleaning up contaminated sites. First, Congress instructed EPA to regulate underground storage tanks (USTs). About 1.1 million active USTs store gasoline, crude oil, heating oil, and hazardous chemicals for residential, commercial, and industrial functions. If these tanks leak, they may contaminate groundwater. In response to the amendments, EPA established technical requirements to prevent and detect releases from new and old USTs. Also, a fund was established to clean up UST leaks. As with Superfund, a combination of legal liability and an excise tax pays for cleanup.

Congress also added the Corrective Action program to RCRA in 1984. This program cleans up sites contaminated as a result of RCRA-regulated activities. Under regulations promulgated in 1990, EPA outlined study, decisionmaking, and cleanup phases that closely match the stages for Superfund sites discussed later (see "The Superfund Process"). However, the program is still in a state of flux. EPA finalized few of the 1990 regulations (even though the proposed regulations serve as guidelines for cleanups that have been conducted in the interim) and issued a general strategy for a revised program in 1996.[8]

As one might expect, the RCRA Corrective Action program encounters many of the same dilemmas as Superfund. In particular, RCRA Corrective Action sites share the "How clean is clean?" debate that rages over Superfund sites. Large sums may be at stake, as under Superfund. EPA estimates that RCRA Corrective Action will cost $18.7 billion dollars (in discounted 1992 dollars) over the life of the program (U.S. EPA 1993).

Enforcement. The effects of regulations depend in part on the enforcement threats that lie behind them. Although few environmental regulations are easy to monitor and enforce, hazardous waste regulations pose a particular problem (Hammitt and Reuter 1988). Waste generators can take waste offsite for illicit disposal ("midnight dumping"), making violations of waste laws especially easy to hide.

The RCRA regulations raise the costs of legal hazardous waste management and thus provide significant incentives for waste generators deliberately to dispose of waste illegally. Consider the experience of a New York police sting operation:

> When undercover detectives, posing as illegal dumpers, went into the business of disposing toxic waste from small businesses for $40 a barrel, no questions asked, they found competition [from other illegal operators] so fierce that they had to lower their price.... One business asked for a payment plan; another insisted it was accustomed to receiving a forged disposal manifest from other illegal dumpers ... (Barbanel 1992, B5).

The illegal dumpers charged less than $40 for illegal waste disposal, whereas the cost of legal disposal for the wastes that the detectives collected averaged $568 per barrel—a dramatic cost difference. The detectives' experience suggests that illegal disposal services are not only inexpensive but also readily available and convenient.

Empirical research demonstrates that these cost differences cause an increase in midnight dumping. In recent research, I found that public policies that raise legal waste management costs for used oil substantially increased the frequency with which this waste was dumped illegally (Sigman 1998b). Because illegal dumping damages the environment much more severely than legal waste management, the regulations may actually provoke greater environmental damage than they prevent.

In an attempt to reduce illegal disposal, the original RCRA law included a system for tracking hazardous wastes. A waste manifest must accompany all hazardous wastes sent off-site for treatment, recycling, storage, or disposal. When the waste reaches its final destination, this facility must send a copy of the manifest back to the waste generator and to state authorities.

However, waste manifests do not seem to have helped enforce hazardous waste laws. For one thing, most states do not have the resources to deal with the paper trail that is created. Only a few states computerize the manifests and thus have any chance of catching inconsistencies. A study of state enforcement programs in the early 1980s found no instances in which manifests tipped off authorities to violations, although the use of manifests may have improved subsequently (U.S. GAO 1985). Illegal operators also can easily beat the system with simple forgery, as the New York police sting encountered.

In addition to the waste manifest system, RCRA has direct monitoring and enforcement. The overall effectiveness of these activities in reducing illegal disposal is unknown, but some data suggest their limitations. In fiscal year 1995, state and federal authorities conducted 20,764 RCRA facility inspections (U.S. EPA 1996a). These inspections might effectively monitor the 1,718 permitted treatment, storage, and disposal facilities and possibly the 19,908 large-quantity generators in that year. However, small-quantity generators (which numbered 113,000 in 1984) probably face little risk of detection for RCRA violations.[9]

Trends in Hazardous Waste

In this section, I describe the recent trends in hazardous waste management and attempt to assess the impacts of RCRA on these activities.

Prices for Hazardous Waste Management. Although RCRA does not directly limit the amount of waste that facilities can generate, it could affect waste generators' decisions if it changes the costs of managing waste. Table 7-3 reports the prices charged by commercial hazardous waste management facilities. The average of low-end and high-end prices quoted by major firms in 1981, 1987, and the early 1990s are shown for various management methods. Although a lot of waste is managed at noncommercial facilities, the costs for commercial services indicate trends for waste management as a whole.

Some kinds of waste management became much more expensive during the 1980s (Table 7-3). For example, incineration prices appear to have increased dramatically. In 1981, incinerators actually paid waste generators for clean liquid wastes with high energy content; by 1987, incinerators charged at least $400 per metric ton for processing these wastes. Landfilling also increased in cost, but less dramatically than incineration.

RCRA's requirements for treatment, storage, and disposal facilities may have caused part of the price increase for landfills and incineration after 1981. Most facilities first faced federal hazardous waste regulation during this period. After issuing the first permanent regulations in 1982, EPA began to phase in performance standards and financial responsibility requirements. In addition, RCRA's land disposal restrictions increased demand for treatment in the mid-1980s. It took time to get new capacity on-line to respond to this demand.[10] The availability of expanded capacity may explain why by 1994, incineration prices had fallen to levels well below their peak in the mid-1980s. Prices in the 1990s for both incineration and landfilling remained above their 1981 levels, perhaps reflecting the costs of RCRA regulation.

However, even absent the RCRA regulations, waste management might have become more expensive in the 1980s for two reasons. First, communities became more aggressive in fighting construction and expansion of local waste management facilities. This community action—sometimes called not-

TABLE 7-3. Commercial Hazardous Waste Management Prices (in 1996 dollars per metric ton, except as indicated).

Management method and type of waste	*1981*		*1987*		*1993 or 1994*[a]	
	Low	*High*	*Low*	*High*	*Low*	*High*
Landfill						
55-gallon drum (per drum)	46	65	80	231	70	140
Bulk wastes	72	108	133	227	114	260
Incineration						
Liquids	69	309	437	1111	58	585
Solids	514	1,031	1,775	2,813	1,169	2,104
Clean liquids, high energy	–16	69	444	969	—	—
Highly toxic liquids	514	1,031	776	1426	—	—
Chemical treatment						
Highly toxic wastes	172	1,101	—	—	—	—
Inorganic liquids	—	—	85	394	—	—
Inorganic solids and sludges	—	—	145	1,153	—	—
Resource recovery						
All organics	86	344	—	—	—	—
Aqueous organics	—	—	131	329	—	—
Nonaqueous organics	—	—	125	789	—	—
Deep well injection						
Oily wastewaters	21	52	30	164	—	—
Other (toxic liquids)	172	344	49	207	—	—
Transportation (metric tons/mile)	0.21		0.29		—	

[a] 1993 price for landfills, 1994 price for incineration.

Sources: Data for 1981 and 1987 are from ICF, Inc. 1985; 1988. Prices for 1993 landfills are from Peretz and Solomon 1995. Prices for 1994 incineration are from a survey by Environmental Information, Ltd. (Krukowski 1995).

in-my-backyard, or NIMBY, politics—probably made expanding waste management capacity more costly.

Second, management facilities began to face a greater threat of lawsuits for environmental contamination. The increase resulted partly from federal regulations: RCRA's Corrective Action program and Superfund allow the government to seek large sums for cleanup of contaminated sites and for any natural resource damages. Lawsuits may also occur under state policies for contaminated sites, which sometimes authorize private parties to sue for damages from contaminated sites (Environmental Law Institute 1996).

With the increase in litigation, the cost of insurance against environmental liability claims has risen dramatically, leading to higher waste management costs. In addition, some facilities cannot find insurers willing to write environmental liability policies at all. In 1991, 87% of treatment, storage, and disposal facilities reported that obtaining adequate pollution liability insurance was very difficult or nearly impossible.[11] When waste treatment facilities cannot purchase insurance for liability claims, they must bear the liability risk themselves, which also raises management costs.

Table 7-3 also gives an idea of the difference in costs among the various waste management methods. Even with the RCRA standards for landfills in place by 1987, land disposal was much cheaper than many alternative management methods. Thus, by precluding this option, the land disposal restrictions dramatically increased waste management costs for many facilities.

Waste Management. Figure 7-1 shows trends in land disposal since the early 1980s. Except for 1981 and 1986, the data on the quantity of waste (bars) are

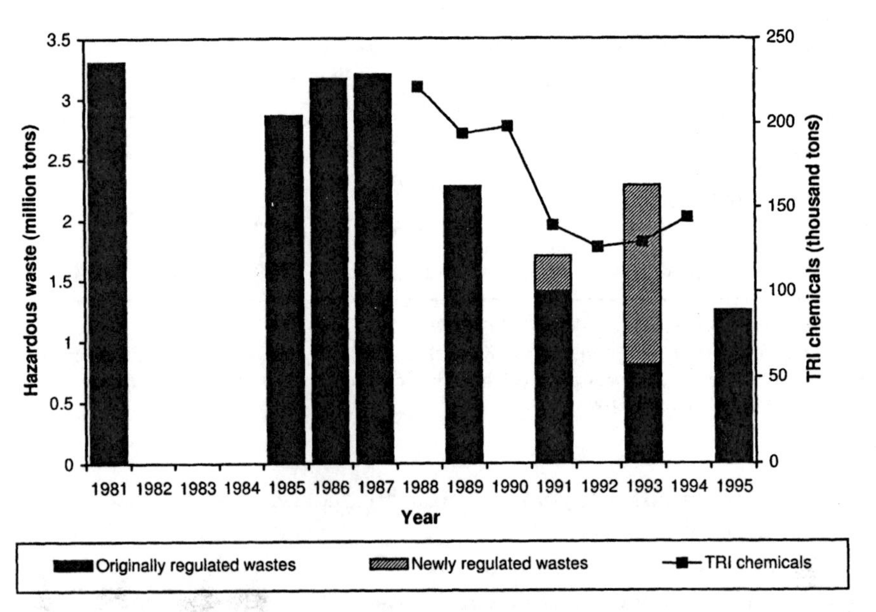

FIGURE 7-1. Disposal of Hazardous Waste in Landfills, 1981–95.

Notes: Bars represent the total quantity of hazardous waste, including media (such as water) generated by large-quantity generators and managed at RCRA-permitted facilities. Points represent the hazardous chemicals in waste reported to the TRI.

Sources: Bars for 1981 and 1986 are from EPA surveys; bars for 1987–95 from the author's calculations based on RCRA biennial reports; point data from author's calculations based on the TRI.

from biennial reports required by RCRA. These data are not ideal for studying trends. First, the states are responsible for collecting the data and may not collect data consistently. EPA stepped up its attempts to standardize data nationally in 1989, so the 1985 and 1987 biennial report values are subject to more uncertainty than the later data. Second, EPA has expanded the set of regulated wastes over time, masking changes for a given set of wastes. Figure 7-1 attempts to distinguish newly regulated wastes from waste regulated originally, but it is not possible to draw this distinction precisely. Most of the newly regulated wastes result from a change in the toxicity test in 1990 and a number of new wastes listed between 1989 and 1991.

The bars in Figure 7-1 show the quantities of hazardous waste, including the media (such as water) that contain hazardous substances. These quantities might decrease without environmental improvements if facilities move toward waste with more concentrated hazardous substances. To show trends in hazardous substances alone, Figure 7-1 also includes data on the quantity of chemicals in waste from the Toxics Release Inventory (TRI), 1988–94. Facilities report their environmental releases and off-site transfers of many toxic substances to the TRI.[12] The universe of TRI chemicals does not include all hazardous constituents of RCRA wastes or vice versa, so the two series represent somewhat different sets of waste.

Although time trends alone cannot be definitive, Figure 7-1 contains a few hints about RCRA's environmental effects. Facilities disposed of about as much waste in landfills in 1987 as they did earlier in that decade, suggesting that RCRA's technological standards did not greatly discourage land disposal. Disposal in landfills did decline dramatically after the RCRA land disposal restrictions began in late 1986.[13] Between 1987 and 1993, landfill disposal of the originally regulated wastes may have declined as much as 75%. TRI data suggest that land disposal of hazardous waste constituents also declined between 1988 and 1992.

The trends at the end of the period are more complicated. The amount of originally regulated wastes placed in landfills appears to have increased somewhat between 1993 and 1995. Similarly, TRI data suggest an increase in land disposal of hazardous substances after 1992. However, the newly listed wastes began to come under land disposal restrictions in 1992; their land disposal fell nearly to zero between 1993 and 1995.

Waste Generation. In addition to its intended effects on land disposal, RCRA may also have had indirect effects on total waste generation. The program may have increased costs of hazardous waste management both because standards raised costs for given management methods and because the land disposal restrictions ruled out cheaper management options. Such cost increases would provide facilities with incentives to reduce the amounts of waste they generate, which they might accomplish through several means.[14] Facilities might change their manufacturing process to generate less waste or to recycle their wastes. For example, a facility that uses solvents to clean manufactured metal parts might have a variety of options to reduce the amount of used solvent waste it generates. It may handle parts differently to keep them cleaner, substitute water cleaning, or purchase equipment for on-site solvent recovery. Waste generators might also scale back their output of those goods that generate a lot of waste. Finally, industries may become more aggressive in seeking out and developing new technologies that allow them to produce less waste.

Did waste generation in fact decline since RCRA went into effect? Figure 7-2 shows trends in this quantity.[15] The biennial report data in Figure 7-2 show a decline in waste generation beginning in the mid-1980s. Figure 7-2 also reports a measure of TRI chemicals in wastes beginning in 1991. Unlike the biennial report data, the quantity of TRI chemicals in waste does not decline in the 1990s. The difference suggests that hazardous chemicals in waste may have become more concentrated, accounting for some of the decline in waste generation.

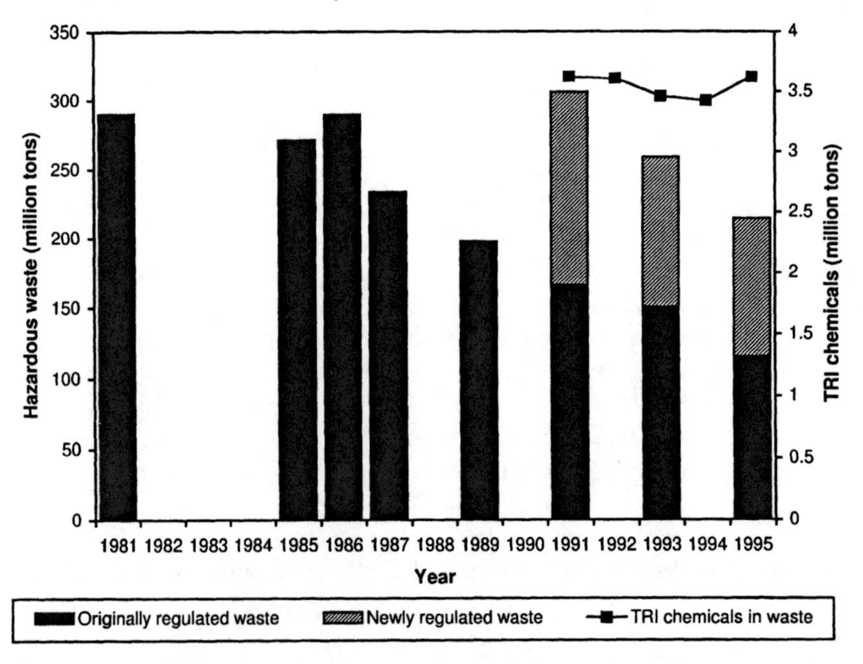

FIGURE 7-2. Generation of Hazardous Waste, 1981–95.

Sources: Same as Figure 7-1.

It is purely speculative to attribute any portion of the 1980s decline to RCRA. Changes in waste management could have a variety of sources. They could result from the cost increases shown in Table 7-3, but RCRA is only one reason for these cost increases. Facilities might also have reduced their waste generation without increased costs of waste management. They may have switched production techniques or decreased their output for reasons unrelated to RCRA.

Economic Evaluation of RCRA

Despite the possible environmental accomplishments described above, we should not consider the RCRA program a success unless it improves the environment at an acceptable cost. Comparing relative costs and benefits of RCRA is difficult because we know little about its effect on the environment. RCRA may have reduced environmental contamination of air, groundwater, surface water, and soils. However, there are no measures of these changes, let alone evaluations of how human health and the environment benefited from reduced contamination. Perhaps because of the lack of information about benefits, no one has attempted an overall assessment of the benefits and costs of RCRA.

Although no overall cost–benefit comparisons are available, there are some evaluations of individual RCRA regulations. Since the early 1980s, EPA must conduct "regulatory impact analyses" of new regulations that have large potential costs. These analyses contain information about the expected costs of the regulation and its environmental benefits. Table 7-4 is a summary of the results of these analyses for the land disposal restrictions, which probably account for the majority of RCRA costs.[16] As described earlier, the land disposal restrictions took place in several stages for different waste types. Table 7-4 shows the estimated costs and health benefits separately for each group of wastes that became subject to the regulations at the same time. In addition, the costs of regulations for surface disposal (which includes landfills and surface impoundments) are reported separately from underground injection, which was regulated later.

The cost and benefit estimates in Table 7-4 have substantial flaws. The estimates may exaggerate the costs of the policies. For example, they assume the preregulation quantity and mix of wastes. Thus, they do not anticipate flexibility on the part of waste generators, who experienced higher waste management costs as a result of the regulations. To avoid these costs, firms may have modified their production processes to reduce the quantity of waste they generated or to make it more easily treatable. Such adjustments would have made compliance with the regulations less costly than reported.

TABLE 7-4. Estimated Costs and Health Effects from Land Disposal Rules from Regulatory Impact Analyses.

Wastes	*Date*	*Annual costs*[a]	*Reported health benefits*
Surface disposal rules			
Solvents	11/86	186	116 cases of cancer over 70 years
Dioxins	11/86	6	Insignificant
California list wastes	7/87	119	2,298 cases of cancer, fetal toxicity, decreased reproductive capacity over 70 years
"First third" wastes	8/88	1,224	360 cancer cases over 70 years; 422 exposures to noncarcinogenic toxic chemicals over reference dose
"Second third" wastes	6/89	39	0.07 cancer cases over 70 years; 555 exposures to noncarcinogenic toxic chemicals over reference dose
"Third third" wastes	8/90	402	316 cancer cases over 70 years; 5,400 exposures to noncarcinogenic toxic chemicals over reference dose
Underground injection rules			
Solvents and dioxins	7/88	80	NA
California list and "first third" wastes	8/88	15	NA
"Second third" wastes	6/89	5	Insignificant
"Third third" wastes	6/89	66	Insignificant

Note: NA = data not available. [a] Millions of 1996 dollars.

Sources: For surface disposal, *Federal Register* 1986; 1987; 1988c; 1989; 1990. For underground injection: *Federal Register* 1988a; 1988b; 1989; 1990.

In addition, the strict regulations could have prompted research and development of alternative treatment technologies, potentially resulting in new, lower-cost treatments.

The land disposal restrictions also give rise to other benefits beyond those in Table 7-4. When land disposal of hazardous waste contaminates groundwater, it may damage ecosystems if the groundwater eventually flows into surface water. The table does not consider any benefits from avoiding ecological damage.

In addition, the people who live around disposal facilities may worry about the sites, even if experts believe the risks are slight (see McClelland and others 1990). Many studies examine the amount that people are willing to pay to avoid living near waste disposal sites and typically find that it is quite substantial.[17] Perhaps the benefits of the land disposal restrictions should include pure distaste for these facilities.

Despite the limitations of the estimates, we still might ask whether the health risk reductions in Table 7-4 merit the estimated costs. This assessment requires comparing health risk reductions to dollar costs, which we can do by looking at the decisions that people make about risk in their own lives. Although few people make life-or-death decisions on the basis of cost, they often trade off money against small amounts of risk, for example, in choosing whether to accept a more dangerous job for a higher wage. Scaling these private trade-offs up to a 100% probability of death provides the value of a "statistical" life. Although neither the public policies nor the private trade-offs actually create a certainty of death for any individual, scaling to a 100% probability is convenient for comparing the trade-offs across different risk levels. Typical estimates for this value range from between $3 million and $8 million per statistical life (in 1996 dollars) (Viscusi 1993).

Compared with Table 7-4, these values suggest that some of the surface disposal restrictions might have benefits to merit their costs. If we assume that the health consequences of land disposal occur continuously over time, the surface disposal restrictions on California list waste cost about $3.6 million per adverse outcome avoided.[18] Some of these adverse health effects are not fatal and thus probably are worth less than reducing the risk of death. Nonetheless, the benefits of this regulation may exceed its costs. It is not surprising that the regulations on this group of wastes look the most favorable. Wastes on the California list are those whose land disposal California chose to restrict before the federal government did. California selected these wastes to regulate because they are hazardous and easy to treat.

By contrast, the surface disposal regulations on the remaining wastes may have high costs for their benefits. As discussed earlier, EPA had divided wastes that did not fall into the first categories of restricted wastes (solvents, dioxin-bearing wastes, and California list wastes) into three groups, referred to as thirds. The $1.2 billion spent on surface disposal restrictions for "first third" wastes amounts to about $240 million per cancer case avoided. In addition, the regulatory impact analysis mentions 455 reduced exposures annually to toxic chemicals above the reference dose (a maximum daily exposure to a substance "likely to be without an appreciable risk of deleterious effects during a lifetime," even for sensitive populations). Given the conservatism of this definition, the chemical will actually harm very few of the people who have exposures above the reference dose. Thus, the benefits of the "first third" regulations are unlikely to justify their costs. The other "thirds" have similarly high costs for their risk reductions.

The difference between these regulations indicates the advantages that might result from balancing costs and benefits in designing regulations. The restrictions applied to many different wastes, with varying damage to health and treatment costs. For example, the analyses sometimes attributed all the reported negative health effects to a few of the hundreds of waste streams addressed by the "thirds" regulations. Because the regulations required that facilities treat each waste stream with the best demonstrated available technology, EPA did not vary regulations based on risks and costs. If it had, EPA might have achieved the same health improvements at much lower cost, by loosening the regulations for wastes with a high cost per cancer case avoided and tightening the regulations for wastes with a low cost per cancer case avoided.

The land disposal restrictions are not the only aspect of RCRA's hazardous waste regulation that does not balance costs and benefits. RCRA's general structure makes it difficult to target the worst risks. For example, RCRA does not distinguish degrees of hazard among wastes. Any waste categorized as hazardous faces the same regulatory requirements, although wastes may vary greatly in the nature and extent of the dangers they pose. Thus, the regulations do not reserve the most expensive treatment for the waste that presents the highest risk. The program might have substantially lower cost if it permitted standards to vary with hazards.

Equity and Hazardous Waste Management

In addition to the overall levels of costs and benefits from hazardous waste management, their distribution across households is important. In particular, policymakers worry that poor and minority communities bear an unfair share of environmental damage. Reflecting these concerns, in 1994, President Clinton issued Executive Order 12898, which calls on federal agencies to develop strategies to ensure "environmental justice."

Table 7-5 presents characteristics for communities with and without commercial waste-management facilities. Derived from a study by Been and Gupta (1997) that used the 1990 census, the table shows the basis for concerns about environmental justice. On average, communities with waste management facilities have higher proportions of minorities, especially Hispanics, than the remaining communities. Communities with waste facilities also have more households with incomes below the poverty line and lower average incomes across the board.

Residents of disadvantaged communities might have more exposure to hazardous waste management for several reasons. Companies trying to choose sites for new or expanded hazardous waste management facilities may disproportionately select poor communities because land is least expensive

TABLE 7-5. Comparisons of Communities with and without Commercial Hazardous Waste Management Facilities in 1990.

	Average values	
Population characteristics	*Census tracts with a facility*	*Census tracts without a facility*
African Americans	14.39%	13.46%
Hispanics	10.34%	7.83%
Minorities (all nonwhite races and Hispanic whites)	27.21%	24.17%
Below poverty line	15.69%	14.59%
Median family income	$31,602	$34,586

Source: Been and Gupta 1997, 54.

there. Residents of disadvantaged communities may also have less access to the political system and therefore less ability to oppose the siting of new facilities than residents of other communities. In addition, racism could play a role in firms' choices or public authorities' oversight of site selection. Finally, disadvantaged communities may develop near noxious facilities rather than vice versa: Poor households may move to the neighborhoods around waste facilities because the housing there is inexpensive.

Several empirical studies try to distinguish the roles of these different factors.[19] In particular, studies ask whether facilities locate disproportionately in minority communities, when the study accounts for the effects of differences in communities' income. If so, this pattern would seem to rule out explanations based on land and housing prices (which should be blind to racial and ethnic differences) and suggest a failure in the political system or direct racism. The literature includes too many studies with different approaches to summarize here, except to say that the results are mixed. Some studies conclude that hazardous waste facilities disproportionately locate in minority communities, whereas others find no effect after they account for income or other characteristics of the areas.

Public Policy for Contaminated Sites

Unlike many other pollutants, certain hazardous substances may persist in the environment long after their initial release. RCRA's regulations regarding current waste management do not solve the problem of contaminated sites that already existed when the law came into effect.

This gap in the law came to public attention with a few incidents in the late 1970s. In perhaps the most dramatic incident, residents of Love Canal (a suburb of Niagara Falls, New York) noticed discoloration on their lawns and basement walls. It turned out that the Hooker Chemical Company had used the area to dispose of chemical wastes from 1942 to 1952. The company then sold the property for a token sum to the community, which built a subdivision and school there, despite being warned of the contamination. By 1978, the plight of Love Canal residents had captured national attention, prompted in part by early reports of high incidence of illness in the community.

In response to the public outcry from this incident, Congress hurriedly passed the Comprehensive Environmental Response, Compensation, and Liability Act (CERCLA) in December 1980. Under CERCLA, EPA identifies sites with hazardous contamination, evaluates the dangers present, and ensures that either the government or private parties clean up the contamination. Because Congress established a large trust fund to pay for site cleanup, the program became known as Superfund. Congress amended the program when it was reauthorized in 1986. As with the RCRA amendments of about the same time, the amendments to Superfund attempted to increase the speed and stringency with which EPA implemented the program.

The Superfund Process

Although abandoned hazardous waste sites originally motivated CERCLA, the program addresses many kinds of contaminated sites. Superfund sites can include not only waste management facilities but also industrial sites, mines, and areas with groundwater contamination of unknown origin. They also include federal facilities, such as defense-related sites with radioactive contamination. EPA has compiled a large inventory of contaminated sites brought to its attention by state and local governments and other concerned groups. By the end of 1996, the inventory included more than 40,000 sites (Table 7-6).

EPA investigates each identified site to determine whether it poses a hazard and whether it falls under CERCLA's jurisdiction. If not, it can designate the site as No Further Remedial Action Planned at this stage. A total of almost 32,000 sites, 79% of the inventory, had received this designation by the end of 1996. Such sites may still receive cleanup under state superfund programs or other federal programs, such as RCRA's Corrective Action.[20]

EPA can respond in two ways to CERCLA sites that are determined to pose environmental risk. First, if the site poses an immediate risk, EPA can conduct (or order liable parties to conduct) a rapid cleanup, called a Removal Action. Removal Actions can cost no more than $2 million if

TABLE 7-6. Progress at Superfund Sites through the End of 1996.

Stage of progress	*Number of sites*
Contaminated sites identified	40,537
Listed as No Further Remedial Action Planned	31,929
At least one Removal Action completed	3,197
NPL activities	
Proposed for the NPL	1,475
Listed on the NPL	1,345
First Record of Decision signed	1,119
First Remedial Design begun	955
First Remedial Action begun	822
First Remedial Action complete	577
Construction Complete	415
Proposed for deletion from NPL	133
Deleted from NPL	129

Note: NPL = National Priorities List.

Sources: Comprehensive Environmental Response, Compensation, and Liability Information System (CERCLIS) and Construction Completion List.

financed by the federal government. They often involve containment of hazardous substances and disposal off-site. At least one Removal Action had occurred at 3,197 sites by the end of 1996 (Table 7-6). Often, a Removal Action completes the necessary cleanup at the site; EPA then may designate the site as No Further Remedial Action Planned.

Federal funds may pay for large-scale cleanup, known as a Remedial Action, only if EPA places the site on the National Priorities List (NPL). To consider a site for the NPL, EPA evaluates it using the Hazard Ranking System, which combines into a single number information about the exposed population and the volume, toxicity, and mobility of contaminants. Although this score is intended to be a first-pass assessment of the site's risk, critics have charged that the score can fail to capture the long-term health risks accurately and may be manipulated for political ends.[21] A site is eligible for the NPL if its Hazard Ranking System score exceeds an arbitrary cutoff value.[22]

The number of sites proposed for the NPL had grown to 1,475 by the end of 1996, and 1,345 were listed on the final NPL (Table 7-6). Almost all sites proposed for the NPL ultimately become final NPL sites. If the rules stay the same in the future, this list is just the beginning: The NPL could grow to more than 8,000 sites in the next few decades, but values below 3,000 may be more plausible (U.S. GAO 1994b). However, proposed legislation would cap the NPL at near its current size and thus begin phasing out Superfund.

After a site is proposed for the NPL, the process of selecting and implementing a remedy begins. EPA can decide to conduct this process once for the entire site or to divide the site up into "operable units." An operable unit corresponds to either an area of contamination or an environmental medium (for example, a contaminated building or surface water). The majority of sites have only one operable unit, but a few sites have as many as twenty.

For each operable unit at the site, detailed studies assess the hazards of the site and evaluate several remedies in terms of cost and effectiveness in reducing contamination. A Record of Decision summarizes the results of the studies, presents a remedy choice, and explains its basis. About 1,100 sites had at least one Record of Decision by the end of 1996 (Table 7-6).

After the Record of Decision, engineering design of the remedy (Remedial Design) begins. Then, Remedial Action finally occurs. After Remedial Action, the site may have further Remedial Actions on other operable units or on the same unit if the first Remedial Action is not successful. When no further cleanup is necessary (or only maintenance or continued groundwater pumping is required), the site is classified as Construction Complete; 415 sites had reached this stage by the end of 1996 (Table 7-6). Finally, EPA proposes to delete the site from the NPL and deletes the site after a comment period. Only 129 sites had completed the process and been deleted through 1996.

Unless Congress ends the Superfund program, this process ultimately will cost a lot. The Congressional Budget Office estimated the present value of Superfund cleanup costs at $51 billion (in 1996 dollars), with a possible range of $29 to $85 billion.[23] These costs exclude federal facilities, where cleanup may be even more expensive. A 1995 study estimated the present value of future costs for federal facilities between $240 and $396 billion.[24] A handful of very contaminated U.S. Department of Energy sites account for most of federal facility cleanup costs.

The money for Superfund comes from several sources. Private parties or government agencies associated with a site pay for most cleanup. Parties that EPA may hold liable are called potentially responsible parties (PRPs). They may include the past and present owners of the site, waste generators, and parties who originally transported toxic substances to a site.[25] PRPs often undertake cleanup themselves under agreements with EPA. Alternatively, EPA can pay for cleanup using the Superfund Trust Fund and then sue the PRPs to recover its costs. In addition to money from PRPs, EPA's Superfund resources come from an excise tax on chemical inputs into petroleum and chemical production, a corporate profits tax, and general government revenues.

How Clean Is Clean?

Each NPL site has its own story. Sites have been used for many different industrial, commercial, and military purposes. Sites that once served as landfills or surface impoundments may contain a witch's caldron of industrial chemicals. Physical attributes of the sites, such as annual rainfall and proximity to surface water and groundwater, also vary greatly. Given this diversity, EPA must choose remedies on a site-by-site basis.

Many cleanup options are available for the sites. Source control remedies address the origin of the contamination, whether it be contaminated soil at an old industrial facility or an abandoned surface impoundment. Table 7-7 shows that these remedies are common but not universal: EPA classified 65% of remedies chosen in Records of Decision signed in fiscal year 1992 as source control remedies. By contrast, groundwater remedies address contamination that has reached underground aquifers. Table 7-7 indicates that 16% of the fiscal year 1992 decisions involved exclusively groundwater remedies; another 45% of remedy decisions in that year had groundwater remedies in addition to source control remedies.

For each kind of contamination, EPA can choose from options that vary in cost and scope. Some remedies treat contaminants. Most source control remedies involve some treatment (Table 7-7). Treatments vary with the chemicals at the site and may include stabilizing, extracting, or destroying contaminants. Groundwater remedies mostly pump out contaminated groundwater and then treat it in some way. A second kind of remedy involves containment, that is, preventing migration of contaminated material. These remedies include constructing retaining walls and installing cement caps over

TABLE 7-7. Summary of Remedies Selected at NPL Sites, Fiscal Year 1992.

Remedy type	*Number of decisions*	*Percentage of total decisions*
Source control remedies		
Principally treatment	50	29%
Treatment and containment in different areas	36	21%
Principally containment	26	15%
Other remedies	7	4%
Only groundwater remedy	28	16%
No action	25	15%
Total Records of Decision	172	100%

Source: U.S. EPA 1994b, 4.

contaminants. Source control remedies commonly rely on containment, but only an occasional groundwater remedy uses this approach. In addition to treatment and containment, EPA can choose approaches that only reduce human exposure, such as fencing the site and providing alternative water supplies (rather than contaminated groundwater). Finally, it can decide to take no further actions at a site or operable unit. EPA opts for no action in a fair number of cases, 15% in 1992.

The choice among these options is difficult. On one hand, EPA would like to choose a remedy that yields permanent protection for human health and the environment under foreseeable contingencies. On the other hand, permanent remedies may not be feasible at some sites or may require great expenditures for small risk reductions. The debate that has developed asks, "How clean is clean?"

The initial Superfund statute in 1980 did not provide much guidance to EPA in selecting remedies. However, in its 1986 amendments, Congress instructed EPA to favor permanent remedies, where permanence implies treatment over containment options. Congress also specified that the chosen remedies should achieve a level of cleanup that meets "applicable or relevant and appropriate requirements" (ARARs).

ARARs include requirements specified in other federal regulations (including RCRA, the Clean Water Act, the Clean Air Act, and the Safe Drinking Water Act) and state requirements, if they exceed federal ones. The EPA site manager must identify ARARs for the site, using professional judgment. A survey found that maximum contaminant levels from the federal Safe Drinking Water Act most frequently serve as ARARs, followed by state drinking water or groundwater standards (Walker and others 1995). These standards can be excessively strict. Using a drinking water standard, for example, means that water coming out of landfill must be as clean as tap water.

In addition to meeting these standards, the cleanup also must achieve acceptable risk levels at the site. EPA now requires that all sites have site-specific risk assessments. The assessments estimate lifetime risks for cancer and noncancer health effects for various scenarios. For example, a scenario might consider a household drinking groundwater that it has pumped, or a person touching contaminated soil. The regulations specify an acceptable risk range for a continuously exposed individual of between one in a million and one in ten thousand (10^{-6} to 10^{-4}). Risks above this range require some remedy; risks below this range do not.

Although ARARs and risk assessments give remedy selection the aura of precise quantitative goals, their application is somewhat haphazard in practice. The roles of ARARs and risk assessments—and the relationship between them—in determining cleanup priorities are often unclear. Not only does the site manager decide which standards should constitute ARARs, but he or she also can waive those that do for several reasons. Similarly, the application of the risk assessments is vague. Although the risk assessments may dictate the need to do something about a particular pathway, they do not determine what to do. At most sites, estimates of how much each of the various remedy options would reduce risks are not available, so the risk assessments cannot affect the choice among possible remedies.[26]

A second flaw in the remedy selection process is that it does not address exposure to risks realistically. To achieve the greatest benefits, the cleanup should be greatest at sites whose risks affect the largest number of people. However, neither the formal risk analysis nor the ARARs consider the number of people actually exposed at the site.[27]

In addition, for many scenarios, people who will be harmed will not be exposed to the site until some time in the future. To determine these health effects, risk analysts must forecast the future use of the site. Critics charge that they too often forecast that people will live at the site and thus future exposures will be high. It is not only unlikely that people will actually want to move to NPL sites but also is something the government can influence. For example, property deeds for NPL sites might prohibit future residential use (or other uses that cause high exposure).[28]

A third problem with remedy selection at NPL sites is the weight given to costs. EPA's regulations allow remedy selection to include the costs of a selected remedy among other considerations, such as the permanence of the remedy. However, it is unclear that this instruction yields a systematic tradeoff of costs against other goals. Gupta and others (1996) examined the decisions that EPA makes to determine the role of costs in practice. They found that EPA systematically trades off the permanence of the remedy against its costs.

Nonetheless, EPA's decisions may not always yield remedies whose benefits justify their costs. Hamilton and Viscusi (1999b) analyzed the cost per cancer case avoided at a sample of 150 NPL sites. They used cancer risk assessments from the official site studies coupled with their own data on exposed populations. Comparing the number of cancer cases avoided to the anticipated cost of the remedy, they concluded that the average cost per cancer case averted was $3 million (in 1993 dollars). This cost falls near the lower end of the range of commonly accepted values for reducing the risk of death. Thus, Hamilton and Viscusi's results suggest that Superfund benefits justify their costs on average.

However, the story became more complicated when Hamilton and Viscusi analyzed the distribution of costs and benefits. The cost per cancer case averted was very low at a few sites. But, at most sites, they found the costs unjustified in terms of the number of cancer cases prevented. At 70% of sites, the cost of reducing cancer was more than $100 million per case, more than an order of magnitude higher than the value of this risk reduction. Thus, although well-targeted Superfund spending may be very desirable, often this spending is not tailored to the cancer risks it eliminates. However, it is possible that considering noncancer benefits of cleanup, such as reducing other health and ecological damage, might justify more of the costs.

A fourth concern about remedy selection is its fairness. High-income communities with Superfund sites may use their political clout to urge especially extensive cleanup for their sites. Superfund attempts to involve communities in decisionmaking, for example, by granting communities funds to hire experts on their behalf. Disadvantaged communities may not get involved in this oversight process. A few studies have looked for inequities across sites in EPA decisionmaking, with mixed conclusions.[29]

Paying for Superfund

Determining who should pay for Superfund cleanup is also controversial.[30] As mentioned earlier, Superfund gets money from taxes and from legal liability imposed on parties who owned a site or contributed contaminants to it.

Two interesting features distinguish Superfund's liability rule. First, Superfund liability is retroactive, which means that EPA may sue PRPs for activities that preceded the beginning of the program. Although critics complain that retroactive liability unfairly changes the rules in the middle of the game, supporters view it as necessary to be sure polluters pay for environmental cleanup. Second, Superfund liability is "joint and several": Any PRP may be required to pay for the entire cost, even if it only contributed a share of the contamination. PRPs who pay all costs then can sue other PRPs for compensation. Again, critics view joint and several liability as unfair, but it does increase the probability that EPA will recover some costs for cleaning up the site.[31]

A debate has begun regarding the wisdom of using legal liability to fund cleanup. One alternative would be to fund Superfund like a traditional public works program: Raise taxes and use them to finance the cleanup. Liability funding has several possible advantages over such a tax-based policy.

First, liability engages PRPs in the study and cleanup at the site, which may result in faster and more efficient cleanups. Many PRPs agree to carry out activities at the site with EPA or state oversight. Table 7-8 summarizes the extent of PRP activity (as a share of nonfederal facility sites to have reached each stage). Through 1996, PRPs had funded at least one Remedial Action at 75% of sites that had experienced a Remedial Action.

TABLE 7-8. Participation of PRPs at NPL Sites.

PRP involvement (through 1996)	*Percentage of sites where PRP financed*[a]
Remedy selection studies (Remedial Investigation/ Feasibility Studies)	60%
Remedial Design	75%
Remedial Action	75%
PRP funding (through FY 1996)	*Value (nominal)*
Estimated value of settlements for remedial activities	$11.9 billion
Cost recovery settlements	$2.1 billion
Cost recoveries collected	$1.4 billion

Note: PRPs = potentially responsible parties.

[a] Percentage of sites to have had at least one such activity financed by PRPs as a share of non-federal facility sites that have reached each stage.

Sources: For percent PRP financed, CERCLIS. For funding levels, U.S. EPA 1997b.

Under an "enforcement first" policy begun in 1989, EPA uses its own funds for remedies only if cannot reach agreements with the PRPs. It then can attempt to recover its costs from the PRPs. EPA estimates that it had reached agreements for Remedial Action cumulatively worth over $11.9 billion (in nominal dollars) through September 1996 (Table 7-8). EPA has received much less money from PRPs in cost recoveries; it had reached agreements for $2.1 billion and collected $1.4 billion. For comparison, EPA spent about $9 billion from taxes and general revenues on cleanup responses over the same period.[32]

PRP participation in the remedy process may improve its outcome.[33] PRPs may have private information about the nature of the contamination at the site that they can use in designing remedies. In addition, private parties may have stronger incentives to control costs than government agencies and thus undertake study and cleanup more efficiently.

A second advantage of liability funding is that it may create incentives for increased precaution by firms that handle toxic substances. If firms expect to bear cleanup costs for future contamination, they may generate less waste and use more permanent treatment and disposal than they would in the absence of liability. Even Superfund's retroactive liability (which concerns activities that already took place and therefore cannot directly change behavior) may create some incentives for current actions if it puts polluters on notice that current behavior may later be judged by even stricter standards.

Finally, proponents of liability funding often argue that it reflects a "polluter pays principle": The beneficiaries of pollution should bear the costs of cleaning it up. Although this goal does seem appealing, it is difficult to implement in practice. The beneficiaries of pollution may include not only the owners of polluting firms but also consumers who purchased their products. However, those consumers who bought products in the past do not necessarily bear the costs of Superfund liability (Fullerton and Tsang 1996).

In addition, liability falls on some parties that one might not think of as polluters. In particular, many municipalities have found themselves the target of Superfund liability because they sent ordinary garbage to landfills that also accepted hazardous waste or because they owned such mixed-waste landfills.[34] Thus, at best, legal liability imperfectly accomplishes the goal of making the beneficiaries of pollution pay for cleanup.

Several arguments speak against liability funding. Many observers are dissatisfied with the speed of progress at Superfund sites and point to the small number of sites deleted from the NPL. Liability funding may contribute to these delays (Sigman forthcoming). According to EPA's site managers, negotiation with PRPs is among the most common sources of delay (Beider 1994a).

Superfund also has a well-deserved reputation as "a full employment act for lawyers." The program has generated an enormous amount of litigation. Not only does EPA sue PRPs, but PRPs also sue one another over their shares of cleanup costs. PRPs also sue their insurance companies over whether insurance policies that the PRPs held before Superfund cover cleanup costs.

Estimates of the cost of all this litigation vary, but most analysts agree that PRPs have spent large sums. Researchers at RAND asked PRPs and their insurers to estimate the share of their Superfund costs that consist of transaction costs such as legal and administrative costs (Dixon 1995). Because many transaction costs come early in the process, before the actual cleanup begins, transaction costs may initially seem to be a higher share of costs than they will be by the end. The RAND researchers forecast the final transaction cost share to be 19–27% of PRPs' total Superfund costs and 69% of their insurers' Superfund costs. Putting these values together, they estimate that private parties' transaction costs will amount to 23–31% of their Superfund liability-related expenditures. The federal government also has legal costs in pursuing private payment, which account for about 10% of its Superfund budget (U.S. EPA 1994a).

Superfund liability may also give rise to costs elsewhere in the economy. In particular, it may discourage the sale and development of land because of potential contamination and thus cause the development of pristine land as a substitute. This phenomenon is called the "brownfield" problem (to contrast with "greenfield" investment, which refers to starting a plant on a new site).

The fact that land ownership may entail Superfund liability does not necessarily discourage sales. Sellers must lower their price by the amount of future Superfund costs, but trades should go through in the end. However, lack of information about contamination may cause Superfund liability to discourage land sales.[35] Differences in buyers' and sellers' knowledge about contamination can give rise to a "lemons" problem: Owners of contaminated land try to sell it to unsuspecting buyers, and buyers become very suspicious and unwilling to buy. In addition, landowners who suspect their land is contaminated may choose not to sell or develop it to avoid inspections that could confirm the contamination and alert authorities.

In recent years, EPA has attempted to address brownfield sites in several ways. Prospective site purchasers can reach agreements with EPA to conduct cleanup in exchange for an assurance that EPA will not sue them. Seven states can offer similar assurances in exchange for voluntary site cleanup (U.S. OTA 1995; U.S. GAO 1997). EPA also funds projects that clean up and develop brownfield sites. However, these programs have affected a limited number of sites thus far.

Given the controversy over Superfund's liability system, Congress has debated restricting the program's scope. For example, proposed changes would eliminate liability for contamination before the passage of Superfund or use taxes to pay for liability shares of PRPs who are no longer financially viable. However, the basic reliance on liability likely will remain in place because political support remains strong for it as a "polluter pays" approach.

Regulating the Use of Toxic Substances

The regulatory programs under RCRA and Superfund address a specific set of environmental releases of hazardous substances: those related to hazardous waste management or other land contamination. Other policies aim to control general exposure to toxic substances. In this section, I discuss three of these policies. The Toxic Substances Control Act (TSCA) provides a general regulatory framework for regulation of toxic substances; the Federal Insecticide, Fungicide, and Rodenticide Act (FIFRA) regulates a specific group of chemicals; and the Emergency Planning and Community Right-to-Know Act (EPCRA) addresses the environmental release of chemicals by providing the public with information about these activities.

Toxic Substances Control Act

As modern environmental regulation began to take form, concern arose that the division of regulation along environmental media lines (air, water, and waste) left no room for a unified policy toward toxic substances. Congress passed TSCA (often pronounced "tos-ca") in 1976 to fill this role. TSCA has two principal functions: to develop information about the toxicity of substances and then, if warranted, regulate those substances.

Under TSCA, EPA has compiled an inventory of about 62,000 chemicals that were produced or imported when the act passed in the mid-1970s. Since then, about 10,000 new chemicals have entered commerce in the United States.[36] Scientists know little about the toxicity of most of these chemicals. The National Research Council (1984) estimated that no data existed on the toxicity of 78% of the most commonly used substances (that is, those with more than 1 million pounds produced per year).

TSCA gives EPA authority to require that manufacturers test chemicals to determine their effects on human health and the environment. Given the large number of chemicals already in commerce and a flow of 1,000–2,000 new chemicals per year, EPA is selective in choosing the chemicals for which it pursues testing. The agency relies principally on structural analyses to compare the chemical with substances of known toxicity and thus determine whether a chemical deserves priority for testing.

TSCA creates separate processes for testing existing chemicals (those already in commerce) and newly introduced chemicals. To require testing of an existing chemical, EPA must go through a full regulatory rulemaking—a long and costly process. The chemical continues to be used during this process.

For new chemicals, more of the burden is on the firm. Firms planning to import or produce a new chemical must submit a Premanufacture Notice (PMN) that provides information about the chemical, the quantities involved, and any known toxicity data. If the information does not satisfy EPA, the agency may delay production pending tests of new chemicals, even without a formal rule.[37] The ability to impose such delay gives EPA more power over new chemicals than existing chemicals. This difference in stringency could be detrimental. It might deter firms from introducing new chemicals that, although hazardous, are less risky (or otherwise better) than existing chemicals. Such distortions from stricter rules for new hazards sometimes are called new source bias.

If tests convince EPA that a new or existing chemical is indeed hazardous, then the agency has two options. First, most exposures to hazardous substances could fall under the jurisdiction of another regulatory program—pollution control, occupational health, or consumer product safety regulations—and EPA may formally or informally refer the chemical to this program. Although there is debate about whether the statute actually requires EPA to defer regulation to other programs, EPA has consistently interpreted the act in this way. Thus, except for the chemical testing program, TSCA largely acts as a last resort, filling in gaps between other regulations.

When EPA does regulate substances under TSCA, it may choose from many types of controls, ranging from labeling requirements and workplace safety precautions to an outright ban on the substance. Unlike many other environmental statutes, TSCA explicitly allows EPA to consider costs in choosing its strategy.

Very little of the testing and regulatory activity under TSCA has gone through the formal regulatory process. Instead, EPA has increasingly relied on agreements with chemical manufacturers. Manufacturers may agree to conduct tests and implement controls on the chemical, such as occupational exposure protection. The threat of regulation under TSCA gives industry an incentive to negotiate such agreements with EPA.

TSCA's accomplishments are hard to pin down. The program may have had some success in generating information about toxic chemicals; about 540 chemicals were tested under the auspices of the program between 1979 and 1996 (U.S. EPA 1996c). However, counting the number of chemicals tested is a poor measure of the benefits of the program. Many chemicals have similar structures and thus similar toxicity, so the absolute number of tested substances does not reflect the degree of information available about chemical risks. More important, information is not an end in itself. The value of this information depends on the extent to which it has permitted better decision-making about exposure to the chemicals.

TSCA has had limited influence on the use of chemicals. EPA has issued formal rules for only five existing and four new chemicals. However, these numbers understate the impact of the regulation because EPA has negotiated agreements with many firms that may include some controls on use of these chemicals. In addition to explicit agreements, the threat of regulation may create incentives for industries to introduce safer chemicals to the market or to produce only small quantities of hazardous chemicals to avoid regulation. Thus, the environmental benefits of TSCA may exceed the count of chemicals regulated. Nonetheless, given EPA's limited ability to regulate chemicals under TSCA, it is unlikely that its incentives are strong enough to alter chemical use significantly.

Federal Insecticide, Fungicide, and Rodenticide Act

FIFRA predates modern environmental legislation. The act began by certifying the efficacy of pesticides in 1947 but shifted focus in the 1960s. With rising concern about DDT and other pesticides, amendments to the act directed its focus toward the protection of human health and environment. When EPA was created in 1970, it took over administration of FIFRA.

FIFRA is much more stringent than TSCA. Whereas TSCA requires testing of only selected chemicals, all pesticides and herbicides regulated under

FIFRA must be tested and approved. The difference between the two statutes results from the great potential risk posed by pesticides. Unlike most industrial chemicals, pesticides are biologically active by design. In addition, pesticides are applied to food products and deliberately released into the environment, so both people and ecosystems have considerable exposure to them.

New pesticides require EPA approval before they can begin production. To obtain this approval (which is called registration), the manufacturer must submit information about the chemical structure and toxicity of the product. In addition to mandating the registration of new pesticides, the 1972 amendments to FIFRA required EPA to reregister all pesticides already in use. This is a substantial task. Currently, about 21,000 pesticide products are formulated from some 860 different active ingredients. Amendments in 1978 simplified the task by requiring reregistration of active ingredients only. As of October 1, 1996, EPA had issued decisions for 148 active ingredients. These approved active ingredients account for about 60% of the 4.2 billion pounds of active ingredients used annually in the United States (U.S. EPA 1997a).

EPA can take several kinds of actions when it reviews a pesticide. If the agency decides to allow continued use of the pesticide, it typically mandates some restrictions designed to protect people who mix or apply pesticides (such as protective clothing). Alternatively, EPA may cancel the registration, thus prohibiting further sale of the pesticide in the United States (however, export for sale abroad is not affected). Cancellation frequently involves lengthy battles that may last from four to eight years.

In choosing among these actions, the FIFRA statute directs EPA to apply a standard of "unreasonable adverse effect." As under TSCA, this standard allows EPA to balance benefits and costs explicitly. EPA can consider factors such as the availability, cost, and toxicity of substitutes for the pesticide. For example, the absence of readily available substitutes for the pesticide might prevent its cancellation.

Although EPA may balance costs and benefits in deciding whether to cancel a pesticide in principle, one might ask whether it does so in practice. To answer this question, Cropper and others (1992) examined pesticide cancellation decisions from 1975 to 1988. They found that EPA's decisions did trade off costs and benefits: The pesticides EPA chose to cancel had fewer benefits relative to their health costs than pesticides allowed to remain on the market. However, Cropper and others also found that EPA's decisions appeared to attribute a value of $35 million per statistical life saved. As discussed earlier, economists find that Americans' personal risk-taking behavior indicates that people place a considerably lower value on avoiding risk of death. Thus, pesticide cancellation decisions under FIFRA may provide reduced health risks at a higher cost than people would choose for themselves.

Toxics Release Inventory

In the 1980s, providing information about pollution through "right-to-know" laws became an instrument of public policies for hazardous substance management. The most important application of this approach in federal environmental policy is the Toxics Release Inventory (TRI), which began with EPCRA in 1986.[38] Beginning in 1987, large manufacturing facilities had to file reports on more than 300 hazardous substances. Facilities must report releases (quantities emitted into the air, discharged to surface water, and land disposed on-site) and transfers (quantities transferred off-site for recycling, treatment, and disposal). They also must report on-site waste management. Table 7-9 is a summary of the TRI data reported in 1997. Point-source air

TABLE 7-9. Toxics Release Inventory Releases, Transfers, and Waste Management in 1997.

Activity	*Quantity (million pounds)*
Releases	
On-site releases	
Air emissions	
Fugitive air emissions	317
Point source air emissions	1,034
Surface water discharge	218
Underground injection	220
Other land disposal	346
Off-site transfers for release	
Landfills	252
Other off-site releases	199
Management	
On-site management	
Recycling	7,987
Energy recovery	3,806
Treatment	7,013
Off-site transfers for management	
Recycling	2,381
Energy recovery	508
Treatment	259
Publicly owned treatment works (wastewater)	267

Source: U.S. EPA 1999a.

releases dominate the releases to the environment. Large amounts of these hazardous substances are recycled, both on- and off-site.

EPA has broadened the scope of the TRI over time. The agency expanded TRI reporting in 1991 to include off-site recycling and reuse (in addition to treatment and disposal) and began to require that facilities report their efforts to reduce pollution at its source. EPA also expanded the set of chemicals included in the TRI from about 300 chemicals initially to more than 600 chemicals by 1995. Beginning in 1998, some nonmanufacturing facilities—such as electric utilities and hazardous waste management facilities—report to the TRI. A proposed expansion would require reporting of chemical use—not only releases, transfers, and waste management.

The TRI could reduce pollution through several effects. As the title of EPCRA suggests, one impetus for the TRI was to provide communities with information about the toxic chemicals released by local plants. In a seminal paper, Ronald Coase (1960) argued that private bargaining between polluters and those exposed to pollution could result in efficient pollution control.[39] However, differences in the information available to the parties can hinder such bargaining. One possible role for the TRI is to improve the flow of information and thus facilitate negotiations between polluters and their communities.

The TRI also might harness the power of public relations and firms' desires to maintain a "green" reputation. Potential customers, employees, and shareholders may value the firm's environmental reputation, which could result in higher profits for firms with low TRI reports. Firms also may cultivate the goodwill of the public at large if this goodwill improves their ability to influence the political process.

EPA tried to strengthen the public relations pressures from the TRI with the 33/50 program. Firms that participated in this program voluntarily agreed to reduce their TRI releases of seventeen common chemicals from 1988 levels 33% by 1992 and 50% by 1995. In exchange, firms could tout their participation in the program. More than 1,300 firms participated in the program.[40] Their reported releases of these chemicals fell 56% by 1995, surpassing the goals of the program (U.S. EPA 1997c).

A final possible effect of the TRI is that it forces firms to collect information about their own releases that may change their behavior. Some firms claim that the exercise has made them aware of new opportunities for pollution reduction.

Overall, TRI releases and transfers have declined dramatically since the program began. Figure 7-3 shows the trends between 1988 and 1994 in releases and transfers, for those chemicals and types of activity that had constant reporting requirements throughout the period. Except for surface water

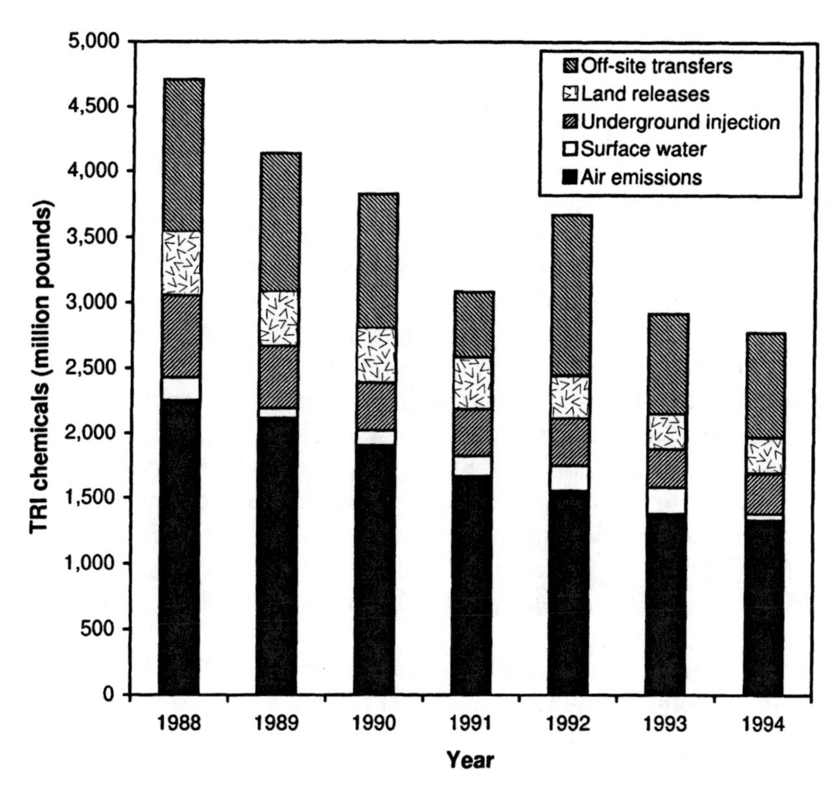

FIGURE 7-3. Trends in TRI Releases and Transfers, 1988–94.
Source: Author's calculations based on the TRI.

discharges and treatment and disposal off-site, large declines appear in all categories of releases. The declines are so dramatic that a skeptic might even wonder whether firms overreported their early releases to show improvements in later years.

Has the TRI itself contributed to the observed decline in toxic chemical releases? Economists have addressed the question by looking for evidence that large reported emissions under the TRI reduce a firm's stock market value. If the market accurately values expected future profits, then a drop in the share price indicates that high TRI emissions reduce a firm's profits. Because studies have found these declines, they support the view that the TRI provides the desired incentives (Hamilton 1995a; Konar and Cohen 1997).

Although most discussion emphasizes the effects of the TRI on polluters, it also may have other beneficial effects. People concerned about pollution may find reassurance, or at least a greater sense of control, in the information that the TRI provides. In addition, if information about the location of toxic releases encourages people to avoid areas with high levels of contaminants, then the TRI could reduce the number of people exposed to these chemicals and hence the damage to health from pollution.[41]

Right-to-know laws sometimes are considered incentive-based environmental policies. Rather than requiring firms to reduce pollution, they simply ask them to report discharges and thus create incentives for firms to gain goodwill by lowering pollution. Polluting facilities have the flexibility to achieve their reductions by using the least costly technologies or to continue to pollute if reductions are expensive. Thus, right-to-know policies might have some of the advantages of incentive-based environmental policies.

However, the right-to-know approach has significant limitations, especially when compared with incentive policies such as emissions taxes and marketable permits. First, the government has little power over the amount of pollution under right-to-know laws. Firms may experience little incentive to reduce emissions if most people are apathetic or tend to "free ride" on other people's efforts to protect the environment. Alternatively, firms may experience too much pressure to reduce emissions if the public is unnecessarily fearful.

Second, right-to-know laws may not target resources to the most hazardous pollution. Both government and media reports tend to summarize the TRI information, reporting total releases to all environmental media for all chemicals. Thus, firms may not receive any greater reward for reducing releases of very toxic chemicals than for reducing releases of much less harmful chemicals. By contrast, a well-designed pollution tax might vary with the hazards posed by the chemical.

Finally, the right-to-know approach could threaten trade secrets. If publicly released data permit competitors to "back out" information about a facility's production technology, this disclosure could weaken incentives to innovate. Right-to-know programs can allow firms to suppress reports as confidential business information, but this secrecy undermines the purpose of the program. Programs that try to protect confidential business information protection have mixed results. Under TSCA, secrecy claims may be too common. A 1992 EPA study found that 90% of TSCA Premanufacture Notices claimed confidential business information and that many of these claims had little basis when challenged (U.S. GAO 1994c). By contrast, firms rarely use this privilege under the TRI, which creates greater hurdles for secrecy claims.

As mentioned earlier, EPA has proposed expanding the TRI to include information about the quantities of chemicals used by plants.[42] This expansion would permit assessment of industrial pollution prevention—that is, changing production processes to avoid using substances that ultimately become pollution. However, the advantage of this additional reporting is unclear, because firms' use of chemicals does not affect the environment unless it results in releases (which are already reported). The affected firms oppose this expansion because they believe it will make the TRI disclose information that is much more sensitive than the information it currently provides.

Conclusions

Policies for hazardous wastes and toxic substances in the United States rely on various strategies. A few of the programs discussed in this chapter—RCRA, TSCA, and FIFRA—have traditional regulatory structures. Under these programs, EPA has the authority to choose the appropriate behavior and enforce industry compliance. However, the economic and environmental effects of these programs extend beyond the direct restrictions in the policies. For example, by raising waste management costs, RCRA may create desirable incentives for reduced waste generation as well as detrimental incentives for increased illegal disposal. Similarly, both TSCA and FIFRA could affect the characteristics of the products that manufacturers decide to bring to market, as well as the innovative activity in which they engage.

In addition to these indirect influences, some of these public policies intentionally provide incentives to polluters. Legal liability plays an important role, especially in the control of hazardous wastes. Superfund and RCRA (through its Corrective Action program and financial responsibility requirements) create incentives for firms to manage wastes in ways that avoid future cleanup costs.

Several policies also provide information as an incentive device. The TRI is the purest application of this approach, but other policies have similar intentions. For example, TSCA aims to gather information about the toxicity characteristics of chemicals and to make this information available to the public.

This hybrid of direct regulations and indirect incentives heightens the need to focus on the environmental impacts of these public policies. Greater attention to the precise damages caused by the regulated pollutants would help in designing programs. Setting policy targets that differ from the true environmental goal can result in unintended effects and unnecessary costs. Focusing on the environmental ends can help us choose policies that successfully achieve environmental benefits but are consistent with other social goals.

Notes

1. Although the states play important roles in public policies for managing hazardous wastes and toxic substances, I focus on federal policies here. I mention state policies in passing, with references for the interested reader.

2. The values in Table 7-1 understate the amount of recycling and reuse as fuel. The biennial report data that form the basis for this table include only waste managed at facilities with a Resource Conservation and Recovery Act (RCRA) permit. Energy and material recovery facilities often do not require RCRA permits.

3. The tortuous process of writing these Resource Conservation and Recovery Act regulations was reviewed by Landy and others (1990).

4. For a perspective from one of the congressional leaders in this Resource Conservation and Recovery Act reauthorization, see Florio 1986.

5. Resource Conservation and Recovery Act, Section 1004(5).

6. Some caution is necessary in interpreting Table 7-2 because a waste generator may not test additional criteria if the waste is found hazardous on the basis of the first criterion (although many do). Therefore, a larger percentage of wastes may be toxic than reported in the table.

7. For state-by-state hazardous waste program information, see Environmental Information, Ltd. 1990.

8. The U.S. Environmental Protection Agency issued an advance notice of proposed rulemaking in May 1996 (*Federal Register* 1996a).

9. They may take advantage of lax enforcement: A survey of small-quantity generators in the San Francisco Bay area found that 57% relied on some form of illegal waste management (Russell and Meiorin 1985).

10. Capacity grew rapidly in the 1980s, but growth slowed by the early 1990s. Industry sources complain of excess capacity in the 1990s (*The Hazardous Waste Consultant* 1997).

11. Most firms in the survey (71%) reported that the principal difficulty in obtaining insurance is that insurance companies simply do not offer coverage (U.S. GAO 1994a). Economists argue that such insurance markets fail for two reasons. First, high-risk buyers demand the most insurance, raising costs and pricing low-risk buyers out of the market. Second, insured parties may behave in an excessively risky fashion because they know that any resulting damages will be covered.

12. A few changes in the Toxics Release Inventory (TRI) over time affect its use to show time trends. Figures 7-1 and 7-2 exclude chemicals with changes in reporting requirements, so the trends are meaningful. In addition, facilities that manufactured or processed between 25,000 and 50,000 pounds of TRI chemicals did not begin reporting until 1989, so the 1988 data understate the relevant quantities.

13. The land disposal restrictions also prohibited placing untreated wastes in surface impoundments and underground injection wells. The trends in underground injection look similar to those for landfills shown in Figure 7-1. The quantity remained stable (or maybe slightly increased in the early 1980s) and then fell after 1987. By the end of the period, almost no originally regulated wastes were managed

by underground injection. It is hard to tell what the trends have been in surface impoundment because the early surveys differ in their classification of this management method.

14. For an econometric study of the sensitivity of waste generation to management costs, see Sigman 1996.

15. The Toxics Release Inventory began to require reporting of recycling and energy recovery in 1991, but Figure 7-2 excludes these categories for consistency with the biennial report, which is incomplete in its recording of recycled wastes.

16. The total costs for the land disposal restrictions in Table 7-4 are $2.1 billion per year. For comparison, manufacturing firms reported spending $2.3 billion dollars (in 1996 dollars) on solid and contained-liquid hazardous waste management in 1994 (U.S. Bureau of the Census 1996). Although the numbers are not directly comparable, they suggest that the land disposal restrictions account for a very large percentage of the costs of the Resource Conservation and Recovery Act.

17. Although many studies estimate willingness to pay to avoid hazardous waste sites, they mostly focus on Superfund sites rather than permitted Resource Conservation and Recovery Act (RCRA) land disposal facilities (which typically may be cleaner). Two studies that examine RCRA facilities are U.S. EPA 1993 and Smith and Desvouges 1986.

18. Assuming that the benefits occur continuously over time probably exaggerates the benefits relative to the costs by moving the benefits forward in time. Many of the health consequences, especially cancer, emerge later in peoples' lives, and delayed health improvements may have lower value than current health improvements.

19. For recent studies of hazardous waste sites and environmental equity, see Hamilton 1995b; Baden and Coursey 1997; Been 1997.

20. For information about state cleanup programs, see U.S. GAO 1989.

21. For a criticism of the Hazard Ranking System, see U.S. OTA 1989.

22. The U.S. Environmental Protection Agency also can add a site to the National Priorities List by state request (each state permitted two sites) and by request of the Agency for Toxic Substances and Disease Registry (a division of the Department of Health and Human Services). However, these mechanisms are rarely used.

23. The study also estimates other costs, such as private-party transactions, government legal expenditures, and Superfund-related research. With all these costs included, the Congressional Budget Office's central estimate for total costs is $79 billion, with a range of $45 billion to $128 billion (Beider 1994b).

24. These figures from the U.S. Council on Environmental Quality (1995) have been converted into 1996 dollars. For details of the estimates for the U.S. Department of Energy sites, see U.S. DOE 1997. For other estimates of the costs of federal and nonfederal facilities, see Russell and others 1992.

25. Initially, banks that lent money to the owners of the site could be liable. The U.S. Environmental Protection Agency (EPA) promulgated a rule to eliminate this liability under the Comprehensive Environmental Response, Liability, and Compensation Act (CERCLA) in 1992, but a court overturned EPA's rule in 1994. However, in 1996, Congress amended Superfund substantially to protect lenders from liability.

26. In one study, only 12% of Records of Decision included assessments of risks after proposed remedies (Doty and Travis 1989).

27. Nonetheless, the U.S. Environmental Protection Agency seems to take some consideration of exposure in practice (Viscusi and Hamilton 1999).

28. Estimates by Hamilton and Viscusi (1997) suggest that these restrictions would greatly improve Superfund decisionmaking. However, Probst and others (1997) are more skeptical.

29. Gupta and others (1996) found no evidence that sites with higher minority populations or lower income in their ZIP code experience less thorough cleanup than other sites. However, Hamilton and Viscusi (1999a) found that sites with a large percentage of minorities within a one-mile radius have less extensive and less costly remedies selected than sites in communities with lower minority populations, even when the authors controlled for the community's income and voter turnout (see also Hird 1994).

30. For an analysis of current Superfund financing and several potential reforms, see Probst and others 1995.

31. For an economic analysis of Superfund's joint and several liability, see Kornhauser and Revesz 1995; Chang and Sigman 2000.

32. Total Superfund Trust Fund spending was $16.4 billion through fiscal year 1996. The U.S. Environmental Protection Agency has spent 62% of the trust fund on cleanup responses rather than other activities, such as research and enforcement (U.S. EPA 1996b). Subtracting cost recoveries from the cleanup share of costs yields $8.8 billion in taxes and general appropriations. This figure includes some spending on federal facility sites. In comparing government and potentially responsible party (PRP) spending, one should note that the PRP settlements reflect future commitments rather than expenditures that have occurred to date.

33. For evidence that potentially responsible parties influence remedy selection, see Sigman 1998a.

34. Although the U.S. Environmental Protection Agency can avoid suing municipalities, it cannot keep other potentially responsible parties from suing the municipalities for a share of costs. Proposals for Superfund reauthorization currently in Congress would reduce or eliminate municipalities' Superfund liability.

35. For theoretical analyses of the brownfield problem, see Boyd and others 1996; Segerson 1993.

36. The Toxic Substances Control Act (TSCA) exempts pesticides, tobacco, nuclear material, firearms and ammunition, food, food additives, drugs, and cosmetics. Other federal laws regulate most of these substances.

37. In addition to new chemicals, if the U.S. Environmental Protection Agency (EPA) judges a chemical to have a "significant new use" it can require the chemical in this use to go through the same review as a new chemical. EPA almost never used this authority until 1989, when it established an expedited process that has greatly expanded its use (U.S. GAO 1994c).

38. In addition to the Toxics Release Inventory, the Emergency Planning and Community Right-to-Know Act also contains requirements that states develop emergency preparedness plans for accidental releases of toxic chemicals and pesticides.

39. For an attempt to test whether the Toxics Release Inventory has a Coasean role, see Hamilton 1999.

40. For a study of determinants of participation in the 33/50 program, see Arora and Cason 1995.

41. For evidence of the relationship between the Toxics Release Inventory releases and housing prices, see Bui and Mayer 1999.

42. The U.S. Environmental Protection Agency issued an "Advance notice of proposed rulemaking" for this expansion of the Toxics Release Inventory in the *Federal Register* (1996b). A similar expansion was included in H.R. 1636 in the 105th Congress. For a discussion, see Hearne 1996.

References

Abt Associates. 1985. *National Small Quantity Hazardous Waste Generator Survey.* Washington, DC: U.S. EPA.

Arora, Seema, and Timothy Cason. 1995. An Experiment in Voluntary Environmental Regulation: Participation in EPA's 33/50 Program. *Journal of Environmental Economics and Management* 28(3):271–86.

Baden, Brett, and Don Coursey. 1997. The Locality of Waste Sites within the City of Chicago: A Demographic, Social, and Economic Analysis. Discussion paper. Chicago, IL: University of Chicago, Harris School of Public Policy.

Barbanel, Josh. 1992. Elaborate Sting Operation Brings Arrests in Illegal Dumping of Toxic Wastes by Businesses. *New York Times,* May 13, B5.

Been, Vicki, with Francis Gupta. 1997. Coming to the Nuisance or Going to the Barrios? A Longitudinal Analysis of Environmental Justice Claims. *Environmental Law Quarterly* 24(1):1–56.

Beider, Perry. 1994a. *Analyzing the Duration of Cleanup at Sites on Superfund's National Priorities List.* Congressional Budget Office Memorandum (series). Washington, DC: Congressional Budget Office.

———. 1994b. *The Total Cost of Cleaning Up at Nonfederal Superfund Sites.* Washington, DC: Congressional Budget Office.

Boyd, James. 1997. "Green Money" in the Bank: Firm Responses to Environmental Financial Responsibility Rules. *Managerial and Decision Economics* 18(6):491–506.

Boyd, James, Winston Harrington, and Molly Macauley. 1996. The Effects of Environmental Liability on Industrial Real Estate Development. *Journal of Real Estate Finance and Economics* 12:37–58.

Bui, Linda T. M., and Christopher J. Mayer. 1999. Capitalization and Regulation of Environmental Amenities: Evidence from the Toxics Release Inventory in Massachusetts. Working paper. Boston, MA: Boston University.

Carson, Rachel L. 1962. *Silent Spring.* Boston, MA: Houghton Mifflin.

Chang, Howard F., and Hilary Sigman. 2000. Incentives to Settle under Joint and Several Liability: An Empirical Analysis of Superfund Litigation. *Journal of Legal Studies* 29(1):205–36.

Coase, Ronald. 1960. The Problem of Social Cost. *Journal of Law and Economics* 3:1–44.
Cropper, Maureen L., William N. Evans, Stephen J. Berard, Maria M. Ducla-Soares, and Paul R. Portney. 1992. The Determinants of Pesticide Regulation: A Statistical Analysis of EPA Decision-Making. *Journal of Political Economy* 100:175–97.
Dixon, Lloyd S. 1995. The Transactions Costs Generated by Superfund's Liability Approach. In *Analyzing Superfund: Economics, Science, and Law*, edited by Richard L. Revesz and Richard B. Stewart. Washington, DC: Resources for the Future, 171–85.
Doty, Carolyn B., and Curtis C. Travis. 1989. The Superfund Remedial Action Decision Process: A Review of Fifty Records of Decision. *Journal of the Air Pollution Control Association* 39:1535–43.
Environmental Information, Ltd. 1990. *Industrial and Hazardous Waste Management Firms*. Minneapolis, MN: Environmental Information, Ltd.
Environmental Law Institute. 1996. *An Analysis of State Superfund Programs: 50 State Study, 1995 Update*. Washington, DC: Environmental Law Institute.
Federal Register. 1986. 51:40572
——. 1987. 52:25760.
——. 1988a. 53:28118.
——. 1988b. 53:30908.
——. 1988c. 53:31138.
——. 1989. 54:26594.
——. 1990. 55:22520.
——. 1996a. 61:19432.
——. 1996b. 61:51321.
Florio, James J. 1986. Congress as Reluctant Regulator: Hazardous Waste Policy in the 1980s. *Yale Journal on Regulation* 3(2):351–82.
Fullerton, Don, and Seng-Su Tsang. 1996. Should Environmental Costs Be Paid by the Polluter or the Beneficiary? The Case of CERCLA and Superfund. *Public Economics Review* 1:85–127.
Gupta, Shreekant, George Van Houtven, and Maureen L. Cropper. 1996. Paying for Permanence: An Economic Analysis of EPA's Cleanup Decisions at Superfund Sites. *RAND Journal of Economics* 27:563–82.
Hamilton, James T. 1995a. Pollution as News: Media and Stock Market Reactions to Toxics Release Inventory Data. *Journal of Environmental Economics and Management* 28(1):98–113.
——. 1995b. Testing for Environmental Racism: Prejudice, Profits, Political Power? *Journal of Policy Analysis and Management* 14(1):107–32.
——. 1999. Exercising Property Rights to Pollute: Do Cancer Risks and Politics Affect Plant Emission Reductions? *Journal of Risk and Uncertainty* 18(2):105–24.
Hamilton, James T., and W. Kip Viscusi. 1997. The Benefits and Costs of Regulatory Reforms for Superfund. *Stanford Environmental Law Journal* 16(2):159–98.
——. 1999a. Environmental Equity at Superfund Sites. In *Calculating Risks: The Spatial and Political Dimensions of Hazardous Waste Policy*. Cambridge, MA: MIT Press, 157–88.
——. 1999b. How Costly Is Clean? An Analysis of the Benefits and Costs of Superfund Site Remediations. *Journal of Policy Analysis and Management* 18(1):2–27.

Hammitt, James K., and Peter Reuter. 1988. *Measuring and Deterring Illegal Disposal of Hazardous Waste: A Preliminary Assessment.* Santa Monica, CA: RAND Corp.

Hearne, Shelley A. 1996. Tracking Toxics: Chemical Use and the Public's "Right-to-Know." *Environment* 38 (6):4ff and the associated industry response.

Hird, John A. 1994. *Superfund: The Political Economy of Environmental Risk.* Baltimore, MD: The Johns Hopkins University Press.

ICF, Inc. 1985. *Survey of Selected Firms in the Commercial Hazardous Waste Management Industry.* Washington, DC: U.S. EPA.

——. 1988. *Survey of Selected Firms in the Commercial Hazardous Waste Management Industry.* Washington, DC: U.S. EPA.

Konar, Shameek, and Mark A. Cohen. 1997. Information as Regulation: The Effect of Community Right-to-Know Laws on Toxic Emissions. *Journal of Environmental Economics and Management* 32(1):109–24.

Kornhauser, Lewis A., and Richard L. Revesz. 1995. Evaluating the Effects of Alternative Superfund Liability Rules. In *Analyzing Superfund: Economics, Science, and Law,* edited by Richard L. Revesz and Richard B. Stewart. Washington, DC: Resources for the Future, 115–44.

Krukowski, John. 1995. Survey: Thermal Prices End Free-Fall? *Pollution Engineering* 27(2).

Landy, Marc K., Marc J. Roberts, and Stephen R. Thomas. 1990. *The Environmental Protection Agency: Asking the Wrong Questions.* New York: Oxford University Press.

McClelland, Gary H., William D. Schulze, and Brian Hurd. 1990. The Effect of Risk Beliefs on Property Values: A Case Study of a Hazardous Waste Site. *Risk Analysis* 10(4):485–97.

National Research Council. 1984. *Toxicity Testing: Strategies to Determine Needs and Priorities.* Washington, DC: National Academy Press.

Peretz, Jean H., and Jeffrey Solomon. 1995. Hazardous Waste Landfill Costs on Decline, Survey Says. *Environmental Solutions,* April, 21–24.

Probst, Katherine N., Don Fullerton, Robert E. Litan, and Paul R. Portney. 1995. *Footing the Bill for Superfund Cleanups: Who Pays and How?* Washington, DC: Resources for the Future.

Probst, Katherine N., Robert Hersh, Kris Wernstedt, and Jan Mazurek. 1997. *Linking Land Use and Superfund Cleanups: Uncharted Territory.* Washington, DC: Resources for the Future.

Russell, Lorene, and Emy Meiorin. 1985. *The Disposal of Hazardous Wastes by Small Quantity Generators: Magnitude of the Problem.* Oakland, CA: Association of Bay Area Governments.

Russell, Milton, E. William Colglazier, and Bruce E. Tonn. 1992. The United States Hazardous Waste Legacy. *Environment* 34(6):12–15, 34–39.

Segerson, Kathleen. 1993. Liability Transfers: An Economic Analysis of Buyer and Lender Liability. *Journal of Environmental Economics and Management* 25(1): S46–S63.

Sigman, Hilary. 1996. The Effects of Hazardous Waste Taxes on Waste Generation and Disposal. *Journal of Environmental Economics and Management* 30:199–217.

——. 1998a. Liability Funding and Superfund Clean-up Remedies. *Journal of Environmental Economics and Management* 35:205–24.

——. 1998b. Midnight Dumping: Public Policies and Illegal Disposal of Used Oil. *RAND Journal of Economics* 29:157–78.

——. Forthcoming. The Pace of Progress at Superfund Sites. *Journal of Law and Economics.*

Smith, V. Kerry, and William H. Desvouges. 1986. The Value of Avoiding a LULU: Hazardous Waste Disposal Sites. *Review of Economics and Statistics* 68(2):293–99.

The Hazardous Waste Consultant. 1997. United States Survey Highlights and Trends. 15(2): D1–D63.

U.S. Bureau of the Census. 1996. *Pollution Abatement Cost and Expenditures, 1994.* Current Industrial Reports. Washington, DC: U.S. GPO (Government Printing Office).

U.S. Council on Environmental Quality. 1995. *Improving Federal Facilities Cleanup: Report of the Federal Facilities Policy Group.* Washington, DC: Office of Management and Budget.

U.S. DOE (Department of Energy). 1997. *1996 Baseline Environmental Management Report.* Washington, DC: U.S. DOE.

U.S. EPA (Environmental Protection Agency). 1990. *Cancer Risk from Outdoor Exposure to Air Toxics.* Vol. 1. Research Triangle Park, NC: U.S. EPA, Office of Air Quality Planning and Standards.

——. 1993. *Draft Regulatory Impact Analysis for the Final Rulemaking on Corrective Action for Solid Waste Management Units.* Washington, DC: U.S. EPA, Office of Solid Waste.

——. 1994a. *Progress toward Implementing Superfund, Fiscal Year 1994.* Washington, DC: U.S. EPA.

——. 1994b. *ROD Annual Report, FY 1992.* Washington, DC: U.S. EPA.

——. 1996a. *FY 1995 Enforcement and Compliance Assurance Accomplishments.* Washington, DC: U.S. EPA.

——. 1996b. Focus on Cleanup Costs. *Superfund Today,* June, 1–4.

——. 1996c. *1996 Master Testing List.* Washington, DC: U.S. EPA, Office of Prevention, Pesticides, and Toxic Substances.

——. 1997a. *Annual Report for 1996.* Washington, DC: U.S. EPA, Office of Pesticide Programs.

——. 1997b. *FY 1996 Enforcement and Compliance Assurance Accomplishments Report.* Washington, DC: U.S. EPA.

——. 1997c. *1995 Toxics Release Inventory Public Data Release Report.* Washington, DC: U.S. EPA.

——. 1999a. *1997 Toxics Release Inventory Public Data Release Report.* Washington, DC: U.S. EPA.

——. 1999b. *The Biennial RCRA Hazardous Waste Report, 1997: National Analysis.* Washington, DC: U.S. EPA.

U.S. GAO (General Accounting Office). 1985. *Illegal Disposal of Hazardous Waste: Difficult to Detect or Deter.* Washington, DC: U.S. EPA.

——. 1989. *State Cleanup Status and Its Implications for Federal Policy.* Washington, DC: U.S. GAO.

——. 1994a. *Hazardous Waste: An Update on the Cost and Availability of Pollution Insurance.* Washington, DC: U.S. GAO.

——. 1994b. *Superfund: Estimates of Number of Sites Vary.* Washington, DC: U.S. GAO.

——. 1994c. *Toxic Substances Control Act: Legislative Changes Could Make the Act More Effective.* Washington, DC: U.S. GAO.

——. 1997. *Superfund: State Voluntary Programs Provide Incentives to Encourage Cleanups.* Washington, DC: U.S. GAO.

U.S. OTA (Office of Technology Assessment). 1989. *Coming Clean: Superfund's Problems Can Be Solved,* OTA-ITE-433. Washington, DC: U.S. GPO, 115–24.

——. 1995. *State of the States on Brownfields: Programs for Cleanup and Reuse of Contaminated Sites.* Washington, DC: U.S. GPO.

Viscusi, W. Kip. 1993. The Value of Risks to Life and Health. *Journal of Economic Literature* 31(4):1912–46.

Viscusi, W. Kip, and James T. Hamilton. 1999. Are Risk Regulators Rational? Evidence from Hazardous Waste Decisions. *American Economic Review* 89(4):1010–27.

Walker, Katherine D., March Sadowitz, and John D. Graham. 1995. Confronting Superfund Mythology: The Case of Risk Assessment and Management. In *Analyzing Superfund: Economics, Science and Law,,* edited by Richard L. Revesz and Richard B. Stewart. Washington, DC: Resources for the Future, 25–53.

eight

Solid Waste Policy

Molly K. Macauley and Margaret A. Walls*

In the late 1980s, municipal solid waste (MSW) issues became important concerns for policymakers and government officials at the local, state, and national levels. The volume of MSW generated in the United States was rising, many landfills were closing, and siting new landfills and incinerators was becoming increasingly difficult because of local opposition. This combination of factors, along with a surge of environmentalism in the general public, heightened public interest in recycling programs, particularly residential curbside recycling. Although these programs were a popular method of conserving landfill capacity, the greatly increased supply of recyclable materials led to collapses in the markets for many of those materials—particularly used newspapers, plastic bottles, and green glass. This situation led policymakers to look for solutions to this problem and alternative strategies for reducing the amount of solid waste going to landfills.

In 1994, the markets for many recyclables picked up, reducing pressure on state and federal policymakers to take action. Nonetheless, numerous policy options continue to be considered at all levels of government. These

*This research was partially supported by the U.S. Environmental Protection Agency, Office of Policy, Planning, and Evaluation, through a cooperative agreement with Resources for the Future (#R821821-01). We thank Karen Palmer, Richard Porter, and the editors for their comments. We also thank David Edelstein and Jon Vranesh for valuable research assistance.

options include taxes to discourage the use of virgin materials; subsidies to encourage and mandates to require recycling, which include mandatory recycled content standards for products and mandatory recycling rate standards for materials; deposit–refund programs for beverage containers, batteries, and other items; advance disposal fees, which amount to taxes on consumer products; "responsible entity" regulations similar to those in Germany, which attempt to charge the costs of recycling and waste disposal to manufacturers of consumer products; and unit-based pricing of residential trash disposal.

Many of these alternatives have been adopted at the state level, and some have found their way into proposed federal legislation. As of the mid-1990s, thirteen states had established standards for the recycled content for newsprint, for example, and a few states had set such standards for other items.[1] Sixteen states had some form of recycling investment tax credit. Nine states had deposit–refund systems for beverage containers.

At the federal level, several bills were introduced in Congress in the 1990s. Two proposals to establish a virgin materials tax were introduced in the House of Representatives in 1990; one would have imposed a tax of $7.50 per ton on virgin materials such as wood pulp, and one would have imposed a tax on intermediate products that contain virgin materials. In 1994, the National Beverage Container Reuse and Recycling Act, a deposit–refund bill, was introduced in the House of Representatives. That bill would have established a 10-cent deposit on bottles and cans in states that do not recycle at least 70% of such containers. A product tax similar to an advance disposal fee was proposed in the Senate; it would vary inversely with the recycled content of a product and would equal zero for products that meet or exceed established target recycling rates. In October 1993, President Bill Clinton signed Executive Order 12873, Federal Acquisition, Recycling, and Waste Prevention, which required federal agencies to purchase only recycled paper for their photocopiers. In May 1996, the President signed the Mercury-Containing and Rechargeable Battery Management Act, which ushered in a national voluntary take-back system for nickel–cadmium rechargeable batteries. This is one example of the notion of extended product responsibility, whereby producers and others throughout the chain of product manufacture, distribution, and use are made responsible for the disposal and other environmental effects of products. This approach is gaining some momentum in the United States, but it is far more popular in Europe. There, it is more commonly called extended *producer* responsibility, and it is often characterized by product "take-back" requirements such as those in the German "Green Dot" program.

Whether one believes that these policies are good or bad ideas depends on what one believes are the appropriate goals for solid waste policy. Some observers view the goal of solid waste policy as reducing the flow of waste to landfills; others view it as increasing the demand for secondary materials; others see it as conserving resources such as energy, water, or virgin materials; and still others view it as reducing externalities or materials use throughout the life cycle of a product. Some proponents appear to believe that solid waste policy should address all of these concerns.

In this chapter, we identify a set of appropriate goals for solid waste policy and talk about why certain policy options may stand a better chance of achieving these goals than others. We begin by describing kinds of MSW, its generation, and its regulation. Then, we discuss the general rationale for government intervention in the MSW market, identify some specific externalities associated with solid waste, and suggest ways to correct these externalities. We also discuss other potential goals (such as conserving resources, increasing secondary materials demand, and addressing life-cycle externalities) and explain why solid waste policy should not attempt to address these concerns directly. To conclude, we offer some ideas about optimal solid waste policy.

What Is Solid Waste, How Much Is Generated, and How Is It Regulated?

The U.S. Environmental Protection Agency (EPA) estimates that some 55–65% of MSW is generated by households (including multifamily dwellings) and that the remainder comes from commercial establishments and industrial sources. EPA has two ways of categorizing MSW: by material (such as paper, aluminum, glass, and yard waste) and by product (such as packaging materials, appliances, automobile tires, and clothing). By definition, MSW does not include automobile bodies, the sludge from municipal sewage treatment plants, combustion ash, or hazardous waste. The definitions used by EPA and states differ somewhat with respect to categorizing other kinds of waste as MSW. For example, EPA does not include construction and demolition waste as part of MSW, but some states do.

Some of the most concise sources of information about MSW are EPA's annual reports, *Characterization of Municipal Solid Waste in the United States.* Based on this report for 1995 (U.S. EPA 1995), the generation of MSW in the United States that year totaled about 208 million tons, or a little more than 4 pounds per person per day.[2] EPA notes that per-person generation of waste has been increasing; it was about 2.7 pounds per person per day in 1960. Figures 8-1 and 8-2 illustrate the composition of the waste stream by material and by product, respectively. Most of the materials waste stream is paper, and

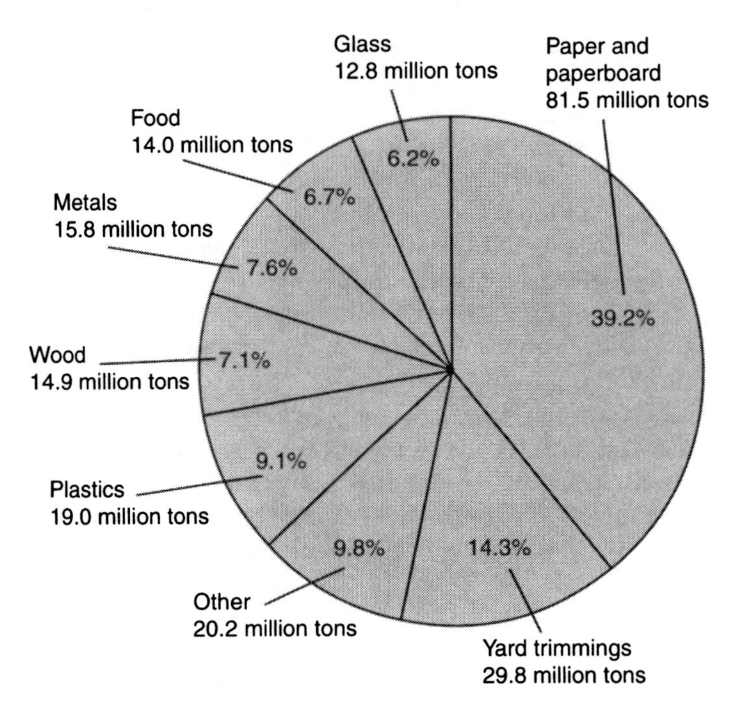

FIGURE 8-1. Materials Generated in MSW by Weight, 1995.

Note: Total weight is 208.0 million tons.

Source: U.S. Environmental Protection Agency 1995.

most of the products are nondurable goods (predominately newspapers, office paper, and magazines), containers, and packaging.

Before the mid-1980s, much MSW was burned, but the use of incineration has declined markedly in recent years. Nationwide in 1995, about 57% of MSW was disposed of in landfills, about 27% was recycled (including composting), and about 16% was incinerated (and the ash generated was disposed of in landfills). Recycling rates vary widely among materials, from about 40% of paper to 5% of plastics. These rates also vary widely among localities, because some areas do not recycle at all and others boast of overall recycling rates of 20–30%.[3]

Because landfilling is the predominant method of waste disposal, it is worth noting that modern landfills are a far cry from landfills twenty years ago, which were merely holes in the ground. As illustrated in Figure 8-3, landfills constructed since the mid-1970s have a series of clay and plastic liners; leachate collection and disposal systems for both liquids and gases; and a host of equipment to monitor air, groundwater, surface water, and soil. In addition, landfill owners and operators must prepare detailed written plans

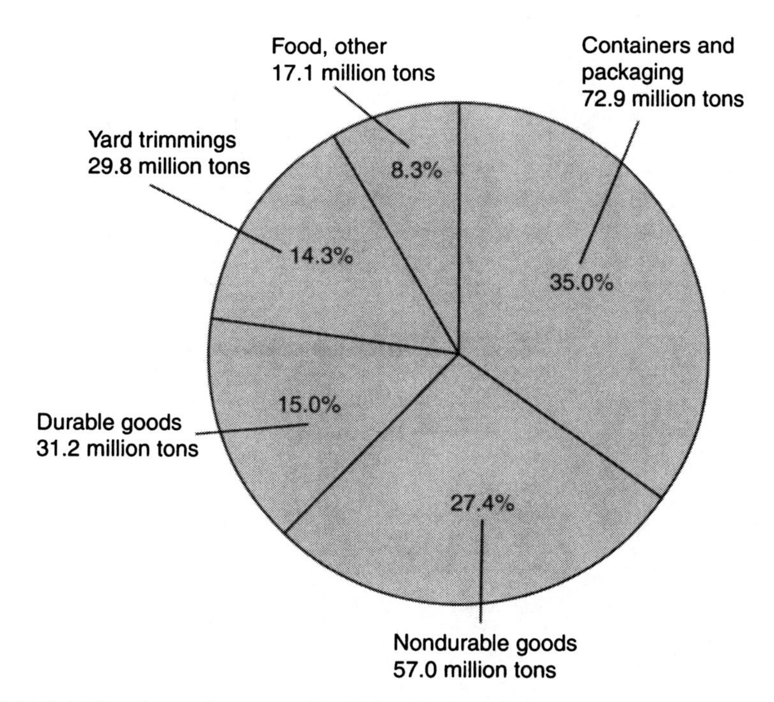

FIGURE 8-2. Products Generated in MSW by Weight, 1995.

Note: Total weight = 208.0 million tons.

Source: U.S. Environmental Protection Agency 1995.

and set aside funding for the care that will be taken when a landfill is closed, including the procedure for covering the fill and monitoring leachate.

In general, federal regulation of MSW has largely vested responsibility for waste management with states and localities.[4] The Solid Waste Disposal Act of 1965 initiated a small program of technical and financial assistance for state and local governments for MSW disposal demonstration projects. The Resource Recovery Act of 1970 established federal authority to issue general guidelines for waste management. Not until the 1976 Resource Conservation and Recovery Act (RCRA) and its 1984 amendments (known as the Hazardous and Solid Waste Amendments) did the federal government assume a direct, although limited, role in management of MSW.

Although most RCRA programs address the management of hazardous waste (so-called Subtitle C waste), the legislation establishes procedures for state development of solid waste management plans (in Part 258, or Subtitle D of the regulations; in fact, MSW is commonly called Subtitle D waste). It is through Subtitle D that the most direct federal influence over MSW has

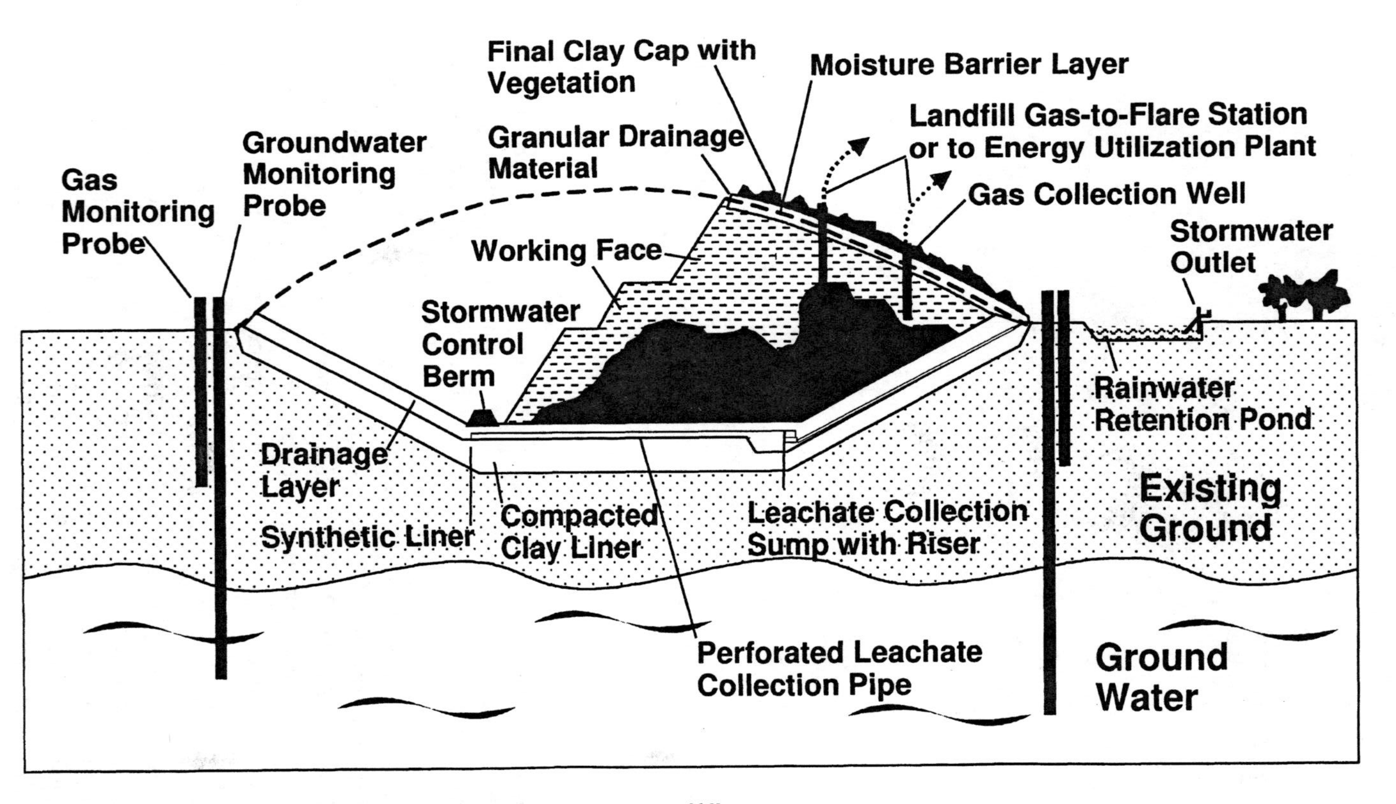

FIGURE 8-3. Cross-Section of a Typical Modern Sanitary Landfill.

Source: Courtesy of Ed Repa, Environmental Industry Associations.

taken place. Subtitle D sets forth criteria that restrict the locations of landfills; establish guidelines for their design and operation; require the monitoring of groundwater near landfills; and establish rules for lining, capping, sealing, and taking other steps when opening and closing landfills. Subtitle D of RCRA required many substandard landfills to either upgrade their operations to meet these standards or close by April 1994. The requirements led to the closing of about 900 landfills that year. Shortly after promulgation of Subtitle D, dire predictions of a pending scarcity of landfill capacity caused an increase in landfill tipping fees (the charge per ton to unload a truck at a disposal facility) and much concern over the future of waste management.[5]

During the 1990s, a few large, state-of-the-art landfills opened, so the nation's overall landfill capacity has remained relatively constant despite the closures of substandard landfills.[6] In addition, a related growing trend in the United States is interstate shipment of MSW (see Ley and others 1996). This trend has alleviated much of the concern about aggregate fill capacity in the United States but has led to a debate about possible new federal government involvement in waste management. Waste used to be discarded in the nearest local disposal facility; now, almost all states routinely import and export some waste, not only to or from a neighboring state but frequently much longer distances (for example, waste from New England may be shipped to Virginia or even the Midwest). The volume of shipments is small—on the order of 5% of all MSW—but the associated transport costs amount to about $500 million annually. The states' willingness to bear these transportation costs arises largely from differences among tipping fees across states. These differences reflect a host of related factors, including land values, landfill closures, and public opposition to expanding capacity at existing landfills or constructing new ones. In the mid-1990s, the per-ton tipping fees ranged from an average of $10 in Nevada and $27 in Ohio to $75 in New Jersey.

Some citizens' groups, environmental organizations, and state legislators have decried the growing trend toward interstate MSW transport. These critics usually express concern about some areas becoming "dumping grounds," the impact of landfill growth on local property values, increased truck traffic in the neighborhood of landfills, and the limited capacity of local landfills. This opposition has led many states to ban, differentially charge (that is, levy taxes on), or impose other restrictions on imported waste. As of 1993, forty-one states had considered or enacted legislation to restrict the flow of waste across their boundaries. Most of these restrictions have been struck down by the courts as violations of the Interstate Commerce Clause of the U.S. Constitution. When state regulations place an undue burden on commerce, including the trade of waste, they are deemed to be unconstitutional. In addition, flow control laws that require local waste to be disposed of in local landfills or incinerators have been struck down because they have the potential to restrict interstate trade in waste. Such laws were put in place to prevent some MSW from being shipped out of state, imperiling the financial health of local waste-to-energy power plants that depend on a steady and low-cost supply of MSW to fuel their boilers. In a landmark decision in May 1994, the U.S. Supreme Court ruled against a municipal flow control ordinance in the state of New York, stating that the ordinance discriminated against interstate commerce (*C&A Carbone v. Clarkstown*, Sup. Ct. 92-1402). Congress has considered numerous bills to allow such controls since the mid-1990s, and in 1995, the Senate passed a bill amending the Solid Waste Disposal Act to permit some types of restrictions on the export and import of waste (Section 534, The Interstate Transportation of Municipal Solid Waste Act of 1995). The House has not yet passed similar legislation.

Rationales for Government Intervention in the Solid Waste Market

The discussion thus far suggests that government at all levels—federal, state, and local—is increasingly involved in solid waste policy, from regulating interstate flows of waste to implementing recycling programs. What is the fundamental reasoning behind this intervention in the solid waste market?

In concept, there are three general rationales for government intervention in private markets:

- when the average costs of producing the good or service decline as more of the good or service is produced—that is, when production is subject to economies of scale over the full range of market output;
- when the good or service is a public good that is not supplied by private markets, such as national defense; and
- when production or use of the good or service results in externalities such as environmental pollution.

Do any of these circumstances characterize the market for solid waste? Yes, particularly the first and third. Some evidence indicates that landfills operate subject to economies of scale, in that their daily operating and other marginal costs tend to be less than the average cost of fill construction. Partly for this reason, and as noted earlier, interstate trade has developed to take advantage of lower costs at large, regional landfills.

As for the third situation, externalities can be associated with solid waste, for instance, when waste is illegally disposed—dumped in an open area, or burned—which can lead to the release of potentially harmful leachate into groundwater systems or undesirable emissions into the air. (Typically, such air emissions include large amounts of methane gas and trace amounts of benzene, hydrogen sulfide, chlorinated hydrocarbons, and other gases.)

Groundwater contamination, air emissions, the buildup and possible explosion risk of methane gas, and neighborhood disamenities (for example, malodors, noise, traffic congestion, and road damage from trucks) may be associated with landfills. The economic consequences of these negative externalities are illustrated in Figure 8-4, which depicts the market for legal solid waste disposal services.[7] The demand for waste disposal (how many tons would be landfilled at various tipping fees) is represented by the demand curve (*D*), and the marginal private cost of providing the disposal services is depicted as *MPC*.[8] However, society also may incur a cost in the form of potential environmental harm. Adding this additional marginal cost per ton of waste disposal to the *MPC* yields the marginal social cost (*MSC*) curve.[9]

If the waste disposal industry faced no outside control on its external damages, it would seek to dispose of *Q* units of waste, where *MPC* is equal to *D*, and it would charge a price per unit of waste disposed equal to *MPC*. In a

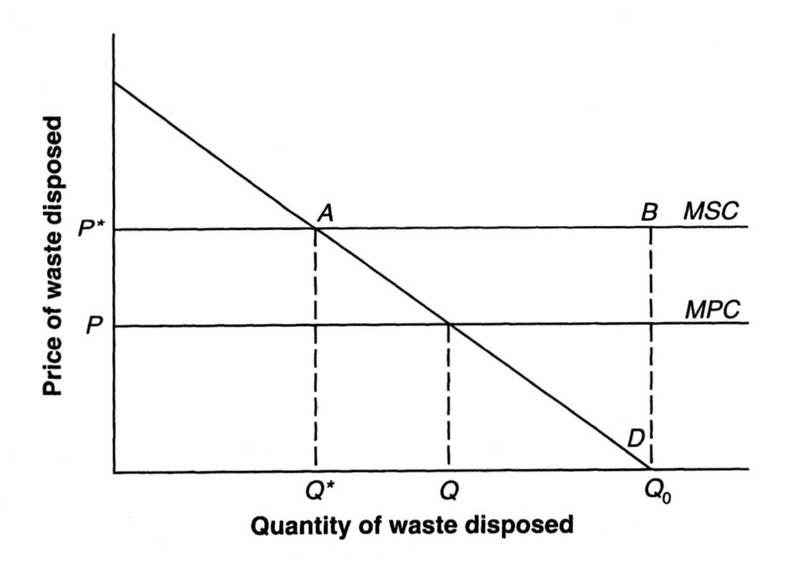

FIGURE 8-4. Legal Waste Disposal Market.

Notes: P^* is the optimal price of waste disposal, P is the price of waste disposal without consideration of *MSC*, *A* is the intersection of demand curve and *MSC*, *B* is *MSC* at Q_0, *MSC* is the marginal social cost of providing disposal services, *MPC* is the marginal private cost of providing disposal services, Q^* is the optimal quantity of waste disposal, Q is the quantity/units of waste, D is demand, and Q_0 is the equilibrium quantity of waste disposed when price is zero. See text and relevant notes for a detailed discussion.

private setting, this choice would maximize the industry's profits. However, this level Q is not the most socially desirable, because each unit generates $\$(P^* - P)$ in environmental costs, where P^* is the correct social price (where *MSC* and *D* intersect) and P is the incorrect private price (where *MPC* and *D* intersect). The net benefit to society is maximized at Q^*, where *MSC* is equal to *D*. At the private market outcome of Q, too much waste is generated, and the environmental damage is too high.

Congress has addressed this problem in part with RCRA Subtitle D which, as explained earlier, specifies operating and postclosure requirements for solid waste landfills. In addition, communities hosting new landfills have increasingly required the payment of host fees from landfill developers as compensation for neighborhood disamenities. For instance, some localities allow construction of new landfills in return for a payment per ton of waste disposed in the landfill. They also may receive free collection and disposal of their own solid waste; construction of new roads, parks, or other community infrastructure; and guaranteed property values for homes adjacent to the landfill. Both the Subtitle D regulations and host fees can operate to close the gap between *MSC* and *MPC*.

However, an important problem still remains. In most communities, the "price" charged to residential customers for trash collection, hauling, and disposal services is close to zero. Rather than charging residents an explicit price, many municipalities historically have provided trash removal service by financing the costs of collection and disposal out of tax revenues, partly to prevent illegal disposal.[10] Recently, we have witnessed a trend toward explicitly charging households for trash services through fees levied on each unit of trash (per bag, per can, or per pound); these kinds of systems are known as user charge, variable rate, unit-based pricing, "pay-as-you-throw," or bag-and-tag pricing policies (see U.S. EPA 1993b).[11] A recent estimate suggests that nearly 4,000 communities in the United States now levy user charges (Miranda and others 1998). But in most locations, zero prices are still the norm.

When the price of discarded MSW is zero, households tend to generate too much waste. In Figure 8-4, the equilibrium quantity of waste disposed when the price is zero is denoted as Q_0. Thus, despite the landfill regulations that seek to close the gap between *MSC* and *MPC*, there is still a deadweight loss to society. A measure of this loss in Figure 8-4 is denoted by the triangle ABQ_0. The loss arises due to the gap between the price that reflects *MSC* and the close-to-zero price that may be levied on households.[12]

Recouping the Deadweight Loss

What are some solutions to this problem? One answer is to get local governments to change the way they pay for MSW collection and disposal by charging each household a price that reflects how much waste it generates. In terms of Figure 8-4, a price of P^*—that is, a price equal to the true costs to society of disposal—should be levied per bag, per can, or per pound. As noted earlier, this system may lead to some illegal disposal; however, if the amount of "midnight dumping" is not significant, then it may not undermine the advantages of charging the price that reflects the true cost of disposal.[13]

Other solutions include measures that would reduce the amount of MSW generated (that is, shift the demand curve down until it intersects the axis at Q^*). Options include the following:

- *Subsidizing alternatives to disposal, such as recycling and composting.* Because this approach makes the alternatives to land disposal more attractive, consumers generate less waste for disposal, moving us back toward Q^* in Figure 8-4.
- *Taxing consumption of final products.* This approach reduces the demand for final products and thus reduces the demand for disposal.[14]
- *Subsidizing alternatives to disposal while taxing consumption.* This kind of system has the same effect on the demand for disposal as either subsidizing alternatives or taxing consumption but will do so at a lower overall cost (Palmer and others 1997).
- *Taxing the use of virgin materials.* Such taxes increase the cost of producing products, raise product prices, reduce the demand for products, and thus reduce demand for land disposal. This approach also increases the demand for recyclables, which, in turn, reduces the demand for land disposal.
- *Imposing standards that require products to incorporate a minimum amount of recycled material.* This alternative increases the demand for recyclable materials and reduces the demand for land disposal.
- *Requiring producers to take back their products or the packaging that comes with their products from final consumers and ensure that it is recycled.* This approach increases the demand for recyclables, which reduces the demand for land disposal. It may also encourage producers to design their products to be more recyclable and lighter in weight, thus further reducing the demand for land disposal.

Choosing among the Options: Cost-effectiveness as a Criterion

All of the options for reducing MSW listed in the previous section have been adopted in one form or another at various levels of government. An important question is, which alternatives perform best? In other words, which alternative achieves the goal of reducing waste at the lowest cost to society, including the costs of administering and enforcing the policy? How does the cost of each alternative compare with unit pricing for a given level of waste reduction? Which policies are the most self-regulating and self-enforcing in providing incentives for households and industries to comply (so-called incentive-compatible policies)? Would combinations of these policies be more effective than any policy operating alone?

These questions relate largely to what economists call the cost-effectiveness of the policies—that is, which option reduces a given amount of waste at a cost lower than all its alternatives. Several studies argue that, in the presence of significant illegal disposal opportunities, a deposit–refund system is the best option for reducing certain kinds of solid waste (Dinan 1993; Sigman 1995; Fullerton and Kinnaman 1995; Palmer and Walls 1997; Palmer and others 1997). A deposit–refund system is, in effect, equivalent to a charge on disposal: The deposit is refunded if the product is returned, so the consumer only bears a cost if the product is discarded. As a result, the deposit–refund system ensures that the least-cost method of reducing disposal is used, whether that method is reducing waste at its source (so-called source reduction) or recycling. By contrast, policies that either subsidize recycling or levy a tax on consumption take advantage of only recycling or only source reduction as an opportunity for waste reduction, but not both.

Palmer and others (1997) empirically assessed the cost of several of these policy options. Relying on estimated elasticities of demand and supply for different materials in the solid waste stream (that is, how responsive demand and supply are to changes in the price of the materials), they developed a simulation model of the markets for disposal and secondary materials and then evaluated the cost of a tax on new products, a recycling subsidy, and a combined tax/subsidy—that is, a deposit–refund system. Using 1990 baseline price and quantity data, they found that for any given percent reduction in waste disposal, a deposit–refund scheme would be the least-cost approach. For example, they estimate that a 10% reduction in waste disposal could be achieved with a deposit–refund of $45 per ton, a product tax equivalent to $85 per ton, or a recycling subsidy of $98 per ton. Even accounting for uncertainty in the elasticity estimates, the authors are reasonably sure that the deposit–refund approach would be the cheapest of the available alternatives.

One outstanding concern about the deposit–refund system is administrative costs. For example, administrative costs in states with bottle bills are high. Under these systems, retailers pay refunds to consumers, sort containers by brand name, and store containers until bottlers collect them. Cost estimates from California, which has a more centralized system in which the state pays handling fees to recycling centers that collect and process all recyclables and do not sort by brand, are apparently much lower (Ackerman and others 1995). Nonetheless, good estimates of administrative costs for different kinds of deposit–refund systems or different kinds of waste materials are not available, and this situation complicates efforts to compare alternative approaches for reducing MSW. Palmer and others (1997) recommend—as have others (see Fullerton and Wolverton 1999)—that both the deposit and refund be placed upstream rather than on consumers of final products. For example, a deposit would be paid on sheets of aluminum purchased by aluminum can manufacturers and the refund paid to those who supply used cans to reprocessors. Imposing the policies at the production rather than the retail level should greatly reduce administrative and transaction costs of the policies, because the number of affected producers and products is small compared with the myriad number of final consumer products.

Other Potential Goals

Environmentalists, legislators and other decisionmakers, waste managers, and the public articulate a host of objectives in managing solid waste besides the goal of cost-effectively reducing solid waste disposal. These objectives include conserving natural resources, offsetting policies that promote virgin material use, reducing greenhouse gas emissions, increasing the demand for secondary materials, meeting state and local recycling goals, and decreasing undesirable environmental and health effects associated with not only waste but the entire life cycle of all products. These objectives represent an amalgam of underlying concerns about safeguarding the environment, natural resources, and human health; they also include some objectives that are not necessarily ends in themselves, such as offsetting virgin materials use or meeting recycling goals—presumably, these goals are means by which to protect the environment and health.

What is the relationship between various waste management policies (specifically, policies used to attain Q^* in Figure 8-4) and these goals? Next, we consider this question in addressing each of these popular objectives.

Conserving Resources. Solid waste generation and management (trash collection and disposal) requires energy, water, and other natural resources. Popular arguments frequently call for policies such as taxes on virgin materials to reduce waste generation and disposal while conserving resources (Environmental Defense Fund 1992; Recycling Advisory Council 1993).

From an economic perspective, the most effective resource conservation policy is one that ensures that the prices of the resources reflect their scarcity. This approach may call for a virgin materials tax, but not as a policy for managing waste or recycling. In fact, several studies have shown the inefficiency of such a tax when the goal is reducing waste disposal (for example, Dinan

1993; Sigman 1995). Fullerton and Kinnaman (1995) show that the tax can be used to address an externality related directly to virgin material use—they use the examples of strip mining or forest clear-cutting, which may impose aesthetic costs on consumers—but they find no useful role for taxes on virgin materials in waste management.

Moreover, the effects of such a tax on waste volumes are unclear, and they could be perverse. For example, about 55% of the fiber used in paper-making comes from wastepaper or waste wood (see Alexander 1993). A tax on virgin timber would reduce the supply of available waste wood and increase the price of paper. It may make wood furniture more expensive but not have much effect on the solid waste stream. It may not reduce significantly the amount of paper discarded in waste, depending on the elasticity of demand for paper (if the demand for paper is fairly price inelastic, increases in price do not lead to large reductions in paper use).

In short, the effect of a virgin materials tax is quite complicated. Because the downstream effects of a tax are difficult to predict and this method has been shown to be less efficient than other options, other policies probably should be used to manage waste and recycling directly. The virgin materials tax probably should be used only to manage externalities associated only with the harvesting and other upstream operations that directly involve these resources.

Offsetting Policies That Promote Virgin Material Use. Another popular argument related to resource conservation and frequently linked with waste management is the desirability of offsetting existing policies that promote use of virgin materials (see Recycling Advisory Council 1993). For instance, advocates of such an approach might argue that an oil depletion allowance that subsidizes petroleum production should be "corrected" by gasoline taxes to reduce consumption. As another example, the sale of timber from public lands, sometimes sold at prices less than the costs of production, is seen by some observers as a practice that markedly favors the use of virgin fiber. Halting below-cost sales, they argue, would help recycled fiber compete with virgin fiber (see Environmental Defense Fund 1992). Similarly, the scrap metal industry historically has claimed that shipping fees set by the former Interstate Commerce Commission (now the Surface Transportation Board) discriminate against the transportation of scrap; the fees charged are a larger proportion of transportation costs for scrap than for virgin materials (see Tietenberg 1992). The industry argues that less discriminatory fees would facilitate the shipping of scrap and, in turn, increase the reuse of scrap instead of virgin materials.[15]

Some use of depletion allowances, discriminatory shipping charges, and other tax and fee treatments that favor the use of virgin materials have been eliminated or reduced through recent tax and regulatory reforms. For instance, since 1975, the percentage depletion allowance has not been available for major integrated oil companies and has been limited to 2,000 barrels per day for independent companies. The 1986 Tax Reform Act further reduced tax privileges for the petroleum industry by eliminating the investment tax credit and instituting the alternative minimum tax.[16] Moreover, some studies have suggested that these "privileges" have had only a small effect in discouraging use and recycling of scrap materials that compete with the use of virgin materials in any case (see U.S. Department of the Treasury 1979; U.S. EPA 1993a). It is unclear what additional reductions in the use of virgin materials would be obtained by attempting to offset existing policies. As suggested earlier, remaining inefficiencies in the use of virgin materials probably should be corrected by reforming the pricing policies for the resources themselves—a "first-best" solution—rather than through more cumbersome indirect approaches.

Reducing Greenhouse Gas Emissions. Another goal related to conservation of resources and discouraging the use of virgin materials is that of reducing greenhouse gas emissions. The Clinton administration's Climate Change Action Plan (Clinton and Gore 1993, 17), for example, states that "increased source reduction and recycling will save energy and money, reduce greenhouse gases, reduce the need for natural resource extraction, and help alleviate disposal problems." A recent U.S. EPA (1997) study quantifies the life-cycle greenhouse gas emission reductions achieved for various products through different solid waste–related policies.

Researchers have examined the greenhouse gas issue more formally by deriving the set of taxes and subsidies than can internalize both an energy-related upstream manufacturing externality, such as global warming, and the downstream disposal problem (see Walls and Palmer 2000). They find that no single policy tool can solve both problems. For example, absent a direct tax on energy, it would be necessary to subsidize all other inputs to production and levy a tax on output. The subsidies to the nonenergy inputs would need to be tailored to each firm to match each firm's production processes (specifically, to match each firm's marginal rate of technical substitution between the input and energy). Walls and Palmer conclude that this option is far inferior to a direct tax on energy, such as a carbon tax. In general, correcting externalities *directly* rather than indirectly through solid waste and recycling policies usually would be the preferred option. Moreover, it is unreasonable to expect policies that address solid waste and recycling to also be able to address global warming concerns. It would be impossible to find a policy that could cost-effectively accomplish both ends.

Reducing Life-Cycle Externalities. Solid waste management frequently is discussed in terms of opportunities to reduce other externalities, such as the air and water pollution that may occur during the life cycle of a product (that is, during production or use of the product in addition to its disposal).

The methodology that has evolved to identify the energy, water, and other resource requirements as well as the environmental impacts associated with every stage in the life of a product is known as product life-cycle analysis (PLCA).[17] PLCA addresses the energy used and the air emissions, water pollution, and solid wastes generated when raw materials are extracted as well as the resources used and the pollution that results during manufacturing, distribution, use, and disposal of the product. In other words, PLCA is similar to environmental full-cost accounting.

PLCA is popular among some manufacturers, environmental and consumer groups, and others as a means of informing consumers about the environmental consequences of producing, using, and disposing of products. Yet the approach has several limitations. One drawback is that interpreting the results of PLCA depends on the relative weights given to various environmental effects. For instance, palm oil–based cleaning agents used in detergents require about half as much energy to manufacture as do petroleum-based cleaners but generate much more solid waste. Another problem is drawing the boundary around the application of PLCA. For example, in addition to the pollution potential of the product itself during its life cycle, should the analyst consider the energy and raw materials used to make the equipment that is used in producing and consuming the product? Should the analyst include the transportation resources used by employees who make the product and by consumers who transport the product from the store to their homes? Finally, consistent use of PLCA would require life-cycle analysis for all products—a daunting task, but one that would be necessary to provide a basis for comparing environmental effects among competing products.

PLCA also can be inconclusive when weighing the impacts of the environmental consequences it identifies. For example, the production and use of one product may generate fewer total life-cycle emissions than another product, but those emissions may affect the environment more detrimentally (for example, if they involve air emissions in densely populated areas or water runoff into major estuaries) than higher levels of another kind of emissions.

Finally, and perhaps most relevant to our discussion in this chapter, PLCA is a highly imperfect tool for addressing solid waste concerns. PLCA can provide a context for comparing the effects of different products on solid waste compared with their effects on other environmental media (air emissions, water quality, energy use). However, it says little about the relative weight that should be given the product on the basis of its solid waste problem per se. In most cases, policies directed toward reducing solid waste will be, at best, blunt instruments for reducing other externalities. As we have emphasized in this chapter, the most efficient policies for reducing externalities associated with, say, the use of energy or water, are policies that focus directly on those activities. Policies such as those that require products to contain a minimum amount of recycled content or that establish a deposit–refund system may very well lead to changes in energy and water use (hence, the possibility of air and water pollution). Reaching the efficient level of those activities *as well as* the efficient level of solid waste generation and recycling through a recycling content standard or a deposit–refund system is probably impossible.

Developing the Concept of Extended Product Responsibility. Concerns about overuse of natural resources, greenhouse gas emissions, and life-cycle externalities are related to the notion of promoting extended product responsibility (EPR) as a principle for waste management and pollution prevention. EPR embodies the notion that individuals along the product chain should share responsibility for the life-cycle environmental impacts of the product (see Davis and others 1997; Fullerton and Wu 1998). The idea originated in Europe, where it is known as "manufacturer responsibility" or "extended *producer* responsibility" because of the focus on making producers responsible for their products, even at the end of the products' useful lives. Some European countries have passed legislation that requires manufacturers to take back the products they make. The German Packaging Ordinance of 1991 is one example.[18] Another is draft legislation under consideration by the European Union to require manufacturers to take sole responsibility for recycling used electronic and electrical equipment. The Mercury-Containing and Rechargeable Battery Management Act, passed by Congress in 1996, is the closest the United States has come to instituting a similar program. The act facilitates a national voluntary take-back system for nickel–cadmium rechargeable batteries.

It is not clear at this point whether policymakers and environmental experts view EPR as a guiding principle by which specific policies should be judged or as a policy itself. It seems to make the most sense to view it as a guiding principle—an umbrella under which different policy tools might sit. However, using EPR as a guiding principle could work at odds with economic efficiency and cost-effectiveness. It is infeasible, in practice, for the myriad manufacturers of consumer products to actually physically collect their products from consumers. To overcome this problem, under some EPR approaches, producers set up a special organization to serve as a centralized collection facility—a producer responsibility organization (PRO)—to collect their products from consumers at the end of the products' useful lives. The

PRO charges member companies a fee for collection and sorting of waste from the products. In this arrangement, take-back requirements look very much like an upstream product tax or advance disposal fee, but, as we have pointed out, this by itself will not be cost-effective at reducing waste disposal since it does not encourage recycling. Moreover, the transaction costs of collecting and hauling, sorting, and administering the operation of a PRO could be quite large. (Additional discussion of EPR is in Palmer and Walls 1999.)

If the goal of solid waste policy is to find the least-cost way to reduce disposal—that is, to internalize the externality we described using Figure 8-4—EPR serves only to confuse the issue. And as we stated earlier, it is impossible to find policies that do it all by reducing disposal and upstream pollution externalities at the same time. Moreover, making sure that responsibility for waste disposal is shared among the market participants might not be cost-effective.

Increasing Demand for Secondary Materials. One of the most important problems involved in recycling, particularly for local governments that operate curbside collection programs, is that supplies of secondary materials collected often become relatively large.[19] Old newspaper, mixed paper, green glass, and some plastics are among the materials that have, at various times, accumulated to amounts that exceeded the market's capacity to absorb them. Reasons for this oversupply include a lack of infrastructure in the recycling industry; general economic conditions; low prices for recovered materials; and, in some cases, lack of uncontaminated or clean supplies of recyclable materials.

A host of recent proposals and policies have sought to increase the demand for secondary materials. These proposals range from requirements that products supplied to federal, state, and local governments contain a minimum recycled content to subsidies to private companies for investing in technologies to make use of recycled materials. For instance, Section 6002 of RCRA requires EPA to recommend practices for federal, state, and local government agencies to follow in purchasing designated products, including standards for the minimum amount of recycled material that the products should contain.[20] Under Section 6002, EPA recently provided detailed guidelines to government procurement agencies for federal purchases of paper and paper products that contain recovered materials (U.S. EPA 1996). By the early 1990s, almost all states had instituted programs for state and municipal procurement of recycled materials. Another section of RCRA, Subtitle E, requires the U.S. Department of Commerce to promote the commercialization of proven recycling technology. In addition, thirty-five states offer grants, loans, or tax incentives to subsidize investment in technologies for using recycled products.[21] Numerous Congressional proposals have considered setting federal standards for minimum recycled content or establishing investment tax credits for recycling technologies.

Although these steps could increase demand for secondary materials, increasing such demand should not be a goal in itself. The problem of oversupply that faces municipalities may be temporary. In fact, periodic gluts and shortages are characteristic of many commodity markets, and recyclables are definitely a commodity. Figure 8-5 illustrates fluctuations in the prices paid for recycled HDPE (high-density polyethylene) plastic and used newspaper during the past few years. With growth in recycling infrastructure and better working markets, some of the oversupply has disappeared. Newspapers are a good example: In 1988, only nine de-inking facilities were located in North America; by 1994, there were twenty-nine (Alexander 1994). Growth in capacity to use old corrugated containers at paper mills was so rapid in the 1990s that observers became worried that demand would outpace the supply of old boxes (McCreery 1994). Likewise, demand for old plastic soda bottles in the mid-1990s was reported to be greater than supply (Miller 1994).

Another problem with recycled content standards is that because they set the required amount of secondary materials as a fraction of overall material use or as a fraction of total production, they do not directly provide incentives to reduce consumption and thus waste. In addition, fairly detailed information is required to specify the right percentage at which to set such standards. Palmer and Walls (1997) suggest that a policymaker would need to know the production technologies of all firms in all industries subject to the standard to be able to get the requirements right.

Meeting Recycling Goals. Another frequent goal of waste management policies is meeting state or municipal recycling goals. According to EPA, as of the mid-1990s, more than forty states had legislatively mandated specific, quantified recycling and/or waste reduction goals (see EPA's Web site, http://www.epa.gov). These goals usually are expressed in terms of percentages of different kinds of materials to be recycled and call for more stringent goals over time. In many cases, the recycling goals themselves appear to be driving other waste management policies—for example, to encourage households to reduce solid waste to meet recycling goals.

We caution that meeting the recycling goals in and of themselves should not be the objective, because to do so can generate additional problems (for example, oversupply in secondary materials markets). Rather, the objective should be to manage the waste stream cost-effectively. It may involve setting recycling goals, but if so, these would be established in the context of overall cost-effectiveness.

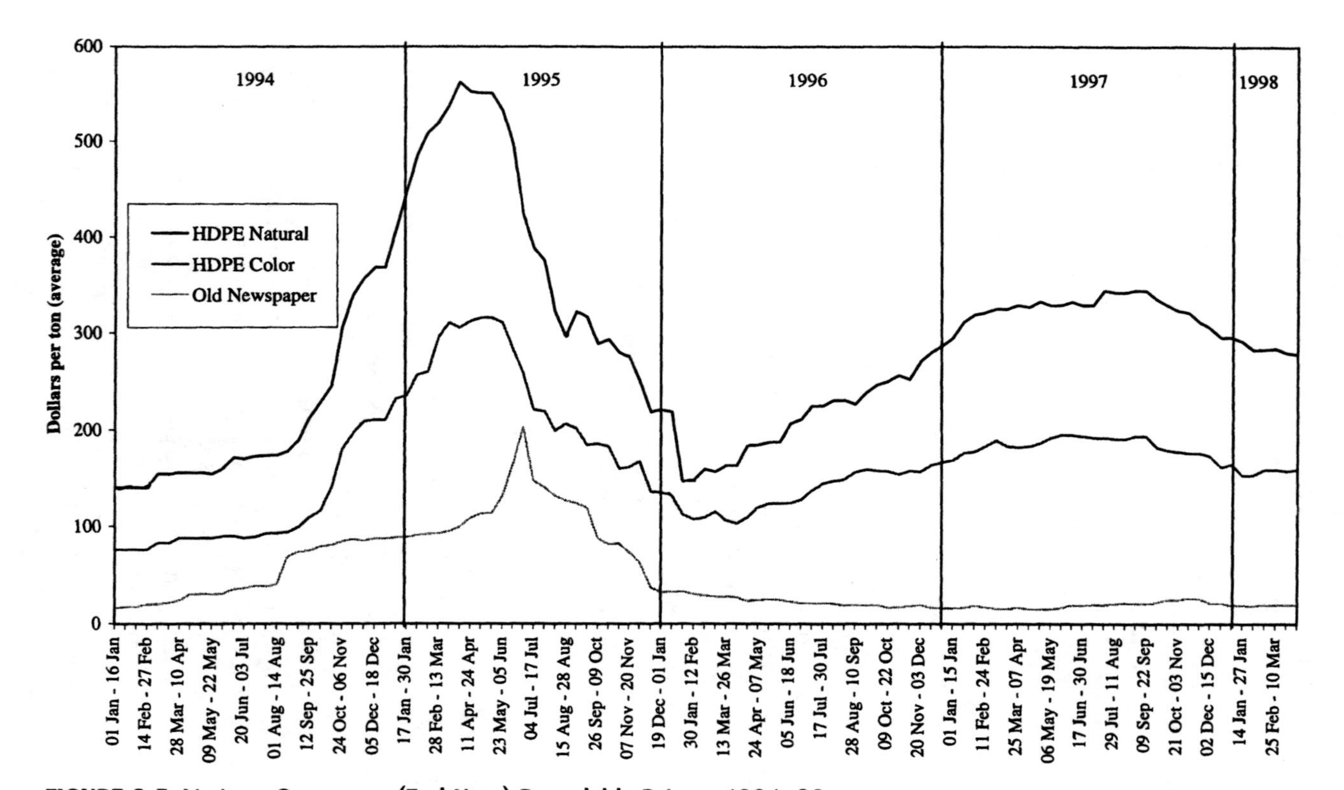

FIGURE 8-5. Various Consumer (End-User) Recyclable Prices, 1994–98.

Note: HDPE = high-density polyethylene.

Conclusions

Because most households in the United States do not pay an easily recognizable price per bag, per can, or per pound to dispose of their trash, the amount of trash generated is more than is socially desirable. In addition, externalities associated with MSW landfills may be not internalized in private decisionmaking. Pricing residential solid waste collection and disposal directly is a first-best solution to these market failures if illegal disposal is not a serious problem. However, in some communities, illegal disposal may become a problem, particularly if the price for legal disposal is set too high.

Of the set of alternative policies that are available to achieve the desired level of solid waste disposal, a combination of a product tax and recycling subsidy—that is, a deposit–refund approach—is the most cost-effective. It encourages both source reduction and recycling and thus reduces waste disposal at the least cost.

Many other potential goals for solid waste policies often are put forth by government officials and environmental advocates. These goals range from resource conservation and correction of externalities in markets for virgin materials to internalizing all externalities throughout the life cycle of a product. Although these concerns are legitimate, it is inappropriate for solid waste and recycling policies to attempt to address them all. Failures in upstream markets—markets for energy or virgin material production—should be addressed at their source. And existing subsidies in those markets, to the extent that they exist, should be eliminated, not somehow corrected for in a second-best way through government intervention in the market for solid waste disposal.

Notes

1. California and Oregon have recycled content standards for glass and plastic containers; Oregon, Maryland, and Connecticut have recycled content standards for telephone directories.

2. The U.S. Environmental Protection Agency (EPA) derives its estimate of municipal solid waste generation by using a method based on the flow of materials and products through the economy. The method begins with the total U.S. domestic production of materials and products, subtracts quantities of materials that are scrapped, adjusts for imports and exports of products, and adjusts for "diversions" from the waste stream (for example, paperboard used in building materials) and for the average lifetime of products. Some experts suggest that this approach may underestimate the amount of waste generated. For example, the trade publication *Biocycle* conducts surveys of states to estimate annual waste generation. Based on this approach, MSW generation in 1993 was 307 million tons, whereas EPA reported 207 million tons for that year (see Steuteville 1995).

3. Some localities, such as Seattle, WA, and Portland, OR, report recycling rates of close to 50% (Heumann 1997).

4. See Hickman 1996, from which much of this discussion of MSW regulation is taken.

5. Some researchers offer a different perspective. Wiseman (1992) estimates that if one year's worth of waste generated in the United States were placed in a single landfill to a depth of 100 yards, it would require an area of about two-thirds of a mile on each side. All municipal solid waste for the next 1,000 years would require less than a 30-mile square, according to his calculations.

6. Additional federal legislation related to municipal solid waste (MSW) includes the Comprehensive Environmental Response, Compensation and Liability Act (CERCLA). CERCLA directly affects localities whose MSW landfills are identified by the U.S. Environmental Protection Agency as Superfund sites. In recent sessions of Congress, bills have been proposed to limit or eliminate local government liability for cleanup costs at sites where the local government disposed of only MSW.

7. We use Figure 8-4 to illustrate the market for legal waste disposal because most of our ensuing discussion involves this kind of disposal. However, external costs also are incurred from illegal disposal.

8. In Figure 8-4, we define quantity along the horizontal axis as units of waste; at this point, we do not specify whether the quantity is measured by weight or volume. Some landfill fees are set by weight, and some are set by volume. A cubic yard of waste may weigh roughly 100 pounds before it is compacted for landfill disposal, at which time it may weigh 800–1,400 pounds on average (see Rathje and Murphy 1992, 48). We also have shown the marginal cost curves to be perfectly elastic in Figure 8-1, but our discussion generalizes to the case where the curves slope upward.

9. Tietenberg (1992) presents an excellent introduction to pollution externalities.

10. The levy typically is buried in other taxes—usually the property tax—or in utility bills. Large and small families usually are charged the same amount, and variations in quantities of waste generated by households are disregarded. In the District of Columbia, trash collection is financed through the property tax. There is no itemized line for the service on the household's tax bill. According to the Department of Real Estate Assessment and Property Tax, the amount that the DC government budget allocates to trash services is $97 per household per year. This amount is about 1–4% of the total property tax bill, depending on the assessed value of the home. In Arlington County, VA, the assessment is included as a separate line item on the household's sewer and water bill.

11. For additional discussion, see Goddard 1975; U.S. EPA 1990; Dobbs 1991; Blume 1991; Repetto and others 1992; Skumatz 1993. Tacoma, WA, apparently was one of the first localities to price solid waste disposal directly and has done so since the 1930s (see Alexander 1993, 148; Goddard 1993, 8).

12. Many local governments collect enough revenues through flat fees or property taxes to cover the costs of collecting and disposing of waste; they do not charge citizens specifically for disposal.

13. Estimates of illegal disposal are difficult to find, but anecdotal evidence suggests that numerous small illegal dumping sites exist, most consisting of "one big

load of one household's garbage," particularly in rural areas (Rathje and Murphy 1992, 85). Researchers point out that illegal dumping may be more likely to occur as disposal fees increase (Alexander 1993). Some anecdotal information about dumping in the wake of levying disposal fees is reported in Blume 1991 and Skumatz 1993; on the basis of this information, it appears that although some illegal dumping occurs, it is not a major concern of localities that have implemented unit fees. Fullerton and Kinnaman (1995, 1996) speculate that the increase in illegal disposal in Charlottesville, VA, after starting a price-per-bag program was fairly substantial. However, their evidence is indirect; they did not measure the amount of dumping that occurs. In the case of used motor oil, Sigman (1998) finds evidence that illegal disposal of this hazardous waste increases as the cost of legal waste management (including disposal and reuse) increases and as the threat of enforcing penalties for illegal disposal increases.

14. Product taxes for the purpose of reducing solid waste often are referred to as advance disposal fees (ADFs). California and Florida have ADFs.

15. Not all policies tend to favor virgin material use. For instance, severance taxes levied by states tend to counteract incentives to use virgin materials by increasing the costs of minerals extraction. Whether severance taxes correct the environmental effects of extraction or are a means of exporting tax liability from one state to another is the topic of much debate. (For a general discussion and references to the literature, see Tietenberg 1992, chapter 8.)

16. The 1990 Omnibus Reconciliation Act reestablished some of these privileges for independent oil companies. The percent depletion allowance was increased for enhanced oil recovery projects undertaken by independent companies, for example, and a tax credit—also for independent companies—was allowed for exploration and development expenditures. The twenty largest companies, which do not receive these benefits, produce approximately 60% of all U.S. crude oil (see American Petroleum Institute 1988).

17. An example of product life-cycle analysis (PLCA) is found in Hocking 1991. Much of the discussion in this section is from Portney 1993–94; see also the references Portney cites for other views of PLCA.

18. Davis and others (1997) present a brief discussion of the German law and other European initiatives.

19. Supply also exceeds demand in other countries. For instance, the nationwide recycling collection system in Germany collected more than three times the tonnage of scrap plastic than the country had capacity to handle (Tanner 1994).

20. For a discussion of federal involvement, see Meyers 1991. For a review of the federal government's role in these activities, see U.S. GAO 1993.

21. Steuteville (1992) surveys several of these state and local initiatives in detail.

References

Ackerman, Frank, Dmitri Cavander, John Stutz, and Brian Zuckerman. 1995. *Preliminary Analysis: The Costs and Benefits of Bottle Bills.* January. Draft report to U.S.

EPA, Office of Solid Waste and Emergency Response. Boston, MA: Tellus Institute.

Alexander, Judd H. 1993. *In Defense of Garbage.* Westport, CT: Praeger.

Alexander, Michael. 1994. Developing Markets for Old Newspapers. *Resource Recycling* 13(July):20–27.

American Petroleum Institute. 1988. *Basic Petroleum Data Book* 8(3).

Blume, Daniel R. 1991. Under What Conditions Should Cities Adopt Volume-Based Pricing for Residential Solid Waste Collection? May. Master's memo study for the Office of Management and Budget, Office of Information and Regulatory Affairs, Natural Resources Branch. Washington, DC: Office of Management and Budget.

Clinton, President William J., and Vice President Albert Gore, Jr. 1993. *The Climate Change Action Plan.* October.

Davis, Gary A., Catherine A. Wilt, and Jack N. Barkenbus. 1997. Extended Product Responsibility: A Tool for a Sustainable Economy. *Environment* 39(7, September):10–18.

Dinan, Terry M. 1993. Economic Efficiency Effects of Alternative Policies for Reducing Waste Disposal. *Journal of Environmental Economics and Management* 25(December):242–56.

Dobbs, Ian M. 1991. Litter and Waste Management: Disposal Taxes versus User Charges. *Canadian Journal of Economics* 24(1):221–27.

Environmental Defense Fund. 1992. *Developing Markets for Recycling Multiple Grades of Residential Paper: A Report to the Northeast Recycling Council and the U.S. Environmental Protection Agency.* July. New York: Environmental Defense Fund.

Fullerton, Don, and Tom Kinnaman. 1995. Garbage, Recycling, and Illegal Burning or Dumping. *Journal of Environmental Economics and Management* 29:78–91.

——. 1996. Household Responses to Pricing Garbage by the Bag. *American Economic Review* 86(4):971–984.

Fullerton, Don, and Ann Wolverton. 1999. The Case for a Two-Part Instrument: Presumptive Tax and Environmental Subsidy. In *Environmental Economics and Public Policy: Essays in Honor of Wallace E. Oates,* edited by Arvint Panagariya, Paul R. Portney and Robert M. Schwab. Northampton, MA: Edward Elgar Publishing Ltd.

Fullerton, Don, and Wenbo Wu. 1998. Policies for Green Design. *Journal of Environmental Economics and Management* 36(2):131–48.

Goddard, Haynes C. 1975. User Charges for Solid Waste Management. *Managing Solid Waste: Economics, Technology, Institutions.* New York: Praeger.

——. 1993. Costs and Benefits of Alternative Waste Management Policies. Paper prepared for Balancing Economic Growth and Environmental Goals, a symposium sponsored by the American Council for Capital Formation, Center for Policy Research, September 29, 1993, Washington, DC.

Heumann, Jenny M. 1997. Most Efficient Municipal Recycling Programs Highlighted by ILSR. *Recycling Times* 9(18, September 1):7.

Hickman, H. Lanier, ed. 1996. *Principles of Municipal Solid Waste Management.* Silver Spring, MD: Solid Waste Association of North America.

Hocking, M. B. 1991. Paper versus Polystyrene: A Complex Choice. *Science* 251(4993):504–506.

Ley, Eduardo, Molly K. Macauley, and Stephen W. Salant. 1996. Spatially and Intertemporally Efficient Waste Management: The Costs of Interstate Flow Control. July. Discussion paper 96-23. Washington, DC: Resources for the Future.

McCreery, Patrick. 1994. New Mills Open, but Feedstock Could Be a Problem. *Recycling Times,* Sept 6, 3.

Meyers, Jonathan Phillip. 1991. Confronting the Garbage Crisis: Increased Federal Involvement as a Means of Addressing Municipal Solid Waste. *Georgetown Law Journal* 79(3):567–90.

Miller, Chaz. 1994. PET: #1 is Number One. *Waste Age* 25(September):49–61.

Miranda, Marie Lynn, Sharon LaPalme, and David Z. Bynum. 1998. *Unit Based Pricing in the United States: A Tally of Communities.* Report to U.S. EPA, July.

Palmer, Karen, and Margaret Walls. 1997. Optimal Policies for Solid Waste Disposal and Recycling: Taxes, Subsidies, and Standards. *Journal of Public Economics* 65(2):193–205.

———. 1999. Extended Product Responsibility: An Economic Assessment of Alternative Policies. Discussion paper 99-12. Washington, DC: Resources for the Future.

Palmer, Karen, Hilary Sigman, and Margaret Walls. 1997. The Cost of Reducing Municipal Solid Waste. *Journal of Environmental Economics and Management* 33(2):128–50.

Portney, Paul R. 1993–94. The Price Is Right: Making Use of Life Cycle Analyses. *Issues in Science and Technology* 10(2, Winter):69–75.

Rathje, William, and Cullen Murphy. 1992. *Rubbish! The Archaeology of Garbage.* New York: Harper Perennial.

Recycling Advisory Council, Market Development Committee. 1993. *Market Structure Policy Options Briefing Book.* April. Washington, DC: Recycling Advisory Council.

Repetto, Robert, Roger C. Dower, Robin Jenkins, and Jacqueline Geoghegan. 1992. *Green Fees: How a Tax Shift Can Work for the Environment and the Economy.* November. Washington, DC: World Resources Institute.

Sigman, Hilary. 1995. A Comparison of Public Policies for Lead Recycling. *RAND Journal of Economics* 26:452–78.

Sigman, Hilary. 1998. Midnight Dumping: Public Policies and Illegal Disposal of Used Oil. *RAND Journal of Economics* 29:157–78.

Skumatz, Lisa. 1993. Variable Rates for Municipal Solid Waste: Implementation, Experience, Economics, and Legislation. June. Policy study no. 160. Los Angeles, CA: Reason Foundation.

Steuteville, Robert. 1992. Economic Development in the Recycling Arena. *BioCycle* 33(8):40–44.

——. 1995. The State of Garbage in America. *BioCycle* 36(4):54–63.

Tanner, Arnold O. 1994. *Materials Recycling.* February. U.S. Department of the Interior, Bureau of Mines. Washington, DC: U.S. Government Printing Office.

Tietenberg, Tom. 1992. *Environmental and Natural Resource Economics.* New York: HarperCollins.

U.S. Department of the Treasury. 1979. *Federal Tax Policy and Recycling of Solid Waste Materials.* Washington, DC: U.S. Department of the Treasury, Office of Tax Analysis.

U.S. EPA (Environmental Protection Agency). 1990. *Charging Households for Waste Collection and Disposals: The Effects of Weight or Volume-Based Pricing on Solid Waste Management.* September. EPA 530-SW-90-047. Washington, DC: U.S. EPA.

——. 1993a. *Federal Disincentives to Recycling.* Washington, DC: U.S. Government Printing Office.

——. 1993b. *Guide to EPA's Unit Pricing Database: Pay-As-You-Throw Municipal Solid Waste Programs in the United States.* April. EPA 230-B-93-002. Washington, DC: U.S. EPA.

——. 1995. *Characterization of Municipal Solid Waste in the United States.* EPA 530-R-96-001. Washington, DC: U.S. EPA, Office of Solid Waste.

——. 1996. Paper Products Recovered Materials Advisory Notice. *Federal Register* 61(104):26986–93, May 29.

——. 1997. Greenhouse Gas Emissions from Municipal Waste Management. March. Draft working paper. EPA 530-R-91-010. Washington, DC: U.S. EPA.

U.S. GAO (General Accounting Office). 1993. *Solid Waste: Federal Program to Buy Products with Recovered Materials Proceeds Slowly.* May. GAO/RCED-93-58. Washington, DC: U.S. GAO.

Walls, Margaret, and Karen Palmer. 2000. Upstream Pollution, Downstream Waste Disposal, and the Design of Comprehensive Environmental Policies. *Journal of Environmental Economics and Management* forthcoming.

Wiseman, A. Clark. 1992. Government and Recycling: Are We Promoting Waste? *The Cato Journal* 12(2):443–60.

Index

CPSIA information can be obtained
at www.ICGtesting.com
Printed in the USA
FFOW01n1036240715
15388FF

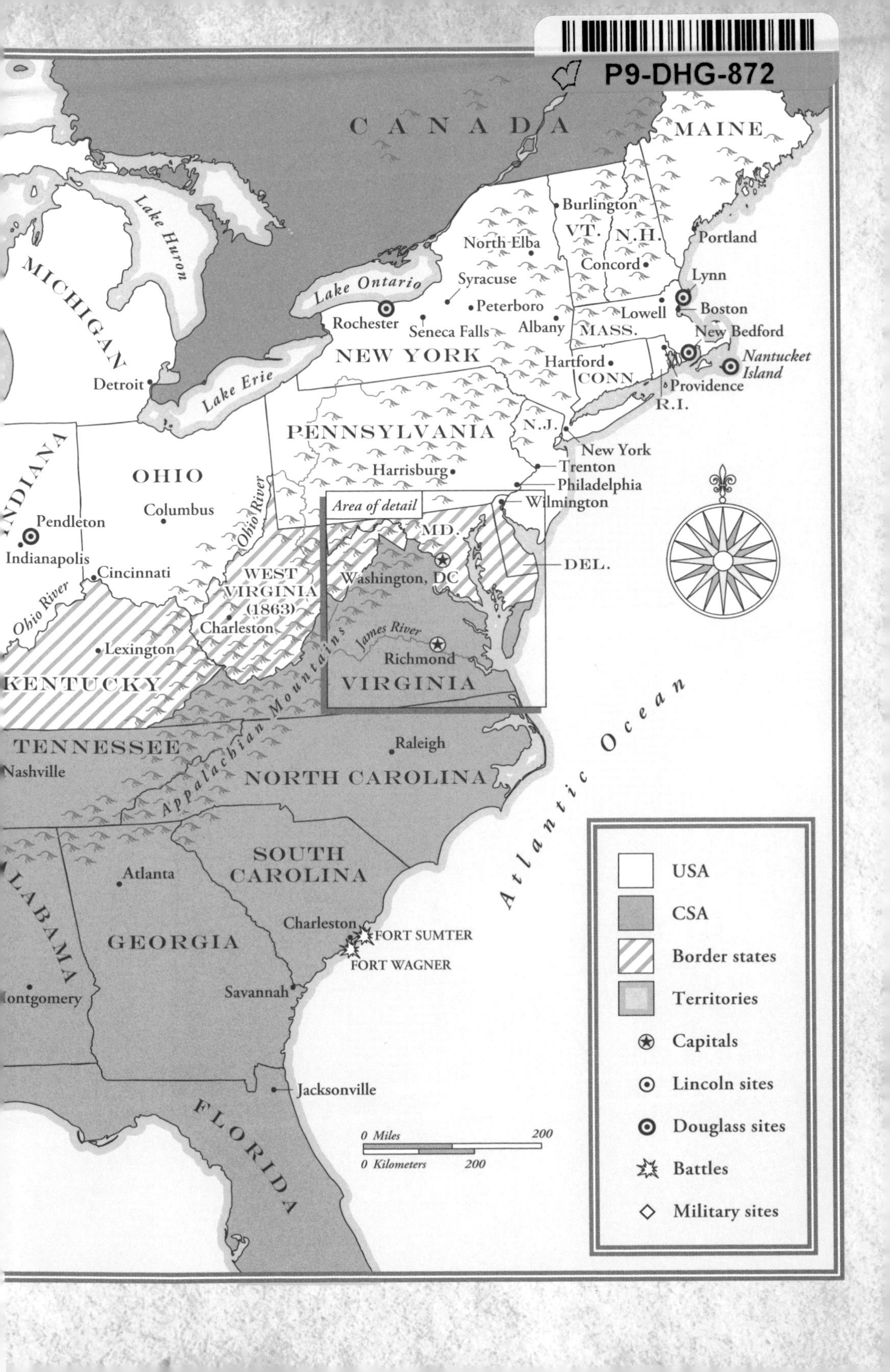

P9-DHG-872
CANADA
MAINE
Lake Huron
MICHIGAN
Lake Ontario
Lake Erie
Burlington
VT.
N.H.
Portland
North Elba
Concord
Syracuse
Lynn
Peterboro
Boston
Rochester
Seneca Falls
Albany
Lowell
MASS.
New Bedford
Nantucket Island
NEW YORK
Hartford
CONN
Providence
R.I.
Detroit
INDIANA
PENNSYLVANIA
N.J.
New York
OHIO
Harrisburg
Trenton
Philadelphia
Ohio River
Area of detail
Wilmington
Columbus
Pendleton
MD.
Indianapolis
Cincinnati
WEST VIRGINIA (1863)
Washington, DC
DEL.
Ohio River
Charleston
James River
Lexington
Richmond
KENTUCKY
VIRGINIA
Appalachian Mountains
Atlantic Ocean
TENNESSEE
Raleigh
Nashville
NORTH CAROLINA
SOUTH CAROLINA
Atlanta
ALABAMA
GEORGIA
Charleston
FORT SUMTER
FORT WAGNER
Savannah
Montgomery
Jacksonville
FLORIDA
0 Miles 200
0 Kilometers 200
USA
CSA
Border states
Territories
Capitals
Lincoln sites
Douglass sites
Battles
Military sites

THE PRESIDENT AND THE FREEDOM FIGHTER

THE PRESIDENT AND THE FREEDOM FIGHTER

ABRAHAM LINCOLN, FREDERICK DOUGLASS, *and* THEIR BATTLE *to* SAVE AMERICA'S SOUL

BRIAN KILMEADE

SENTINEL

SENTINEL
An imprint of Penguin Random House LLC
penguinrandomhouse.com

Most Sentinel books are available at a discount when purchased in quantity for sales promotions or corporate use. Special editions, which include personalized covers, excerpts, and corporate imprints, can be created when purchased in large quantities. For more information, please call (212) 572-2232 or e-mail specialmarkets@penguinrandomhouse.com. Your local bookstore can also assist with discounted bulk purchases using the Penguin Random House corporate Business-to-Business program. For assistance in locating a participating retailer, e-mail B2B@penguinrandomhouse.com.

Owing to limitations of space, image credits may be found on pages 290–292.

Library of Congress Cataloging-in-Publication Data

Names: Kilmeade, Brian, author.
Title: The president and the freedom fighter : Abraham Lincoln, Frederick Douglass, and their battle to save America's soul / Brian Kilmeade.
Identifiers: LCCN 2021024463 (print) | LCCN 2021024464 (ebook) |
ISBN 9780525540571 (hardcover) | ISBN 9780525540601 (ebook)
Subjects: LCSH: Lincoln, Abraham, 1809–1865—Friends and associates. |
Douglass, Frederick, 1818–1895—Friends and associates. | Slavery—Law and legislation—United States—History. | Slaves—Emancipation—United States. | Presidents—United States—Biography. | Abolitionists—United States—Biography. | United States—Politics and government—1849–1877. | United States—History—1849–1877.
Classification: LCC E457.2 .K49 2021 (print) | LCC E457.2 (ebook) | DDC 973.7092—dc23
LC record available at https://lccn.loc.gov/2021024463
LC ebook record available at https://lccn.loc.gov/2021024464

Printed in the United States of America
1 3 5 7 9 10 8 6 4 2

CJKV

BOOK DESIGN BY MEIGHAN CAVANAUGH

For all the teachers who have dedicated their careers to showing young learners that America is a truly exceptional nation—not because we are perfect, but because we try to be.

Liberty has been won. The battle for
Equality is still pending.

CHARLES SUMNER, JUNE 1, 1865

CONTENTS

PREAMBLE

> We hold these truths to be self-evident, that all men are created equal, that they are endowed by their Creator with certain unalienable Rights, that among these are Life, Liberty and the pursuit of Happiness.
>
> —Declaration of Independence, July 4, 1776

In an early draft of the Declaration of Independence, Thomas Jefferson called slavery "a cruel war against human nature itself."[1] James Madison argued that "it would be wrong to admit in the Constitution the idea that there could be property in men."[2] Benjamin Franklin, a former slaveholder, described slavery as "an atrocious debasement of human nature."[3] But in the early days of the republic, slavery remained legal, the law of the land.

The Founders recognized that slavery, which was still practiced on every continent, threatened the new nation's foundation. But the men who possessed the genius to launch a country that changed the world could not agree upon a way to end the institution of human bondage. The closest they came was to acknowledge, as Jefferson confided sadly to a friend, "We have the wolf by the ear, and we can neither hold him, nor safely let him go."

Every generation since—including our own—has grappled with this legacy of racial inequity, bequeathed by our political parents. In 1861, that inheritance threatened to destroy the United States in a bloody civil war. No longer was it possible to ignore the question of whether the Declaration's ideals of "Life, Liberty and the pursuit of Happiness" applied to all people. But how could a practice that was so much a part of the nation's way of life be abolished? And could the United States survive the abolition?

The job of guiding the nation to a fairer future fell to two remarkable Americans, an unexpected pair. One was White, born impoverished on a frontier farm, the other Black, a child of slavery who had risked his life escaping to freedom in the North. Without fancy pedigrees, neither had had an easy path to influence. No one would have expected them to become friends—or to change the country. But Abraham Lincoln and Frederick Douglass believed in their nation's greatness and were determined to make the grand democratic experiment live up to its ideals. Sharing little more than the conviction that slavery was evil, the two men's paths converged, and they would ultimately succeed where the Founders fell short.

This is their story.

ONE

FROM THE BOTTOM UP

I do not remember to have ever met a slave who could tell of his birthday.

—Frederick Douglass, 1845

Abraham Lincoln had a problem. His flatboat, carried by the rush of spring waters, had run aground atop a mill dam in the Sangamon River. The square bow of the eighty-foot-long boat hung over the dam, cantilevered like a diving board. Meanwhile, the stern was sinking lower and lower as it took on water. If Lincoln didn't think of something quickly, the vessel might break apart.

The young man had built the boat with a plan in mind. Along with his cousin, he would take on cargo, travel down the river from central Illinois to New Orleans, and there dismantle the boat, selling both its timber and the cargo on behalf of a man willing to underwrite the venture. Together, he and his cousin had cut down trees for lumber upstream from where they were now marooned. They had built the boat and loaded it with dried pork, corn, and live hogs. All had seemed well when they set off only hours before, but now, on April 19, 1831, far from his intended destination, Lincoln had to do something to save his boat and his cargo.

As goods slid slowly astern in the tilting craft, Lincoln went into action. Removing his boots, hat, and coat, he improvised. First, he and his two-man crew shifted most of the goods to the nearby shore. Next, while he hurriedly bored a large hole with a hand drill, his team began rolling the remaining cargo of heavy barrels forward, thereby shifting the boat's center of gravity.

The strategy worked: As the flatboat's bow began to tilt downward, water poured out the hole. As the boat got lighter, it rose in the water. After plugging the hole, Lincoln and his men, helped by the spring currents, managed to ease the box-like craft clear of the dam.

The crowd of villagers that had gathered to observe the spectacle of a sinking boat was astonished. No one had seen anything like it. But then they had also never met Abraham Lincoln, just two months beyond his twenty-second birthday. At first sight, he was unmistakably a country bumpkin, dressed in ill-fitting clothes that exaggerated his six-foot, four-inch height, with long arms and exposed ankles sticking out of too-short shirts and homespun trousers. He made, said one observer, "a rather Singular grotesque appearance." But the young man who saved the boat possessed a loose-limbed grace that disguised both unexpected strength and a driving ambition to make something of himself. Weighing over two hundred pounds, he could lift great weights and throw a cannonball farther than anyone around.[1] He ran and jumped with the best of his peers.

To the people he met, the young man's appearance quickly became secondary. "When I first [saw] him," reported one New Salemite, "i thought him a Green horn. His Appearance was very od [but] after all this bad Apperance I Soon found [him] to be a very intelligent young man."[2] Lincoln surprised people, who found he was not an illiterate rube but a man with a lively wit and keen intelligence.

He impressed not only that day's onlookers but also the owner of the flatboat. After completing the trip to New Orleans, Lincoln returned to New Salem to accept the man's offer to clerk at a new general store. He would sell foodstuffs, cloth, hardware, tobacco, gunpowder, boots, whiskey, and other goods to the people of New Salem and the local farmers who visited the little market town to sell their grains.

Like "a piece of floating driftwood," as Lincoln later described himself, he accidentally lodged at New Salem.[3] He would establish a new and happy life there, a world apart from his childhood in the backwoods.

"THE SHORT AND SIMPLE ANNALS OF THE POOR"*

Before striking out on his own, as Abraham Lincoln himself would tell the story, he had been under the thumb of his hard-luck father.

Born during the Revolution, Thomas Lincoln was still a boy when his family, following in the footsteps of distant cousin Daniel Boone, left the Shenandoah Valley for the territory known as Kentucky. Barely two years later, in 1786, Thomas's father, Abraham, died when planting corn in a field, shot by a roving Native American war party. His entire estate went to the eldest son, leaving the youngest, Thomas, destitute, "a wandering laboring boy [who] grew up literally without education."[4]

Thomas Lincoln eventually saved enough money to buy a farm in Hardin County and, in 1806 took Nancy Hanks as his wife. The following year they became parents, with the birth of daughter Sarah. Three

* When asked about his childhood, Lincoln explained that his early years weren't worth a close look. His first two decades could be summed up, he said, as "The short and simple annals of the poor," quoting a 1750 poem by Thomas Gray, "Elegy Written in a Country Churchyard."

years later, the little family expanded again with the arrival of a son, Abraham, born on February 12, 1809, and named for his grandfather.

Unfortunately for baby Abraham, the Lincoln family's stability was short-lived. Before he turned two, poor soils forced the family to abandon their first log cabin for more fertile ground a few miles away. Just five years later, after a title dispute over land ownership, the Lincolns started yet again, this time moving across the Ohio River to a crude, dirt-floored home in Indiana. The family lived a life of long days of work, brutally cold winter nights, and next to no comforts.

Then life got even tougher. Nancy Lincoln fell ill with milk sickness, poisoned by milk from a cow that had eaten snakeroot while grazing in the forest. After a week of watching his mother suffer acute intestinal pains and persistent vomiting, Abraham Lincoln was left motherless at age nine.

Fourteen months later, Thomas Lincoln remarried. The arrival of widow Sarah Bush Lincoln was a bright spot in the boy's hardscrabble childhood. "Mama," as Abraham called his stepmother, brought three children with her. As she blended two families into one, Sarah became the boy's "best friend in the world."[5]

He would never express such affection for his father; the son would remember Thomas Lincoln as a taskmaster. At age eight, Lincoln later said, his father "[put] an axe put into his hands . . . and [until] his twenty-third year, he was almost constantly handling that most useful instrument."[6] Virtually all the work on the farm was handwork, the tools crude, and the chores many, with livestock to care for, fields to plant, wood to cut and split, and gardens to tend. Abraham attended school only at the rare times when there wasn't other work to be done, and Thomas always seemed to have more tasks for his growing son. By law, boys at that time were effectively indentured servants obliged to work for their fathers until they came of age, and Thomas took full advantage. He hired Abraham out to other farmers and kept the wages his son earned. Even when Abraham hired himself out as a boatman in his late teens, the wages he earned belonged to his father.

Thomas was known to beat Abraham, and he kept him at work on the farm so much that the boy got less than one full year of formal schooling in his entire childhood, never attending classes after he turned fifteen. But Abraham would not be held back. Few as they were, his school days inspired in him an intense love of learning. He was eager for knowledge, looking for it wherever he could. "He must understand Every thing—even to the smallest thing—Minutely and Exactly," Sarah would remember years later. "He would then repeat it over to himself & again—sometimes in one form and then in another & when it was fixed in his mind to suit him he . . . never lost that fact or his understanding of it."[7]

Thomas Lincoln could barely scratch out his own name, and Sarah could neither read nor write. But Abraham was a fast learner who mastered reading despite his irregular visits to the schoolroom. He found a new world in books and was desperate to read all he could. Although books were rare in rural Indiana, Sarah had brought some of her late husband's from Kentucky and encouraged Abraham to read them. He read her family Bible and John Bunyan's *Pilgrim's Progress*, and he reread *Aesop's Fables* so often he could recite the moral stories from memory. Daniel Defoe's novel *Robinson Crusoe* and biographies of Benjamin Franklin and George Washington were other favorites. According to a cousin who lived with the Lincolns, Abe (some called him that, though he preferred his full first name) was a "Constant and voracious reader."[8]

Thomas had no patience for a son who "fool[ed] hisself with eddication."[9] Nonetheless, Abraham read widely and began traveling downriver, his imagination filling with thoughts of the day when he could leave behind both his father and the hard and monotonous labor of frontier farming.

Finally, in February 1830, Abraham turned twenty-one. The next month he served his father one last time, helping the Lincolns make another move, this time to Illinois. Abe led an ox team on the long trek. He helped build a log cabin "at the junction of the timber-land and prairie." He aided in planting a corn crop and split enough rails to fence ten acres of ground.[10] But that would be the last growing season for Abe Lincoln. Now, at last an adult, he left the farm behind and set out looking for a life of his own.

BORN TO SLAVERY

Abraham Lincoln's father permitted him few school days. The child given the birth name Frederick Augustus Washington Bailey got none at all.

Born on a farm on Maryland's Eastern Shore, he never knew the exact date of his birth. He spent his early childhood in the care of his grandmother after his mother was hired out to another farm miles away. Although she managed a few nighttime visits, Harriet Bailey was just a vague memory to the son who had no recollection of ever seeing her in daylight.

The child knew nothing of his father beyond the rumor that the White man who impregnated his mother might have been her enslaver, Aaron Anthony, a man he feared and hated. He once watched Anthony whip an aunt "till she was literally covered with blood." The very young boy would carry the harsh recollection forever, a "horrible exhibition," a "terrible spectacle," and a glimpse of "the blood-stained gate, the entrance to the hell of slavery."[11]

At age six, Fred Bailey was relocated by his owner to a grand, eighteenth-century mansion house on the Wye River. Put to work in the house, he was a playmate and companion to twelve-year-old Daniel, youngest son of Edward Lloyd V, a former governor of Maryland.

Young "Mas' Daniel" took a liking to his boy, acting as his protector. He shared bread with him, too, when the usual fare for the plantation's Black inhabitants was a daily portion of boiled cornmeal mush. Daniel's kindness led Fred to observe many years later that, "The equality of nature is strongly asserted in childhood . . . *Color* makes no difference with a child."[12]

In the eyes of no one else, however, were the boys anything like equal. On school days Daniel disappeared into the plantation's schoolroom, a place where no Black face was welcome. Still, the enslaved child benefited from the tutoring Daniel received. The Lloyds spoke correct English, a very different language from the broken, half-African dialect spoken on the plantation. The precocious Bailey, who stowed everything he heard in his prodigious memory, was left curious, hungry to know more.

In March 1826, at age eight, he was sent to the next stop in his enslavement. With no close family ties to sever, he left behind only the "hardship, whipping and nakedness" of the plantation.[13] Fred Bailey stepped aboard the sloop *Sally Lloyd*, headed for Baltimore and servitude in the home of his owner's brother. After years spent wearing nothing more than a crude linen shirt that hung to his knees, he wore his first pair of trousers, an outfit better suited for the city that would be his home. On its overnight sail, the vessel and its cargo of sheep crossed the Chesapeake Bay; to a farm boy who had always lived upriver, it seemed an unfathomably large expanse of water. When the city came into sight, young Fred was equally amazed by the tall church spires and five-story buildings. The *Sally Lloyd* entered a harbor full of three-masted seagoing vessels, warehouses, and steamships belonging to merchants in what had become one of the country's busiest ports.

A crewman escorted the eight-year-old to Aliceanna Street, an easy walk from the harbor, and the home of an aspiring shipbuilder named Hugh Auld. Auld and his wife, Sophia, met him at the door, together with their son, Thomas. "Little Tommy" was told the new arrival was to be "his Freddy." The dark-skinned boy would be caring for the younger White one.

Little did Freddy Bailey know that he was about to get the unexpected gift of the ABCs, the instrument that would permit him to elevate himself to one of the great men of the century.

A HUNGER FOR KNOWLEDGE

The city differed in a thousand ways from the rural life Freddy had lived. On Baltimore's paved streets, the noise was constant, but there were fewer of the "country cruelties" he had known on the plantation. In his new home, he slept on a mattress of straw, his first bed after the cold, damp dirt floors on which he had always slept. He wore clean clothes and ate better than ever before. But the kindness of Sophia Auld was the biggest shock of all.

As wholly owned property, he previously "had been treated as a *pig* on the plantation." But Miss Sopha, as he called his new mistress, "made me something like [Tommy's] half-brother in her affections."[14] A regular churchgoer, Mrs. Auld continually surprised him. Part of the explanation was that she had not grown up in a slaveholding family. Not having been hardened to the barbarity of owning other people, she didn't expect the cowering servility demanded by the previous White adults Freddy had known. Miss Sopha actively encouraged him to look her in the eye, and he "scarcely knew how to behave towards her."[15]

Though his primary job was to keep Tommy out of harm's way, he also ran errands for Mistress Auld. Moving freely on the streets of Baltimore, he explored a city where a minority of Whites were slaveholders and a majority of the Black inhabitants were freemen. As a "city slave," he had become, in comparison to his status on the plantation, "almost a free citizen."[16] That wasn't quite true, of course; in the night he could hear his less fortunate brothers and sisters being walked to the docks in chains. But looking back later, he described how important was his arrival in a cosmopolitan city of eighty thousand people. "Going to live at Baltimore," he would write at age twenty-five, "laid the foundation, and opened the gateway."[17]

The revelation that truly changed his life occurred in the Auld household. Listening to Miss Sopha read the Bible aloud, Freddy began to recognize that words had power. When he asked if he, too, might learn to read, she agreed without hesitation. An apt pupil, he quickly mastered the alphabet. Teacher and student moved on to sounding out simple words of three and four letters. Taking pride in his progress, Mrs. Auld told her husband how well Freddy was doing, confiding that she hoped he might one day be able to read the Bible.

Hugh Auld was flabbergasted. He was appalled that his wife didn't understand that teaching the Black child "the A B C" was irresponsible, unwise, and unlawful. She was violating the basic code of slaveholding society, which demanded total compliance on the part of the enslaved. Slavery and education were incompatible, he told her; tutoring Freddy was an invitation for him to violate Auld's property rights. "He should know nothing but to obey his master—to do as he is told to do."[18]

The reading lessons ended abruptly, and Miss Sopha hardened her heart toward the boy she had previously treated with affection. Yet her reaction proved less significant than the bright light of understanding that Auld's words lit in the mind of Fred Bailey.

Although directed at his wife, Auld's rebukes had been a revelation to the wary nine-year-old, who had stood by, watching and listening, as Hugh chastised Sophia. He saw for the first time a fundamental truth about slavery. "Learning would do him no good," Auld had said, "but probably a great deal of harm—making him disconsolate and unhappy.... If you learn him now to read, he'll want to know how to write; and, this accomplished, he'll be running away with himself."[19]

In the coming years the enslaved child, by then a man, would quote Auld's words often. They had sparked the crucial understanding that reading and knowledge were the keys to his freedom. Young as he was, Freddy saw that ignorance was a weapon in the hands of the slaveholder, one that was far more dangerous than the whip.

Mistress Auld no longer tutored him, but Hugh Auld's words inspired a new determination to learn. Seeking instruction elsewhere, Fred deciphered words on signs. He bartered for reading lessons with little White boys he befriended when doing errands in the town. In return for biscuits he pocketed at the Auld home, these "hungry little urchins" gave him "the more valuable bread of knowledge."[20]

Tommy grew up and Fred, no longer charged with child-minding duties, spent his days performing small tasks in the shipyard of Auld & Harrison. As Hugh Auld had predicted, Fred taught himself to write, laboriously copying the letters that ships' carpenters incised on beams. He worked secretly at night with a dog-eared Webster's speller. He filled in the blanks in forgotten copybooks from Tommy's early schooling.

Nothing the boy learned would mean more to him than the word *abolition.* When he first heard Auld and other slaveholders speak the term *abolitionist,* Freddy observed that they spat out the syllables with anger, disgust, and even fear. The term's meaning was unknown to him, but he heard other variations, such as *abolition movement* and *abolitionism.*

The child did not yet know the idea of abolition had a growing currency among the educated, especially in the North. While national politicians had argued for decades about a gradual end to slavery by various means, including colonization, in which the enslaved would be resettled in Africa, an outspoken Boston printer and newspaperman named William Lloyd Garrison had founded, in 1831, an anti-slavery newspaper called

The Liberator. In its pages he advocated for the immediate, unconditional, and total abolition of slavery. He believed the entire Constitution itself was a corrupt pro-slavery document, and some years later he would burn a copy of the founding document in public, proclaiming it "a Covenant with Death: an Agreement with Hell."[21] As a member of Congress, former president John Quincy Adams had begun talking about slavery as a matter of national conscience. Although not as radical in his beliefs as Garrison, Adams wondered aloud "whether a population spread over an immense territory, consisting of one great division all freemen, and another of master and slaves, could exist permanently together as members of one community or not."[22]

In the idea of abolition, Freddy saw hope for his future—a future that he decided would be a relentless quest for freedom.

TWO

A FIGHTING CHANCE

I knew Mr. L well . . . he was a great man long years since. . . . [I] Knew he was a Rising man.

—Elizabeth Edwards

After returning from New Orleans in July 1831, Abraham Lincoln settled in New Salem, Illinois. The river town was no metropolis; just two years before it had been an uninhabited ridge overlooking the Sangamon River. Now, with a gristmill and sawmill at the river's edge, New Salem had grown into a village much larger than any community Lincoln had ever called home, with a population of more than a hundred men, women, and children.

Its founders had wanted their frontier settlement to become the market town for the surrounding rural hamlets and farms. That had happened, and the village's streets were already lined with houses and shops for a blacksmith, cooper, hatmaker, and other tradesmen, along with two general stores, a tavern, and two doctor's offices. This seemed the perfect place for Lincoln, a young man with little experience in the wider world but great aspirations. He spent his days at his employer's store behind the counter, his nights sleeping in a cot in the attic above.

The townspeople immediately took to the young clerk, who had a gift for telling stories and entertaining an audience. They were impressed with the size and strength of this stranger, but that also meant that Lincoln would meet with an unexpected test after one of his admirers bragged that the tall and sinewy young man "could outrun, whip, or throw down any man in Sangamon country." Word of the boast reached Jack Armstrong, a local man who headed a pack of neighborhood rowdies, and he responded by challenging Lincoln to a fight, as he did whenever a new neighbor with a reputation for strength came to town.[1] As the town schoolteacher, Mentor Graham, explained, taking on the local bully "was an ordeal through which all Comers had to pass."[2]

Lincoln felt he had no choice but to accept the challenge, but only if he could dictate the terms. Young as he was, Lincoln was also savvy. He knew from experience that not all frontier fighting was the same. Word traveled around town; this would be a public event, and Lincoln, wanting to avoid a violent, anything-goes brawl, in which eye gouging and ear biting were accepted tactics, let it be known he wouldn't be "tussled & scuffled."[3] He insisted his rite of initiation must be less animal attack and more a test of skill and strength. A wrestling match was agreed upon.

Wrestling had rules, and a match typically started with an "Indian hug," in which the wrestlers wrapped their arms around each other's torso. What followed would involve practiced holds and other maneuvers, with the goal being to throw the opponent, not pin him.

On the appointed day a crowd of local men gathered to watch the spectacle. People said Jack Armstrong was the best fighter in Sangamon County, "a man in the prime of life, square built, muscular and strong as an ox."[4] But this match might be a fight worth watching, as Lincoln was an experienced wrestler. His height and strength suited him to the sport, and his intelligence was an added asset. As one of his New Salem neighbors put it, Lincoln was "a scientific wrestler."[5]

Lincoln's employer at the store put ten dollars on his young clerk. His bet was matched by a tavern keeper who favored Jack Armstrong. Men of lesser means bet knives and whiskey on the outcome. The fight would take place in front of the store where a relaxed Lincoln, according to one witness, stood with his back against the wall waiting for the fight to begin, "seaming undaunted and fearless."[6]

Once the fight was under way, neither man gained a clear early advantage. "They wrestled for a long time," recalled one New Salem man, "withough either being able to throw the other."[7] Armstrong was accustomed to winning, both by thrashing his opponents and taking their money in wagers, but he soon realized that this new man in town wasn't about to give up—and that Lincoln might even beat him.

Deciding that winning was more important than the agreed-upon rules, Armstrong broke his hold and dove for Lincoln's legs. But Lincoln countered with a rapid maneuver of his own. As twelve-year-old Robert Rutledge reported, "Mr. Lincoln seized him by the throat and thrust him at arms length from him."[8]

Suddenly, the match was at an impasse. Although everyone had seen Armstrong commit a foul, his supporters, caught up in the overheated moment, were ready to jump in to defend their champion. The match could have degenerated into a big brawl, with neighbors taking on neighbors. But Lincoln didn't let that happen.

"Jack, let's quit," he said to his opponent. "I can't throw you—you can't throw me."[9]

To Lincoln, the outcome of a wrestling match was much less important than gaining friends in a town where he wanted very much to fit in. He gave the other man an out and Armstrong agreed to call the match a draw. They left the bettors to argue about what exactly had happened,

An illustration from an early twentieth-century children's biography depicting Lincoln's wresting match with Jack Armstrong.

launching a debate that continues to this day since the many accounts of the fight—from friends of both men, bystanders, and others—agree on little. One thing is clear, however: A tough fight ended with the two men sharing a bond of mutual respect and friendship that would last their lifetimes.

As his authorized biographers put it many years later, Lincoln gained something important in the eyes of New Salem. "He became from that moment, in a certain sense, a personage, with a name and standing of his own."[10] The farmer boy from nowhere was now a known and respected man.

LIKABLE LINCOLN

Lincoln stood out in New Salem. In the words of Dr. Jason Duncan, a Vermont native who practiced medicine in New Salem, the young man had "intelligence far beyond the generality of youth of his age and opportunities."[11] Unlike many in the two-saloon town where whiskey drinking was a dawn-to-dark pastime, Lincoln never drank alcohol. Yet he wore his smarts and sobriety lightly, and his customers liked him.

Lincoln's boss soon rewarded his young clerk by giving him the added responsibility of managing the town's gristmill. The young man was moving up the ranks of the town, but he had something on his mind beyond promotions: more education. Whenever and wherever he could, he read. He read in bed. He read sitting astride a log by a stream. One friend spotted him reading in the woods, lying on the ground with his legs extended upward along a tree trunk. He read while walking, so absorbed in his text that he would sometimes stop, oblivious to anything but the words on the page, before continuing on, never having lifted his gaze. Neighbors lent him books from their libraries, and Lincoln began to memorize long passages from Shakespeare (*Macbeth* became his particular favorite). He talked poetry with the town blacksmith and committed to memory poems by Robert Burns, which he then recited in a proper Scots dialect.

If he was to rise in society, Lincoln decided, he needed to talk like he was somebody. He recognized his own unpolished way of speaking was different from the speech of his more educated customers, and he set his mind to learning proper English. He asked around and borrowed a grammar book, which he memorized. As he taught himself how to speak correctly—to use plural verbs with plural nouns and such niceties as when to use the word *shall*—he also made it his business to learn how to speak effectively since he would need to persuade people to believe not only in him but in his ideas. During his first winter in New Salem, he spent many, many hours studying and "practicing polemics" from a little volume called *The Columbian Orator.*

The Columbian Orator offered advice on how to use cadence, emphasis, pauses, demeanor, and even gesticulations to enhance oral expression. But the bulk of the book consisted of writings by George Washington, Napoleon, Ben Franklin, Cicero, Socrates, and dozens of other historic figures remembered for their speeches. Lincoln copied the ones he liked most into his copybook word for word. Some he could recite verbatim. He absorbed the structure, syntax, and sentiment of the great men's words. The man managing the gristmill was clearly hungry for something bigger.

Lincoln remained in New Salem for almost six years, during which he struggled to make a living. The store failed as New Salem's growth slowed and Lincoln's employer demonstrated his true colors—he was, observed schoolteacher Graham, "an unsteady . . . rattle brained man, wild & unprovidential."[12]

Lincoln's sudden unemployment provided him an opportunity. In the spring of 1832, Lincoln tried soldiering, signing up for the local militia when a conflict with a local Sauk chief named Black Hawk prompted the Illinois governor to gather a fighting force. To his great satisfaction, Lincoln was elected captain of his volunteer company, which consisted largely of men he knew from New Salem. After three months' service, he returned from the little conflict in northern Illinois, remembered as the Black Hawk War. His only bloody battles, he later joked, were with mosquitoes.

Lincoln would spend the next few months picking up enough odd jobs in and around New Salem to sustain himself, but the ambitious lad wasn't looking for just a job. He was looking for a calling. With his local popularity running high, he decided to try his hand at politics, and that autumn he took his first step; at the urging of his neighbors, he ran for the Illinois General Assembly. In his first public speech, he made plain the simplicity of his motives:

> I am young and unknown to many of you; I was born and have ever remained in the most humble walks of life. I have no

> wealthy or popular relations or friends to recommend me. My case is thrown exclusively upon the independent voters of the county, and if elected, they will have conferred a favour upon me for which I shall be unremitting in my labours to compensate. But if the good people in their wisdom shall see fit to keep me in the background, I have been too familiar with disappointments to be very much chagrined.[13]

Lincoln's early political priorities were equally simple. Adopting the thinking of Henry Clay, leader of the national Whig Party, he ran on a platform of improving roads and waterways—improvements that would mean a lot to the voters of rural Illinois. He promised to make education a priority, too. Every man, he insisted, should "receive at least, a moderate education."[14] No other boys would have to fight for an education the way he had if he could help it.

Due most likely to a lack of name recognition in the outlying districts, Lincoln lost the election. In his home precinct, though, he won 277 of the 300 votes. Those who did know him clearly saw that he was "a rising man."[15]

Seemingly unfazed by his political defeat, Lincoln bided his time—and got to know more people. He and a partner opened a store of their own. In 1833, Lincoln was appointed the village postmaster. He pored over newspapers that arrived from near and far, reading out his favorite bits to his customers. The townspeople saw him as a reliable source of news and opinions. His stockpile of off-color stories didn't hurt.

When that store failed, too, Lincoln was left with heavy debts, and his postmaster's salary wasn't even sufficient to pay his room and board. The county surveyor offered him work, and Lincoln taught himself metes and bounds from a book. His good reputation continued to grow. "When any dispute arose among the Settlers," one neighbor remembered, "Mr. Lincolns Compass and chain always settled the matter satisfactorily."[16] The work also got him into the surrounding countryside, enlarging his circle of acquaintances. But he aimed higher than surveying, setting his sights on becoming a lawyer.

Even as a teenager in Indiana, he had attended lawsuits conducted before the local justice, and in New Salem he added law books to his reading list, among them Blackstone's *Commentaries*, a basic work on British common law. Using forms that he found in one reference work, he drafted deeds, wills, and contracts for friends and neighbors, "never charging one cent for his trouble."[17]

In a village with no lawyers, it was inevitable that the talkative and competent young man would be asked to act the advocate. In his first trial, he was charged with defending the honor of a young woman in a case brought against a suitor who had refused to marry her. With no formal training, Lincoln drew upon common sense in his pleading. Dressed in the workaday clothing of a surveyor, he made a strong case that the girl deserved damages. His surveying assistant was in attendance and told the story.

"[Lincoln] made a comparison [and] cald the young man a white dress, the young lady a glass bottle," the man recalled. "He said you could soil the dress, it cold be made to look well again, but strik a blow at the bottle and it was gon."[18] In his debut at the plaintiff's table, Lincoln with his plainspoken argument won the day. His client was awarded a hundred dollars in damages.

In August 1836, he won something else. He ran again for a seat in the Illinois legislature, this time running what he called "a hand-shaking campaign" to make sure he won over voters outside New Salem.[19] The strategy worked, and his victory catapulted the twenty-seven-year-old to his first elective office.

Looking to leave behind his country-boy appearance, Lincoln bought himself a new suit—it was part of his personal reinvention—before boarding a four-horse stage bound for the state capital. Though larger than New Salem, Vandalia was a modest place, with a population of less than a thousand, its capitol building a dilapidated brick structure of just two stories. When he took his seat as a member of the legislature, the novice politician said little, listened intently, and attended faithfully. He saw these first sessions as part of his education. "He was reserved in manner," a friend reported, but "but learned much."[20]

In 1830s Illinois, few people lost any sleep over the evils of slavery. Lincoln's constituents were almost all White and unworried about the state's small number of enslaved people. Of greater concern were the eastern abolitionists whom they feared as radicals. Lincoln, like most of his neighbors, was of Southern birth. Unlike them, Lincoln was gravely uncomfortable with slavery, due, in part, to his upbringing. Lincoln's father, he recalled, had wanted to leave Kentucky "partly on account of slavery," having often heard the practice condemned from the pulpit of their church.[21] When a resolution to condemn abolitionist societies was raised before the General Assembly during his first term, Lincoln spoke up, going further than many would have, saying on the record that "the institution of slavery is founded on both injustice and bad policy."[22] On the other hand, he also accepted that Congress lacked the power to interfere with slavery in individual states, and many years would be required before he became a friend to abolitionism.

In those days, his law studies, not abolition, were Lincoln's top priority. He'd begun to suspect he'd need to have the power and social status a lawyer had if he was to be a successful legislator. A fellow lawmaker and friend from his militia days, John Todd Stuart, pushed Lincoln to pursue his law studies in earnest, and with more borrowed books—this time from his friend's firm, Stuart & Drummond, in nearby Springfield—Lincoln spent every available minute between legislative sessions studying the law. Although he "studied with nobody" (Lincoln's words), he passed the exams with ease, and on September 9, 1836, he was awarded a license to practice law.[23]

As one friend observed, "[Lincoln's] ambition was a little engine that knew no rest."[24] Now, as a lawyer and legislator—he would serve eight years in the Illinois House of Representatives—he had a way to harness it.

THE TURNING POINT

During the summer of 1834, while Abraham Lincoln was glad-handing neighbors and winning votes with his easy manner and lighthearted stories, Fred Bailey made a dangerous choice.

Bailey was back in the fields after spending eight years in Baltimore. As an enslaved person, he was considered little different than a slaveholder's horses, sheep, and other chattels; the bondsman had no say in what he did or where he went. Ordered back to Maryland's Eastern Shore, Bailey had no choice but to go. Making matters even worse, he was lent out for a year to a farmer who was intent upon erasing all memory of the looser discipline of Bailey's city days. At almost seventeen, he now labored for Edward Covey, a man reputed to be "a first rate hand at breaking young negroes."[25]

On Covey's farm, enslaved workers rose before first light and worked until dark and beyond, even on the coldest and hottest days. Bailey was awkward and unfamiliar with his new tasks in the fields. "The work was simple . . . yet to one entirely unused to such work, it came very hard."[26] He was subjected to brutal whippings for his failures, and in every way was "made to drink the bitterest dregs of slavery."[27]

He tilled fields that lay beside the waters of the Chesapeake Bay, his life a misery. But as he glimpsed the boats sailing by, he could not help but sense intimations of freedom in the schooners carrying their passengers to destinations forbidden to the enslaved. More than once he thought to himself, "One hundred miles straight north, and I am free!"

Bailey could not resign himself to a life in bondage—"It cannot be that I shall live and die a slave"—but Covey did everything he could to break him.[28]

One scorching August day, Fred Bailey collapsed while harvesting wheat. Felled by sunstroke, he crawled to a nearby fence seeking shelter from the sun. When Covey found him, he kicked Bailey savagely and ordered him to rise. Though the seventeen-year-old managed to stand, he collapsed again. The slave breaker then delivered a blow to his head with a hickory club, gashing open a large wound.

With blood oozing from his skull, Bailey managed to make his way into the woods. Eluding pursuit, he stumbled for miles until he reached his home farm hours later. He told his enslaver that Covey aimed to kill him, and at first, Auld seemed affected by the story. Soon, however, as the boy recalled the events in his autobiography years later, Captain Auld "became cold as iron." This was a transformation that Bailey was coming to understand: As he watched, the man's "humanity fell before the systematic tyranny of slavery." After a few minutes of pacing, apparently arguing with his own conscience, Auld decided to take Covey's side. "My dizziness was laziness," Douglass later reported the slaveholder told him, "and Covey did right to flog me."[29] Auld let Bailey stay overnight on his home farm, but ordered him to return to Covey before first light the next day.

Bailey managed to avoid facing the slave breaker until Monday morning. He was in the hayloft working before sunrise when Covey quietly snuck into the barn. As Bailey came down the ladder, Covey emerged from the shadows to grab him from below, attempting to bind his legs with a long rope, prepared to give him a beating. Instinctively, Fred jumped free. In the instant that he crashed to the stable floor, he understood that he was less afraid for his life than he was for the death of his freedom. This time, he resolved, he would not submit and take his whipping like an animal with no will of its own. He would defend himself.

He seized Covey by the throat and, holding him in an iron grip, rose to his feet. The two men grappled, but Fred carefully parried the blows sent his way, making no attempt to injure Covey. But the farmer, shocked that Bailey was actually fighting back, demanded, "Are you going to resist, you scoundrel?"

"*Yes, sir,*" was Bailey's polite reply, his gaze eye to eye with Covey's.

Covey called for help. When another bondsman came to Covey's aid, Fred incapacitated him with a wicked kick to the ribs. The fight continued, one man against the other.

Covey attempted to drag his opponent to the stable door, but Fred floored him. They fought on. For almost two hours the confrontation continued, and with the sun up and its rays lighting the barnyard, Covey finally realized this was a fight he could not win. Both men were exhausted, and Covey, who was "huffing and puffing at a great rate," brought the contest to a close.[30]

"Go to work," he ordered, adding, "I would not have whipped you half so much as I have had you not resisted."

In truth, Fred Bailey realized, Covey had drawn no blood. Nor would he summon the constable, despite the fact that Bailey had raised his hand to a White man, a crime that would ordinarily be punished by a beating at the whipping post in the town square—or worse. He would pay no penalty because the man's cherished reputation as a slave breaker would have suffered if word of his failure to discipline Bailey got out.

Not once during Bailey's four remaining months in his service would Covey again lay so much as a finger on him, but Bailey gained something much greater than a reprieve from whippings. By risking everything, he shifted the balance of power. His unlikely victory, he later said, was "the turning point in my '*life as a slave.*'" Until their fight, Covey had been on the verge of breaking his spirit. But resistance had "rekindled in my breast the smouldering embers of liberty; it brought up my Baltimore dreams, and revived a sense of my own manhood. I was a changed being after that fight. I was *nothing* before; I WAS A MAN NOW. It . . . inspired me with a renewed determination to be A FREEMAN."[31]

HOPE AND FAILURE

Like Abraham Lincoln, Frederick Bailey owned a well-thumbed copy of *The Columbian Orator.* He had purchased it at age thirteen, for a precious fifty cents, from a bookshop on Baltimore's Thames Street. He brought the "gem of a book" from the city to the country, and as the child in bondage had grown into a man, *The Columbian Orator* had been more than an introduction to public speaking.[32]

The book aimed to educate young White minds to the principles of freedom and the responsibilities of citizenship, but its editor had been strongly opposed to slavery. Bailey found himself drawn again and again to the text of an imaginary dialogue, in which a slaveholder berates an enslaved man, calling him an "ungrateful rascal" for attempting to escape. "You have been comfortably fed and lodged, not over-worked," insists the angry enslaver, "and attended with the most humane care when you were sick." Why would he want to run away?

The response is a deeper question, one that rejects the logic of slavery. "What have you done, what can you do for me that will compensate for the liberty that you have taken away?"

But, the slaveholder counters, the enslaved man had been "fairly purchased." Again, the response is a question: "Did I give my consent . . . when I was treacherously kidnapped in my own country . . . put in chains [and] brought hither . . . like a beast in the market[?]" A slave is not a beast, he insists. "Look at these limbs; are they not those of a man?"

In the end, the debate is resolved in favor of the bondsman when, recognizing that his argument for enslavement has been demolished, the slaveholder relents and acknowledges the other man's humanity, granting him freedom.[33]

The dialogue wasn't a transcript of a real moment; it never happened. Yet for a Black man with a growing passion for words and ideas, as well as deep desire for freedom, the dialogue had the power of revelation. "Slaveholders," Bailey thought, "are only a band of successful robbers, who left their homes and went into Africa for the purpose of stealing and reducing my people to slavery."[34] *The Columbian Orator* showed him how words and reason might sway minds, and the dialogue dovetailed with his own experience. Above all, it embodied a most deep-seated hope: "I could not help feeling that the day might come, when the well-directed answers made by the slave to the master, in this instance, would find their counterpart in myself."[35] Freedom seemed closer, perhaps even possible, as he committed the dialogue to memory.

When his year of service to Edward Covey had ended, Bailey was sent to work on a neighboring farm on the Eastern Shore. He found William Freeland wasn't the brutal taskmaster Covey was, and the possibility of freedom seemed constantly in his thoughts. Freeland's more permissive rules also meant that Bailey was able to quietly run a Sabbath school.

On Sundays, behind a barn or at the home of a freeman, he tutored as many as forty students in the rudiments of reading and writing. Bailey brought his Webster's spelling book and, as a teacher standing before his first flock, he began to find his voice, to shape his arguments, to gain confidence in the power of his words. Most of his pupils kept mum about their learning and their charismatic instructor.

Bailey also plotted an escape. He and five others planned to steal a canoe, paddle north on the Chesapeake, and follow the North Star to freedom. They chose a date—Easter eve 1836—and he laboriously forged traveling papers. The plan might have worked, but one of the coconspirators betrayed them. Just hours before their planned departure, Bailey and the others were rounded up and thrown in jail. The rest were released after a week, but, identified by his jailers as the mastermind, Bailey spent another week within the walls of the stone prison, worried that he might be handed over to slave traders and "sold south" to the much harder labor of a cotton or sugar plantation, "beyond the remotest hope of emancipation."[36]

To his surprise, however, he was sent back to Baltimore, once again to serve Hugh Auld. There Frederick Bailey would work in the shipyards, and within the relative freedom of the city, he would make a better plan.

ESCAPING TO FREEDOM

In Baltimore, Fred Bailey no longer did errands. He now stood six-one, his solid frame well muscled from his years of toil in the fields. Although he learned the caulker's trade in the shipyards, his wages were not his to keep, despite helping to construct some of the fleetest ships on the sea, the schooners called Baltimore clippers. The unfairness of having to hand over his pay to a slaveholder for doing the same job done by White workers on the same docks fed his simmering desire to be his own man.

He made fast friends among the city's free Black population. His appetite for words led him to become an active member of a debating club, the East Baltimore Mental Improvement Society. And there he met and fell in love with Anna Murray. Five years older than Fred, Anna was a free woman of color.

Like him, she had been born on the Eastern Shore; unlike his, her parents had gained their freedom from servitude. She had come to Baltimore at age seventeen to make a place for herself and had worked as a

Although her parents had been enslaved, Anna Murray was free when she met Frederick Bailey, and she played an essential role in helping him escape bondage.

maid for the postmaster. Despite being illiterate, she could read music and taught Fred to play the violin; soon they were playing duets. She lived in rooms of her own and would underwrite the adventure that was about to unfold, selling one of her two featherbeds and drawing upon her savings.[37]

His months under Hugh Auld's control had darker moments, too. Four White apprentices in one shipyard had attacked Bailey; when a blow with a handspike to the back of his head sent him to the ground, one of them delivered a vicious kick with his boot, nearly blinding Bailey in one eye. Dozens of White men watching the fight had urged his assailants on with cries of "kill him—kill him . . . knock his brains out—he struck a white person!"[38] As he recovered from his wounds, he was more resolved than ever to escape bondage, but the final spur would be a bitter quarrel with Auld, after the slaveholder abruptly withdrew hard-earned privileges.

Over a period of weeks, Fred and Anna meticulously planned his departure to a new life. When the day came—September 3, 1838—Douglass stood alone on the platform, since the plan called for Anna to follow later. Waiting for the Philadelphia, Wilmington and Baltimore train, he wore a red shirt, loosely tied black cravat, and wide-brimmed canvas hat of a sailor, and he carried free papers lent him by a man retired from the sea in case he was stopped by officials or slave catchers and ordered to produce proof he was free. Although Stanley was both darker and much older than he, Bailey planned to bluster his way to the North with Stanley's "sailor's protection," a document emblazoned with an American eagle.[39]

The timing was important, and to avoid suspicion, only when the train was ready to depart did a friend arrive with Bailey's baggage; then, with the train already in motion, Bailey jumped aboard. For his fellow passengers, this was just another day, but this particular Monday felt like it just might be the most important day of Bailey's life.

With the journey underway, his sailor suit and the eagle did their job: The train conductor let him pass. At Havre de Grace, Maryland, the fugitive took a ferry across the Susquehanna and boarded a second train, this one bound for Wilmington, Delaware. Again, his papers passed muster. A steamboat ride to Philadelphia would follow, and he hoped to catch a train from there to New York. But first he had to get out of slave country. Until then, any man could call his bluff, looking to collect the substantial reward for turning in a young and healthy runaway whose market value was perhaps a thousand dollars.

For hours, he had to keep his composure and suppress his rising panic. "The heart of no fox or deer, with hungry hounds on his trail, in full chase, could have beaten more anxiously or noisily than did mine," he later wrote, "from the time I left Baltimore till I reached Philadelphia."[40]

Three times on the journey he encountered men who could have sent him home in chains. When a free Black man he knew seemed to be asking him too many questions, Bailey managed to slip away. When he spied a ship's captain on whose ship he had worked the previous week, Bailey's fear peaked again, but the man failed to recognize him.

The closest call came when a German blacksmith he knew well looked directly and intently in his direction. But the White man proved to be an unlikely ally after Bailey thought he saw a spark of recognition in the man's eye. "I really believe he knew me," he remembered later, but this would prove to be the escapee's lucky day. Despite the fact that Bailey wore an obvious disguise, "[the blacksmith] had no heart to betray me. At any rate he saw me escaping and held his peace."[41]

Frederick Bailey made it to New York at 1:00 A.M. the next day. Good fortune had ridden with him, along with his tattered copy of *The Columbian Orator.* At first he wandered the streets, knowing no one, feeling untethered: "I was free from slavery," he later remembered, "but I was free from home as well."[42] After a night spent sleeping among the barrels on a wharf, he found shelter in the home of an abolitionist named David Ruggles, the editor of *Mirror of Liberty*, the nation's first Black-owned and -operated magazine. He managed to get word of his safe arrival to Anna, and she made haste to New York, where they were married, on September 15, in Ruggles's parlor.

Anna was not the only one to change her name. The couple decided to begin their new lives together in New Bedford, Massachusetts, where Fred hoped to put his shipyard skills to use. They made their way to the Massachusetts port, planning to settle into the town's well-established Black community. "To preserve a sense of my identity," he later wrote, he retained the name Frederick, but because in the eyes of the law he remained someone's property, he needed to take steps to keep slave catchers off his trail. He became thereafter to his fellow citizens in New Bedford and to posterity Frederick Douglass.

THREE

SELF-MADE MEN

He was the insurgent slave, taking hold of the right of speech, and charging on his tyrants the bondage of his race.

—Nathaniel Rogers

Frederick Douglass soon discovered that prejudice was alive and thriving in the North. White workers in the New Bedford, Massachusetts, shipyards refused to work side by side with Black laborers. Yet, for a man still reveling in his newfound freedom, the setback seemed almost unimportant. Douglass was starting a new life, with a new wife, and "no work [was] too hard—none too dirty."[1] He reveled in being a free man.

Douglass felt safe in a town of many Quakers, with their strong antislavery tradition, along with more than a thousand Black faces, three-quarters of whom were free men, the remainder fellow fugitives. He found work when, seeing a pile of coal in front of a minister's manse, he knocked at the back door and asked if he could move the gritty fuel to its storage in the cellar. After he made short work of the job, the lady of the house gave him two silver half-dollars. A new sensation rolled over him as he held the coins in his palm: He was now "not only a freeman but a free-working man." No slaveholder would be taking this pay. "*It was mine* [and] *my hands were my own*."[2]

Within a year, Anna and Frederick welcomed their first child, daughter Rosetta, then a son, Lewis, born in the autumn of 1840. Douglass tended to his growing family and performed whatever menial work he could get, from cutting wood and sweeping chimneys to loading ships and pumping the bellows at a foundry. This last job suited him best. With almost no spare time to improve his mind, he could nail a newspaper to a post near his bellows and read as he worked. Sometimes the publication he studied was William Lloyd Garrison's *Liberator*, to which Douglass now subscribed. He took in the abolitionist's words hungrily. As he absorbed Garrison's compelling arguments, he felt his "heart burning at every true utterance against the slave system."[3]

He looked for a house of worship and, after having been ordered to the balcony at a Methodist church because they were Black, the Douglasses joined the African Methodist Episcopal Zion Church, a small church on Second Street near the docks. There Frederick took an active role, first as sexton and Sunday school teacher, then from the pulpit. Word of the tall, handsome man's powerful sermonizing soon spread beyond Zion's Black congregation. The fugitive, once happily anonymous in the obscurity of a busy whaling port, would soon take a fateful trip to Nantucket Island, where he would begin to speak to a much larger public.

A NEW DESTINY

When the sloop *Telegraph* steamed out of New Bedford harbor on Tuesday, August 10, 1841, the passengers included Frederick and Anna Douglass. Forbidden to go below—Whites only, the ship's notorious captain decreed—the couple and three dozen other abolitionists, both White and Black, rode in solidarity, exposed to a mix of sun, wind, and rain on the upper deck. After a sixty-mile sail, they arrived on Nantucket Island for a three-day convention of abolitionists.

In the three years since his arrival in New England, Douglass had taken no holiday, often working in the foundry "all night as well as all day."[4] He needed a rest but also wished to attend the events at the Nantucket Atheneum, together with other like-minded opponents of slavery, one of whom had invited him to come along. In particular, he looked forward to hearing William Lloyd Garrison, editor of *The Liberator* and founder of the Massachusetts Anti-Slavery Society, the organization sponsoring the convention.

Garrison was largely self-educated, having grown up in poverty in Newburyport, Massachusetts, after his alcoholic father abandoned the family. He went to work as an apprentice printer at age thirteen and quickly mastered both writing and setting type; as publisher of *The Liberator*, he now composed his editorials as he typeset them, never putting pen to paper. Garrison's voracious reading over the years had led him to

A late-in-life photo of William Lloyd Garrison, founder and editor of The Liberator, *and Frederick Douglass's first mentor.*

take a moral stand in favor of racial equality. At first a supporter of colonization, he had rejected the gradual approach to freeing the enslaved, becoming an abolitionist true believer.

Despite a slightly owlish appearance, the bald and bespectacled thirty-five-year-old was fearless, accustomed to the fact that his ideas upset people. He faced down outbursts of violence at his public appearances and ignored regular threats on his life. He was a pacifist, stressed nonviolence, and believed that membership in the society should include women. He rejected alliances with any political party, firm in his belief that the U.S. Constitution was a pro-slavery document. To cautious people, including many who opposed slavery, he was a dangerous radical.

Douglass idolized this White man, who had the perfect certainty of one who knew he was right. Garrison had promised in the first issue of *The Liberator*, "I will be as harsh as Truth, and as uncompromising as Justice. . . . Tell a man whose house is on fire to give a moderate alarm; tell him to moderately rescue his wife from the hands of a ravisher; tell the mother to gradually extricate her babe from the fire into which it has fallen—but urge me not to use moderation. . . . I am in earnest—I will not equivocate—I will not excuse—I will not retreat a single inch—and I will be heard."[5]

More than a thousand people attended the Nantucket convention, many of them off-islanders. The Atheneum's hall was full for the Wednesday evening meeting when Garrison introduced a resolution condemning the complicity of people in the North who "act[ed] as the body-guard of slavery."[6] He invited discussion from the floor.

As Douglass later remembered, "I felt strongly moved to speak."[7] A local Quaker had already urged him to do so, but Douglass hesitated. He had stood in front of audiences in churches and meetings before, but he had never addressed a largely White audience. "The truth was, I felt myself a slave, and the idea of speaking to white people weighed me down." But he rose and, with all eyes on him, moved to the platform.[8]

At first, he stammered. He felt his limbs trembling and apologized for his ignorance, intimidated by the large crowd. Of the countless speeches he delivered over the years, this would be "the only one I ever made," he said later, "of which I do not remember a single connected sentence."[9] He spoke extemporaneously, with no notes before him. As he recounted his life story, he felt his tension ease.

The audience was captivated. Before them stood a runaway slave who, even at that moment, remained the legal property of a White man in Maryland. He was young—just twenty-three—and if his speech was grammatically imperfect, he was obviously highly intelligent. He exhibited a sharp wit as well as an unexpected eloquence. He drew his listeners in, a man speaking not only for himself but for millions of others. To him, slavery was no abstraction. Frederick Douglass, formerly Frederick Bailey, did not have to imagine life in bondage.

"Flinty hearts were pierced," said one listener, "and cold ones melted."[10] His audience listened in a surprised silence, and when Douglass finished, Garrison took back the podium. He was as stunned as everyone else at what they'd just witnessed. Still absorbing the experience they'd shared, Garrison realized, "I never hated slavery so intensely as at that moment."[11]

He asked the audience, "Have we been listening to a thing, a piece of property, or a man?" He spoke to a hall full of Garrisonians, already converts to the anti-slavery cause. Would they allow this "self-emancipated young man" to be returned to slavery?

"*NO!*" was the thunderous and unanimous response.[12]

Would they welcome him as brother man in their state? In response they shouted, *Yes, yes, yes,* they would.

Douglass's performance and his powerful story planted an idea in Garrison's mind. He offered Douglass a job, complete with a contract and a salary. As a paid lecturer, Garrison proposed, Douglass would travel the speaking circuit. The cause had other Black speakers, but

The strikingly handsome Frederick Douglass looks intently at the camera in this daguerreotype image, taken about 1841.

none had been enslaved and none had Douglass's rhetorical gifts. His task would be to open people's minds to the humanity of those in bondage, to galvanize audiences far and wide just as he had that evening in Nantucket. He would be a living embodiment of why abolition was a moral imperative, necessary to mankind.

On hearing the proposal, however, Frederick Douglass again hesitated. He wondered aloud, Garrison reported, whether "he was not adequate to the performance of so great a task."[13] Douglass also worried a new notoriety might lead to his discovery and that he would be risking recapture and a return to slavery. But he also felt an undeniable calling, understanding this work could help lead to freedom for others still bound by the knots of servitude.

By the time he and Anna Douglass boarded the *Telegraph* to return to New Bedford the next day, Frederick Douglass had truly embraced his calling. No longer would he depend upon his calloused hands to put bread on the table before his family. With the guidance of the fatherly

William Lloyd Garrison, Douglass's mind and words, his gifts of self-expression, and his harrowing personal story would take him to a new destiny.

LINCOLN, LEGISLATOR AND LAWYER

During Douglass's New Bedford days, Lincoln, too, relocated, when the state of Illinois, in 1839, removed its capital to Springfield, a rapidly growing town of some 2,500 people near the center of the state. The young legislator found a room over a general store, which he shared with the proprietor, Joshua Speed, who soon became a trusted friend and confidant, and joined the law practice of his friend John Stuart.

Lincoln's personality and persistence, his hard work and goodwill, and his association with the well-liked Stuart gained him acceptance among his Springfield neighbors. It didn't hurt that as a first-term legislator in the Illinois General Assembly, Lincoln had cast a vote in favor of making Springfield the new state capital. He soon was appointed a Springfield trustee, which cemented his local influence, but most of his income came not from public offices but his law practice. Since the court sat in Springfield for a total of just four weeks a year, for two months at a time Lincoln traveled the Eighth Circuit, which encompassed most of central Illinois. He rode an aging and tired horse called Old Tom from one county seat to another, seeking business. This likable man gained a reputation far and wide for his integrity and fairness.

His confidence growing, Lincoln began to think in national terms. He was outraged by outbreaks of violence in Louisiana, where a band of vigilantes lynched alleged gamblers, and in Mississippi, where a mixed-race man accused of murder was burned to death. Another clash closer to home, in Alton, Indiana, made national news when the editor of an abolitionist paper, Elijah P. Lovejoy, was shotgunned by a pro-slavery rabble, his printing press broken up and thrown into the Mississippi. To Lincoln, these mob executions were ominous, and he delivered a speech on the subject at Springfield's Lyceum.

Mob justice was just wrong, he argued, and would "subvert our national freedom." He preached *"a reverence for the constitution and laws."* He embraced neither slavery nor abolitionism; for him, the guiding principle was "reason, cold calculating, unimpassioned reason."[14]

Lincoln played an official role in the national election of 1840 after he was selected as a presidential elector for the Whig Party in Illinois. He was already well known for speeches he gave while running for his legislative seat and his gift for skewering his opposition to the great amusement of the crowd. He traveled across the state in 1840 on behalf of the party's nominee, aging general William Henry Harrison, victor at the Battle of Tippecanoe during the War of 1812, almost thirty years earlier. As Lincoln campaigned for Harrison in Illinois, his chief opponent was another rising Illinois politician, Stephen Douglas, who stumped for the incumbent, Democrat Martin Van Buren.

Lincoln and Douglas repeatedly faced off at campaign events, previewing their later rivalry when they would oppose one another in the U.S. Senate race in 1858. Douglas, a hard-driving young Democrat who stood a foot shorter than Lincoln, could fill a room with his booming voice, large head, bushy eyebrows, and ever-present cigar. Nonetheless, Lincoln's man, Harrison, would win the presidency—and Lincoln got his first write-up in national press when the *National Intelligencer*, the principal newspaper in Washington, D.C., quoted a speech he gave praising Harrison.

A life-changing new passion arrived in Lincoln's life when he met another Harrison supporter, twenty-year-old Mary Todd, at a Christmas party in December 1839. His law partner's cousin, Mary was pretty, pert, and so passionate about politics that a relation had called her a "violent little Whig" when she was just fourteen.[15] Mary had ambitions,

These images of Mr. and Mrs. Abraham Lincoln were taken prior to their move to Washington in 1847, after he won the right to represent Illinois's Seventh District in the House of Representatives; this is the earliest known photograph of the then-obscure frontier lawyer.

too, joking that she planned to marry a man who would become president.

Though both were born in Kentucky, their upbringings could hardly have been more different. The fourth of seven children, she grew up in a slaveholding family that occupied a fourteen-room mansion tended by Black "servants." Her hometown, Lexington, was one of the largest cities west of the Allegheny Mountains. When she met the tall and lank Lincoln—she was visiting a married older sister in Springfield—Mary, a decade younger than he, was a pleasantly plump five-foot-two with blue eyes and a reddish tinge to her brown hair. After more than a dozen years of study at a series of academies, she was well educated, well mannered, and entirely at home in civilized society; Lincoln, inexperienced and shy in the presence of women, was bowled over. Despite their differences, a shared love of poetry and politics helped bring them together.

Lincoln's store of memorized poems charmed Mary, while the fact that she actually knew Henry Clay, a friend of her father's and a man Lincoln idolized, added to her attraction.

Their courtship was on-again, off-again, and after asking for her hand, he broke off their engagement because of doubts that he could support and please her. But three years after meeting, Mary and Abraham married on November 4, 1842. Their marriage would have its ups and downs, with Mary adjusting to the demands of housekeeping in a home without servants. Terrified by thunderstorms and plagued by headaches, she could be willful, while Lincoln was by habit remote and was sometimes depressed ("generally," remarked Joshua Speed, who knew him as well as anyone, "he was a very sad man").[16] Nine months after their marriage, on August 1, 1843, their first son, Robert Todd Lincoln, arrived.

With his new commitment to marriage and family, Lincoln left the legislature, investing his full energies in building his law practice. As a general practitioner, he drafted wills and petitions, handling bankruptcies, divorces, estates, and criminal cases. He prepared diligently. Comfortable in front of audiences, he demonstrated great skills as a courtroom litigator. As his reputation grew, he appeared often before the Illinois Supreme Court, compiling an enviable record.

In 1844, he purchased a cottage in Springfield, the only house he would ever own, on the corner of Eighth and Jackson Streets. The same year he formed a new law firm with him as senior partner. He prospered and their marriage thrived, yet neither he nor Mary would be willing to settle for life in Springfield. His larger political aspirations simmered.

FOUR

ON THE ROAD

We looked up to [Frederick Douglass] almost as we do to the memory of Abraham Lincoln.

—Congressman George W. Murray

Frederick Douglass's reputation grew by the day. After joining Garrison's team of lecturers, he traveled to dozens of towns in Massachusetts, Rhode Island, and New Hampshire, emerging as more than a spokesman for the abolitionist cause. He was its chief exhibit, a living case history, a "graduate of the peculiar institution," who literally bore the scars of slavery, as Douglass said, "*with my diploma written on my back!*"[1]

He moved audiences with stories from his enslaved past, but his time traveling also meant new chapters were added to an already eventful life. During his first autumn on the road, in late September 1841, he was ordered to move to the "Negro car" at a train station in Lynn, Massachusetts. When his White traveling companions demanded to know why, the conductor growled at Douglass, "Because you are black."

The situation deteriorated after the conductor summoned a band of railroad enforcers and they assaulted Douglass. He still refused to leave, gripping his seat with all his strength. Moments later the toughs tore the seat free from its floor bolts and ejected it, along with its occupant, from the car. His luggage was next, landing on Douglass, who lay dazed on the platform.[2]

The incident prompted an outcry in the town, with public protests on behalf of Douglass, threats of a railroad boycott, and outraged accounts in the *Lynn Record.* Douglass responded to the community's support by moving to Lynn, which was convenient to Boston and home to many sympathetic Quakers. With the help of the American Anti-Slavery Society, he bought a home in December, and three months later, on March 3, 1842, Anna gave birth to their third child, Frederick Jr.

Yet the man of the house would be there only rarely. He won over audiences at Boston's famed Faneuil Hall and barnstormed across New York State, following the route of the Erie Canal. He spoke to full houses in New York City and Hartford, Connecticut. In 1843 he embarked on a six-month, hundred-meeting tour with lectures and conventions in Vermont, New York, Ohio, Pennsylvania, and Indiana.

Though he impressed his audiences, Douglass and his abolitionist entourage were often unwelcome. They were regularly refused admission to churches and halls and were heckled and pelted with rotten vegetables and racist epithets. But it wasn't until an event in Pendleton, Indiana, that the violence escalated dangerously out of control.

Abolition was a fightin' word among Indiana settlers, many of whom had moved there from nearby slaveholding states and held to their Southern sympathies. When Douglass and two colleagues arrived for a two-day meeting, they were warned that a mob planned to disrupt the events. Worried about their house of worship, the trustees at Pendleton's Baptist church, where Douglass and company were scheduled to speak, told the abolitionists they were no longer welcome. They attempted to convene outside on the church steps, but loud jeers and a rainstorm intervened and they adjourned.

The next day being pleasant—it was September 15, 1843—they tried again in a nearby wood, having arranged seats and a makeshift platform. During an opening song, one of the lecturers scheduled to speak, William White, scanned the audience. Among the crowd, which numbered more than a hundred people, the young Harvard graduate spotted perhaps a dozen troublemakers. Though a few more drifted in, they quietly departed when the speeches began, melting into the forest. It seemed like maybe the abolitionists were going to be able to conduct their program without harassment.

"In a few minutes we heard a shout," White later reported. Thirty or more men marched two by two into the clearing; coatless, their sleeves rolled up, they carried bricks, stones, and eggs. Their leader, a coonskin cap on his head, ordered the abolitionists to "be off."

None of the abolitionists budged, but members of the crowd rose to exit. As White pleaded with them to stay, the man in the coonskin cap issued an order.

"Surround them!" he yelled.

His mob surged forward, hurling rocks and rotten eggs. The attackers made short work of wrecking the platform, but true to the Garrisonian principle of nonviolence, Douglass stood safely aside. Then, looking around him, he realized he could not catch sight of William White. Fearing White might be in danger, Douglass instinctively grabbed a large stick and entered the fray in search of his friend.

The entrance of the Black man into the melee immediately drew the fury of the mob, and Douglass had to run for his life. A torrent of blows struck him as he fled, one of them smashing his hand. The riot's leader led the pursuit, and a blow to the head knocked Douglass to the ground, where he lay unconscious and defenseless. As the man prepared to strike

This illustration from his 1881 autobiography Life and Times of Frederick Douglass, Written by Himself, *portrays Douglass at the moment when, forced to defend himself in Pendleton, Indiana, his commitment to civil disobedience first wavered.*

again, William White appeared and raced into the melee to save his friend.[3] Despite bleeding from a head wound and having had two teeth knocked from his mouth, he threw himself at Douglass's attacker, preventing what might have been a death blow.[4] When the townspeople finally restored peace, the unconscious Douglass was loaded into a wagon and transported to the nearby farm of a Quaker.

Douglass recovered in time to deliver a lecture the following evening, but the beating in the woods changed him. When Garrison had drafted him into the ranks of the abolitionists, Douglass agreed to the importance of nonviolence; as he later wrote, he had been "a No Resistant till I got to fighting with a mob in Pendleton." When he leapt to defend his friend, he defied Garrison's prohibition on fighting back—doing so seemed necessary, even divine, Douglass felt, and on other days, in other places, he would have no choice but to fight again. He had no regrets, admitting "I cannot feel I did wrong." Yet in picking up a cudgel and entering the fray, he opened a tiny crack that would eventually become a yawning schism separating Douglass and Garrison, the two most essential men of abolitionism.[5]

WRITTEN BY HIMSELF

Douglass's right hand, sorely injured at Pendleton, never recovered its strength and dexterity. Nonetheless, he put that same hand, his writing hand, to work putting his autobiography on paper.

Anna birthed their fourth child, Charles, on October 21, 1844, and with a new baby to help care for, Douglass spent the winter in Lynn. He had something new to prove, too, since people in his audiences had begun to wonder aloud, "How [could] a man, only six years out of bondage, and who had never gone to school a day in life, . . . speak with such eloquence—with such precision of language and power of thought[?]"[6] Douglass answered with his story, this time his whole story.

Over the course of the next six months, seated at a table at the home he shared with his family on Union Street, he wrote out the story of his life as an enslaved person in Talbot County, Maryland, and on the streets of Baltimore. Just as he once found his public voice, knees shaking, speaking to a crowd in Nantucket, he trusted his own truth as he wrote. From the first page, he took the reader on a sometimes hellish journey into his world, one that no White person had known. The title he gave his life story, *Narrative of the Life of Frederick Douglass, an American Slave, Written by Himself*, was itself a rebuttal to his doubters. The book was like no other, its narrative more powerful than any fictional story.

The small volume, consisting of 125 pages plus a preface from William Lloyd Garrison, went on sale in May 1845 for fifty cents. Admiring reviews accumulated, including a gratifying rave in the *Lynn Pioneer,* which called the book "the most thrilling work which the American press ever issued—*and the most important.* If it does not open the eyes of this people, they must be petrified into eternal sleep."[7] By autumn, the book, released by Garrison's Massachusetts Anti-Slavery Society, sold an astonishing 4,500 copies. Editions in English, French, and German were in the works, and before the decade ended, Douglass's *Narrative* sold more than 30,000 copies. In comparison, neither *Walden* by Henry David Thoreau nor Walt Whitman's *Leaves of Grass,* both published in the next decade, would sell more than a few hundred copies in their first year.

Douglass used the written word skillfully, speaking directly to his reader. He was at times sarcastic, at others wistful, but always his words carried an undeniable authenticity. In telling his story he named names:

This song was published in 1845, the same year as Narrative of the Life of Frederick Douglass. *The sheet music cover stated that the abolitionist song was "respectfully dedicated, in token of confident esteem to Frederick Douglass" and identified him as "a graduate from the peculiar institution."*

His readers met Sophia and Hugh Auld and Edward Covey. Douglass also revealed the name Frederick Augustus Washington Bailey as his own. The fiction of Frederick Douglass fell away, thereby exposing him to the risk that slave catchers would appear at his door. According to a recent Supreme Court decision, a fugitive slave could be legally recaptured despite residence in a free state. Even so, Douglass chose to send a copy of *Narrative* to his owner in Maryland. The Aulds were outraged at their portrayal, and Hugh promised to regain his lost property and "place him in the cotton fields of the South."[8]

By the time those words were written, however, Frederick Douglass was well out of reach, having steamed across the Atlantic for a speaking tour of Ireland, Scotland, and England. The trip would stretch to twenty months, and for Douglass the journey abroad was a revelation. For the first time in his life, no one cared his skin was a different color. Not once while overseas, he reported, did he meet with "a single word, look, or gesture, which gave me the slightest reason to think my color was an offense to anybody."[9] On boarding his ship in Liverpool, however, for the trip home, he was reminded that no man of African ancestry should expect to be treated as the equal of the White man. Despite his prepaid ticket for a first-class cabin, he was refused his berth and exiled to segregated quarters in the steamship's stern.

THE NORTH STAR

With a little help from new friends, Douglass returned to America a free man. On his behalf a small band of English Quakers had contacted Hugh Auld, asking to buy his freedom. Auld agreed: In return for payment of £150 (in American currency, $710.96), he filed a bill of sale and a deed of manumission at the Baltimore Court House. The documents formally freed "Frederick Bailey, or Douglass, as he calls himself."[10]

When he arrived in Boston on April 20, 1847, Douglass leapt off the ship and caught the first train to Lynn. There, as he reported to one of his English benefactors, "I was met by my two bright-eyed boys, Lewis and Frederick, running and dancing with joy to meet me."[11] He would depart barely a week later on another lecture tour. But those would be his last days as Garrison's headline performer.

While in England, Douglass hatched a new life plan. Now a renowned world figure—one English newspaper called him that "illustrious transatlantic"—Douglass possessed a new confidence in his ability to run his own public life.[12] He hoped to launch an anti-slavery newspaper, believing that, if done well, a Black-run paper "would be a *telling* act against the American doctrine of natural inferiority, and the inveterate prejudice which so universally prevails in [the United States] against the colored race."[13] His admirers abroad had raised $2,174 to underwrite start-up costs, including the purchase of a printing press.

When William Lloyd Garrison got wind of the idea, he dismissed it. "The land is full of the wreck of such experiments," said Garrison. In fact, a handful of earlier Black-operated papers had failed, among them *Freedom's Journal*, *The Elevator*, and *The Colored American*.[14] Garrison insisted that Douglass could better serve the cause by carrying on as abolitionism's most persuasive spokesman, traveling and pleading the cause directly to audiences.

Taken aback by Garrison's objections, Douglass briefly put his plan on hold. But his desire to prove himself soon won out, and by October, Douglass was purchasing type. In November he moved to Rochester, New York, where he rented office space. He became his own man, confident his proposed publication would make him an even more effective advocate for Black Americans.

He chose Rochester deliberately. The western New York location gave him a healthy distance from Garrison and *The Liberator*, with which he would compete for readers and advertisers. Rochester was also a hotbed of abolitionism, for many years a way station on the Underground Railroad. Rochester had a substantial Black community, and Douglass's family would join him there a few months later. The upstate city became home for Anna and Frederick for the next quarter century.

Douglass christened his paper *The North Star*, echoing the advice—*Go north! Follow the North Star!*—that runaways were traditionally given when they ran for freedom. Issue number 1 went on sale on December 3, 1847, and within weeks Douglass's four-page weekly had a mailing list of seven hundred, a number that would eventually rise to an average of three thousand subscribers.

Still hungry to learn, he found his new paper "the best school possible." As its founding editor and writer, he taught himself to be a journalist and master the rules of grammar. "It obliged me to think and read, it taught me to express my thoughts clearly, and was perhaps better than any other course I could have adopted."[15]

The North Star gave Douglass a broad public platform. In the first issue he attacked Henry Clay, Abraham Lincoln's idol, a colonization man who, while favoring the return of bondsmen to their "ancestral" home in Africa, remained a slaveholder himself. "Do you think that God will hold you guiltless," demanded Douglass in *The North Star*, "if you die with the blood of these fifty slaves clinging to your garments?"[16] His argument echoed the "Moral Persuasion" central to Garrison's thinking, yet by choosing Clay as a target, Douglass also took a step away from his former benefactor. In the first issue of *The North Star*, Douglass opened a political conversation.

Douglass also worked to educate himself concerning the issues of the day. Among them was the Mexican War, which was drawing to a close. Launched by slaveholder president James Polk, the conflict was, in the eye of abolitionists, a transparent attempt by Southern politicians to extend their influence by adding slave states to the Union. To others, it was simply a way to solidify the recent Texas annexation, though along the way it did upset the balance of power between free and slaveholding states. Douglass saw signs that slavery and politics were inseparable everywhere he looked.

In the summer of 1848, he traveled to nearby Seneca Falls to speak at America's first women's rights convention. He was true to his paper's motto, printed on the masthead, which proclaimed, "Right Is of No Sex—Truth Is of No Color." His was the only Black face in attendance, but Elizabeth Cady Stanton and her friends were his allies, all of them anti-slavery advocates. Theirs was a shared pursuit of equality.

MR. LINCOLN GOES TO WASHINGTON

As Douglass gained fame, Lincoln the family man and lawyer decided to resume his public life, this time as a U.S. congressman. After her husband won Illinois's Seventh District seat, Mary refused to be left behind in provincial Springfield. On moving into Mrs. Sprigg's boardinghouse on First Street, in December 1847, they were a family of four, including son Robert, now four years old, and toddler Eddie, a year and a half.

Washington, D.C., was a revelation to the Lincolns, a burgeoning city with a population of more than thirty thousand White and ten thousand Black citizens, including roughly two thousand men and women in bondage. Lincoln witnessed the city's racial inequalities in small ways and large. One day at Mrs. Sprigg's, the quiet of the house was interrupted by the arrival of three officers to seize one of her waiters. Guns drawn, they put irons on his wrists and dragged him away while his wife and the Lincolns watched. The man had been using his wages to buy his freedom and had done nothing wrong. But just sixty dollars short of the purchase price, his owner had changed his mind. The police had been sent to return the man to slavery.[17]

Lincoln was gaining a new awareness of the issue that divided the country, with the practice entirely legal in more than a dozen states in a wide swath that encompassed the entire South. Slaveholders, dependent upon free labor, regarded slavery as a fundamental right; a growing number of people in the North, galvanized by abolitionists, thought human bondage a fundamental wrong. But Lincoln, as a first-year congressman, had no illusions about changing the country's course.

He served diligently on committees and missed almost no roll call votes. With the White House and Senate in Democratic hands, his Whig Party wielded little power, but he eyed the campaign of 1848. Looking to gain an advantage over the opposition, Lincoln, in his first major speech, accused President James Polk of provoking the Mexican War. Calling the war a land grab and a transparent attempt to expand slavery, the freshman congressman dismissed Polk as a "bewildered, confounded, and perplexed man."[18] The speech got Lincoln nowhere since it angered constituents back home for his opposition to a war in which many of them fought. As for Polk, he simply ignored the accusations and never so much as uttered Lincoln's name.

In his second session the following winter, Lincoln tried a different approach. Rather than getting into a dispute about the recent past, he looked to pass legislation that would lead to the abolition of slavery in Washington, D.C.

His attitude toward slavery remained a bundle of contradictions. To the man on the street in Illinois, as in the North in general, slavery had little immediate reality. With just 1 percent of the Black population living in the free states, relatively few White citizens interacted with people of color. Most of Lincoln's own experience with the enslaved had been limited to encounters at the Kentucky homes of friends or with the Todd family, where living conditions bore almost no resemblance to the hardships of bondsmen who worked in the cotton fields or on sugarcane plantations.

By nature, Lincoln was against slavery, as his father had been. But as a lawyer and lawmaker, he also believed the Constitution protected the rights of each state to decide its own policy. Although the "peculiar institution," in the words of powerful pro-slavery senator John Calhoun, wasn't a daily presence in his life, Lincoln had gotten disquieting glimpses of slavery's cruelty. Returning from Kentucky a few years earlier, he had shared a long, meandering steamboat ride with a coffle of "ten or a dozen slaves, shackled together with irons." The sight of the poor wretches had stayed with him, the memory of it a "continual torment."[19] In his Washington days, the business of slavery was an inescapable, day-to-day fact of life.

Foreign visitors to the American capital had often remarked upon the hypocrisy of a country that boasted of its individual freedoms when numerous so-called "slave pens" could be seen on the city's streets. Lincoln overlooked one of them from the Capitol. "[It was] a sort of negro-livery stable," he wrote, "where droves of negroes were collected [and] temporarily kept . . . precisely like droves of horses."[20] These human corrals were way stations for the enslaved, many of whom were headed to the deep South from Virginia's played-out tobacco plantations.

In particular, Lincoln found his growing anger directed at slave traders; he called them "a small, odious, and detested class."[21] He also recognized that an ongoing congressional debate concerning the slave trade in the District of Columbia was dividing not only the nation but his party. In response, he drafted a bill he hoped might stand a chance of passage, incorporating elements calculated to appeal to various constituencies. One required "full cash value" be paid to slaveholders as compensation for those they freed. The law would declare all children born of enslaved women after 1850 free at birth. Another clause would require city authorities to arrest and return fugitives who sought safe harbor in Washington. Finally, a referendum would be mandated in the district to approve (or reject) the legislation, since Lincoln believed abolition was a matter for the people, not Congress, to decide.

When he went public with the laboriously negotiated bill, a barrage of angry criticism led former backers to melt away. Lincoln had tried to craft a compromise with enough sweeteners that his congressional brethren on both sides of the slavery argument could hold their noses, ignore what they did not like, and advance the issue, but the strategy failed. Lincoln would never formally introduce the bill and, instead, reaped only insults, even from abolitionists appalled that Lincoln's proposal allowed for apprehending fugitives within the district for return to their slaveholders. John Calhoun also attacked Lincoln—though not by name, calling him only a "member from Illinois"—and denounced the bill, which Calhoun claimed was a threat to life in the South as he knew it and would lead to "the prostration of the white race."[22]

Looking to please everyone, Lincoln satisfied no one. When Congress adjourned in the early morning hours of Sunday, March 4, 1849, his congressional term ended. After wrapping up his affairs, he headed back to Springfield, where, Lincoln noted later, "he went to the practice of the law with greater earnestness than ever before."[23] Like everyone else, he had found the problem of slavery intractable. Dreams of a career in national politics looked to be fading away.

A NEW MENTOR

As the decade drew to a close, Abraham Lincoln walked away from politics. Meanwhile, Frederick Douglass struggled to keep his paper afloat.

His seed money quickly spent, Douglass mortgaged his new brick home in Rochester to pay his employees. He traveled often to lecture and seek new subscribers, and he looked to his allies for contributions. Among the most important would be a new American friend, an upstate New Yorker named Gerrit Smith, who would contribute more than money. As Garrison's control over his protégé faded, Smith became the single most important influence on Douglass's thinking about how to confront slavery and the slavers.

Smith had introduced himself in a nearly illegible letter that landed on Douglass's desk a few days after the first issue of *The North Star* went on sale. The stranger sent warm greetings and a five-dollar check, good for a two-year subscription. "I welcome you to the State of New York," Douglass read as he deciphered Smith's scrawl. "May you and yours, and your labors of love for your oppressed race, be all greatly blessed of God."[24]

To Douglass's surprise, the envelope also contained a deed for forty acres of land in a place called Timbuctoo. Douglass published the letter in the next issue of *The North Star*, thereby inaugurating a friendship that would be both public and private.

Gerrit Smith was a very rich man, the owner of vast tracts of undeveloped land in upstate New York. His father, Peter, a business partner of America's first multimillionaire, John Jacob Astor, had accumulated more than a million acres of real estate. But Gerrit, despite being raised in a mansion near Syracuse, had seen his share of unhappy moments. On the day after his graduation from Hamilton College in 1818, his mother died; one year later, his wife of seven months passed away unexpectedly. Then the young man abandoned what he called the "laudable pursuit" of literature (he read Greek and Latin and greatly admired the poetry of Lord Byron) when his still-grieving father handed him full responsibility for the massive family estate. It was a task Gerrit had neither asked for nor wanted.[25]

Some years later, Gerrit Smith and his second wife grew interested in issues of social reform. As a budding philanthropist, his favorite charities included the American Temperance Society, which regarded alcohol consumption as a sin that corrupted body, mind, and soul, and the American Colonization Society, an organization devoted to resettling

Gerrit Smith, the very rich New Yorker who provided land to Black homesteaders and influenced Frederick Douglass's thinking on the merits of the Constitution, pictured here in the 1840s.

American Blacks in Africa. Over time, Smith's thinking shifted from colonization, never a practical solution in a country with millions of enslaved people, to abolitionism. He proudly boasted he had become an "antislavery fanatic."

Smith's home in Peterboro, New York, became an important stop on the Underground Railroad, and he personally purchased the freedom of dozens of the enslaved. Peterboro was a rarity for its time, a fully integrated town, and on a typical evening in his own dining room Smith shared his table with fashionable visitors of his own social class, "three or four Indians from the neighborhood," and "a sprinkling of negroes from the sunny South, on their way to Canada."[26] But the Timbuctoo community he founded more than a hundred miles away in the Adirondack Mountains was his most remarkable social experiment.

In North Elba, New York, he set out to create a self-sustaining community. In 1846, he designated 120,000 acres for the purpose, giving forty-acre parcels to each of a hoped-for total of three thousand Black New Yorkers, both freemen and fugitives. Smith believed strongly in political action, and his "Smith Grants" would enable the inhabitants of Timbuctoo to gain access to the ballot box by fulfilling New York State's strict property requirement. Black men had to have assets of $250 to vote, though there was no such requirement for White males.

Smith's first letter to Douglass opened the dialogue between the two men, and Douglass soon visited Smith's Peterboro estate. He found that Smith, unlike Garrison, valued discussion and argument. While Garrison brooked no dissent among his followers, Douglass and his new friend often disagreed, both in their in-person conversations and in the pages of *The North Star*, which in the coming years regularly published Smith's responses to Douglass's editorials.

CONSTITUTIONALLY SPEAKING

Although the two shared a passionate hatred of slavery, they started from fundamentally different positions in thinking about the United States Constitution. Douglass came to the discussion tutored by Garrison to hate the Constitution. The founding document was so flawed, Garrison maintained, that it could only be rejected and reinvented. Though it did not explicitly condone slavery, article 1, section 2 of the Constitution distinguished "free Persons" from "other Persons." Without being called by name, these "other persons"—logically, they could only be the enslaved—were assigned a value for purposes of congressional apportionment of three-fifths that of "free persons." According to Garrison's reading, the Constitution thus enshrined slavery as legal and legitimate, rendering the entire document, in his opinion, corrupt, pro-slavery, and worthless.

Gerrit Smith took a different view. He saw no need to nullify the Constitution. As Lincoln had done, Smith studied Blackstone's *Commentaries*, the foundational work of English law. From its dense pages he extracted the notion of "natural law," which he concluded was incompatible with slavery. Thus, in Smith's interpretation, the Constitution was an aspirational document, a working instrument that could be improved, its flaws fixed. Most significant of all, Smith believed that the Constitution empowered Congress to legislate the abolition of slavery in the Southern states. This had led to a parting of the ways with Garrison, and Smith had founded his own political party, the Liberty Party, in 1840. He had even run for president himself under the Liberty Party banner, but he caused little more than a ripple in the election of 1848, when he got just 2,500 votes out of roughly three million cast. But Douglass found himself drawn to Smith's thinking.

By 1850, Douglass was veering sharply away from Garrison's and toward Smith's interpretation of American law. At first the change was slow, but the passage of the Fugitive Slave Act sped the process.

The Fugitive Slave Act was one of the five laws that made up the Compromise of 1850. Senators Henry Clay and Stephen Douglas had masterminded the Compromise, hoping to unite the representatives of free and slave states and put the slavery debate behind them. One of the laws abolished the slave trade (though not slavery itself) in Washington. Another defined Texas's disputed western and northern boundaries. Other acts established territorial governments in Utah and New Mexico, freshly acquired in the Mexican War along with California, and permitted California to join the Union as a free state. Both sides won something, it seemed. But the fifth law proved incendiary.

The draconian Fugitive Slave Act required federal and local law enforcement officials to arrest and return runaways in all states and territories, including the free states. It specified fines and imprisonment for citizens caught aiding escapees. The Fugitive Slave Act removed runaway cases from Northern courts, stripping away due process. In short, it required Northerners who opposed slavery to betray their principles, in effect to collaborate with the enforcement of slavery.

Some in the North called it the "Kidnapping Act," and for Douglass, this was personal. If the law had been in effect when he escaped a dozen years before, anyone who helped him could have been punished; had he been caught, even in Northern territory, he would have been shackled and returned to bondage. But the passage of the Fugitive Slave Act posed another question, too: If the federal government could enforce slavery, could it not also choose to abolish it?

In May 1851, Douglass, after carefully thinking through his argument, took a public stand against the act at an American Anti-Slavery Society convention meeting in Syracuse. His argument boiled down to this: Consistent with "the noble purposes in its preamble," Douglass believed the Constitution could "be wielded in behalf of emancipation."[27] He wanted not to reject the founding document but to put it to work. In the preamble, its purposes are clearly stated, key among them to "secure the Blessings of Liberty." That was the document's true promise, and in Douglass's reading, it could power abolition. It was an almost complete reversal in his thinking.

When Garrison got wind of Douglass's speech, he was apoplectic. He labeled his protégé an apostate and declared him an avowed enemy; in the years to come, the Garrisonians would do everything they could to undermine Douglass's work. But new ally Gerrit Smith welcomed Douglass to the ranks of political abolitionists and looked for new ways they might work together for the cause. Douglass's change of heart led to the merger of the *Liberty Party Paper*, which Smith backed, and *The North Star*. The first issue of the combined publication, renamed *Frederick Douglass' Paper*, appeared on June 30, 1852. Its motto, suggested by Gerrit Smith, was "All Rights for All."

Douglass, a decade out of slavery, now free of both Hugh Auld and Garrison, had forged his own public identity. During the next decade, he would continue to opine from the columns of his paper about slavery and American politics. He had truly become his own man.

FIVE

WHERE THERE IS SMOKE

One flash from the heart-supplied intellect of Harriet Beecher Stowe could light a million camp fires in front of the embattled host of slavery, which not all the waters of the Mississippi, mingled as they are with blood, could extinguish.

—Frederick Douglass, *My Bondage, My Freedom*, 1855

Douglass and Lincoln were far from the only people drawn to the thorny problem of slavery. After decades of denial and willful ignorance, new voices in the 1850s brought human bondage to people's attention. Among them was a minister's wife in Maine—she would make millions of people cry—and a senator who would find himself lying in a pool of his own blood on the floor of Congress.

In July 1851, Frederick Douglass received a letter from an early subscriber to *The North Star.* Mrs. H. E. B. Stowe wanted Douglass's help in a writing project of her own, which she described as "a series of sketches" about slavery. The first ones had already appeared in a weekly paper, *The National Era*, but she needed to get direct testimony about plantation life. "I wish to be able to make a picture," she told Douglass, "that shall be graphic & true to nature in its details."[1]

A petite woman who stood less than five feet tall, Harriet Beecher Stowe would write a very big book. She reworked her serial sketches into the novel *Uncle Tom's Cabin; or, Life Among the Lowly.* Published in 1852, it found an immense audience, selling more rapidly than any book in the nation's history, three hundred thousand copies in 1852 alone. The next decade would see some two million copies printed in the United States and several times that worldwide.

In reaction to the Fugitive Slave Act, Frederick Douglass had ratcheted up his rhetoric. "I am a peace man," he proclaimed before issuing a stern warning to those who favored slavery and, in particular, slave catchers. He could imagine fugitives and their protectors fighting back.

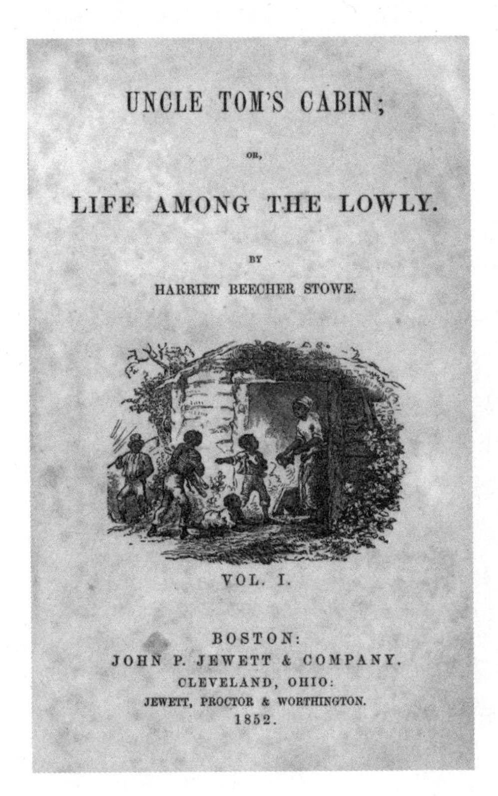

UNCLE TOM'S CABIN;

OR,

LIFE AMONG THE LOWLY.

BY

HARRIET BEECHER STOWE.

VOL. I.

BOSTON:
JOHN P. JEWETT & COMPANY.
CLEVELAND, OHIO:
JEWETT, PROCTOR & WORTHINGTON.
1852.

The title sheet from an early edition of Mrs. Stowe's moving novel.

"I believe that two or three dead slaveholders," said Douglass, "will make this law a dead letter."[2] But Mrs. Stowe's emotionalism in response to the same legislation impacted the slavery debate beyond Douglass's angry words.

Mrs. Stowe wanted her readers see slavery with "new eyes."[3] In a sentimental manner, she told the story of her title character. Though saintly and kind to everyone, Tom was beaten to death by slaveholder Simon Legree and his overseers. A fugitive named Eliza met a better fate when she escaped slavery, her son Harry clutched to her bosom, by leaping onto an ice floe that carried her across the Ohio River. Readers of the melodrama also met the mischievous Topsy, a badly abused enslaved woman, and Cassy, who, having seen two of her children sold, killed her third child.

Stowe could tell a story, and the humanity of her Black characters

Harriet Beecher Stowe in a photograph taken shortly after the publication of Uncle Tom's Cabin.

shone through on every page; their hardships and the perils they experienced left readers in tears. To millions of White people who knew few or no Blacks, the book offered a larger revelation: The enslaved were people, too. They had human voices, feelings, and hopes.

Douglass and Stowe would not meet until early 1853, when she invited him to visit her. The two did not see eye to eye on all matters, since several fugitives at the close of *Uncle Tom's Cabin* were headed for Liberia, the African nation established for formerly enslaved people. Colonization, he said to the "good lady," was no answer to ending slavery. "The truth is, dear madam," Douglass explained, "we are *here,* & here we are likely to remain."[4]

Yet they would be allies and occasional correspondents for many years to come. She tried to mend the rift between Douglass and William Lloyd Garrison, telling Garrison that "[Douglass] holds no opinion which he cannot defend . . . [and] his plans for the elevation of his own race, are manly, sensible, comprehensive."[5] In his speeches and his writings, Douglass often made reference to *Uncle Tom's Cabin* and its raw emotional truths, calling it "a work of marvelous depth and power. . . . No book on the subject of slavery had so generally and favorably touched the American heart."[6]

The success of her book dwarfed the sales of Douglass's *Narrative,* but he was more than happy that *Uncle Tom* made slavery a topic of dinner-table talk across the country. Mrs. Stowe's portrayal of slavery's cruelties outraged many in the South, but more importantly, it appealed to people of conscience of the North, whether they were farmers in Maine, factory workers in eastern cities, or settlers making their way in the West. By making everyone look more closely at slavery and by demanding slaveowners look in the mirror, *Uncle Tom's Cabin* raised the country's emotional temperature a good deal closer to the national boiling point.

LINCOLN RISING

After completing his term in Congress, Lincoln once again pushed politics to the back of his mind.[7] For the next five years, his personal life was eventful, with the death of his three-year-old son, Eddie, taken by tuberculosis, in February 1850, and the birth of William Wallace that December. He and Mary welcomed a last child, Thomas—nicknamed "Tad," short for "Tadpole," because of his very large head—in 1853, so Lincoln spent much of the first half of the 1850s helping to raise his three boys, including the eldest, Robert, who turned ten in 1853.

His Illinois law practice prospered, partly because of his reputation for diligence and truthfulness. The man now widely known in Illinois as "Honest Old Abe" summed up his legal philosophy in a single line. "[I]f in your own judgment, you can not be an honest lawyer," he wrote, "resolve to be honest without being a lawyer."[8]

Then, in 1854, Lincoln heard the call to public life once more.

Signed into law on May 30, 1854, the Kansas–Nebraska Act sent a tremor across the country. The argument over slavery had become a heated debate over which of the western territories could become slave states and which free states. Sponsored by Lincoln's old rival Stephen Douglas, the Kansas–Nebraska legislation attempted to break a legislative logjam, but it did so by repealing a three-decades-old agreement, spelled out in the Missouri Compromise, that established latitude 36°30' as the northern boundary for slavery in territories and new states. Douglas's legislation erased that hard line and replaced it with the principle of "popular sovereignty." That meant residents would vote on permitting or prohibiting slavery in the upper half of the Louisiana Purchase, now the Nebraska and Kansas territories.

By blowing up the central principle of the Missouri Compromise, the Kansas–Nebraska Act ended a fragile truce that had prevailed for years.

Southerners rejoiced because new slave states could be established where once they could not, potentially adding more congressional representation to what was called the "slave power." In the North, opponents of slavery feared the balance of power would now shift permanently into Southern hands.

For Lincoln, it was as if he had been awakened from a slumber by a bucket of water. When the act became law he was "thunderstruck and stunned," and he felt compelled by the sudden change in his country's ground rules to leap into the unfolding national controversy.[9] But before he did, Lincoln, as a man who always lived by the book, went to the library.

For Frederick Douglass, anti-slavery was a cause he carried like a cross every hour of his life. For Abraham Lincoln, a White man who lived in a free state, slavery wasn't an everyday matter. Instinctively he hated the idea of it. He thought slavery was morally wrong and, some said, he had quietly contributed money to help fugitives on the Underground Railroad.[10] As a young politician, he admitted, "the slavery question often bothered me," but as he readied to weigh into the debate in 1854, he felt the need to master its history, economics, and politics.[11] That summer he set about improving his education at the State Library in Springfield.

For weeks he pored over historical and legal texts, reading what the Founders said about slavery. He analyzed the census records and studied earlier congressional debates. Wanting to know everything about the history of slavery in the United States, he assembled a mass of facts and accumulated "scraps of arguments against slavery."[12] At summer's end he emerged with a sheaf of papers and, in late August, gave the first of a dozen speeches around Illinois, speaking and debating as the election of 1854 approached. In Peoria, on October 16, he presented his final argument, one buttressed by a deep and careful reading of the American past.

LINCOLN'S REBUTTAL

Stephen Douglas had spoken earlier, and after a break for dinner, the crowd reassembled to listen to Lincoln in front of the tall facade of Peoria's stone and brick courthouse. At seven o'clock he appeared, one leg first, maneuvering his angular frame through a window to stand atop the roof over the entrance portico. Lincoln looked down at the crowd that filled the town square, their faces lit in the autumn dark by lanterns and by candles in the windows of nearby houses.

He told his listeners he had long believed, along with many of the Founders, that slavery would die a natural death. He pointed out that the Framers omitted all mention of "slave" and "slavery" in the Constitution. "[T]he thing is hid away, in the constitution, just as an afflicted man hides away a wen or a cancer, which he dares not cut out at once, lest he bleed to death." He pointed to subsequent limitations put on the slave trade and noted that the Northwest Ordinance of 1787 forbade slavery in Ohio, Indiana, Illinois, Michigan, and Wisconsin. Lastly, he pointed to the Missouri Compromise, which had been "canonized," he said, "as a sacred thing."

Only now Stephen Douglas's Kansas–Nebraska Act had gutted it, making the spread of slavery seem inevitable.

His speech lasted for three hours, but "perfect silence prevailed" as he laid out his argument.[13] He argued for the restoration of the 36°30' boundary, though he knew that was unlikely. Nor did he have another answer. "If all earthly power were given me, I should not know what to do [though] . . . my first impulse would be to free all the slaves, and send them to Liberia." Despite its appeal—and he would advocate for it in the years to come—colonization was not a practical solution. But neither could Lincoln imagine freeing all Blacks to "make them politically and socially our equals." He did not believe they were.

What he did do was lay down a hard line of his own: a stubborn opposition to the extension of "the monstrous injustice of slavery itself." He hated slavery because it made Americans hypocrites and diminished his country's influence as a representative to the world of republican values. He felt in his bones that the further spread of slavery must be prevented.

The speaker that cool autumn evening in Peoria wasn't Lincoln the amusing storyteller who won over juries and audiences by poking fun at himself and others. Instead he stood as a passionate man who had found a topic that tested his intellect as he spoke from the heart. "Mr. Lincoln's eloquence," one young Chicago journalist observed, "produced conviction in others because of the conviction of the speaker himself. His listeners felt that he believed every word he said."[14]

After delivering his speech, he handed the manuscript to the editors of the *Illinois State Journal.* Its great length, some seventeen thousand words, necessitated serial publication in the single-sheet paper, and in late October the full text appeared piece by piece in seven consecutive issues of the *Journal.* Lincoln was announcing himself to his nation, and for the next few years, his Peoria speech would be his most essential position paper as he found his place in the politics of his state and country. His opposition to slavery—most specifically, to preventing its extension—was a cause equal to his ambitions.

The immediate consequences of the Kansas–Nebraska Act were on his mind, too. As he asked the crowd at Peoria, "Bowie-knives and six-shooters are seen plainly enough [in the territories] but never a glimpse of the ballot-box. . . . Is it not probable that the contest will come to blows, and bloodshed?" Events would prove him prescient: So many violent clashes between slavery and anti-slavery men occurred in the territory in the coming months that *New-York Tribune* editor Horace Greeley nicknamed the place "Bleeding Kansas." But the violence and intimidation would not be limited to Kansas or Nebraska. One bloodletting would soon occur in Washington, D.C., within the hallowed halls of the United States Congress.

A FIGHT ON THE FLOOR

The Senate chamber was unusually quiet. After an early adjournment, the hush of the room was broken only by the voices of a few stragglers talking quietly. For these about-to-be witnesses, May 22, 1856, would be anything but a typical day in the Capitol.

Senator Charles Sumner of Massachusetts sat alone. At work on the correspondence spread across his desk, he was unaware of the arrival of the other player in this two-man drama, Congressman Preston Brooks. Elected from an upland district in South Carolina, Brooks spent most of his time in the House of Representatives at the opposite end of the building, but on this particular Thursday he had a pressing matter to take up with the senior senator from the Bay State.

Earlier that week, Sumner had delivered a long speech. He was a lawyer and a scholar, and at a time when few in Congress would call themselves abolitionists, he was proud to claim that label. He had spent weeks drafting his speech, which he titled "The Crimes Against Kansas." On the floor of the Senate, he made a full-throated argument that Kansas was falling into "the hateful embrace of slavery."[15]

His fears were well-founded. Just as Lincoln and others predicted, Kansas had become a battleground. After passage of the Kansas–Nebraska Act, immigrants on both sides of the slavery argument flooded into the territory, looking to establish residence and gain the right to vote. Towns like Lawrence and Topeka became hotbeds of abolitionism, while Leavenworth and Atchison filled with so-called "border ruffians," mostly pro-slavery Missourians. Widespread voter fraud led to unrest, with Kansans struggling to decide whether their state would be

Gifted orator, brilliant lawyer, linguist, and committed reformer, Senator Charles Sumner of Massachusetts. The photograph dates to the early 1850s.

free or slave. "Beecher's bibles" arrived, too. To avoid capture by the enemy, Sharps rifles, renamed for the Brooklyn-based minister Henry Ward Beecher, had been packaged in crates labeled "Books" to arm abolitionist forces.

Sumner's oration had lasted for hours on each of two consecutive days. Speaking from his Senate desk, he criticized the pro-slavery administration of President Franklin Pierce. He condemned Senator Stephen Douglas, too, and in particular his doctrine of popular sovereignty, which Sumner renamed "Popular Slavery." In Sumner's hot words, popular sovereignty was an evil idea that had emerged from Douglas's mind just as sin had from Satan's. But he had reserved his sharpest barbs for Senator Andrew Butler.

Sumner compared Butler to Don Quixote. "The senator from South Carolina . . . believes himself a chivalrous knight," Sumner began. But rather than taking as his mistress an innocent farm girl, as the title character in Cervantes's novel had done, Butler chose a woman "ugly to others [and] polluted in the sight of the world." She was, said Sumner, "the harlot slavery." Even an indirect allusion to sexuality violated the Senate's rules of decorum, but Sumner blew through that prohibition—and his meaning was clear to every informed listener. He was referencing the widespread accusation that slaveholders were sexual predators who fathered children with the women they owned.

For Preston Brooks, a cousin of Butler, Sumner's words were too much to take. After reading the speech in the newspaper, he set out for the Capitol, walking stick in hand, to complete what he viewed as an errand of honor. Sumner's words had besmirched the honor not only of his cousin, who was absent that week from Washington, but of their state. According to the Southern code of chivalry, that demanded a response.

Arriving at the mostly empty Senate floor, Brooks walked down the aisle, limping noticeably, the result of a gunshot wound sustained in a duel years before. When Brooks reached Sumner's desk, he leaned in.

Congressman Brooks's assault on Charles Sumner shocked the nation. When this illustration was published, the caption read, THE SYMBOL OF THE NORTH IS THE PEN; THE SYMBOL OF THE SOUTH IS THE BLUDGEON.

"Mr. Sumner," he began, his tone calm. He told Sumner he had read his Kansas speech and that he took exception to his slander of "a relative of mine." Even before he finished speaking—as Brooks remembered, his closing words were, "I feel it to be my duty to punish you for it"—he struck Sumner with his walking stick.[16]

The gold head of the heavy cane crashed into the senator's head. Before Sumner could rise, Brooks swung the walking stick again. And again and yet again, delivering a rapid series of blows with such force that the stick cut clean to Sumner's skull, opening deep lacerations in the flesh. With blood pouring down his face, Sumner, a sturdy man of six-foot-four, struggled to stand, but his long legs caught beneath the desk. He teetered, but Brooks, taking Sumner by the lapel, steadied him as he continued to strike blow after blow with the stick clenched in his other hand.

Blinded by the blood running into his eyes, Sumner raised his arms to protect himself. The shaft of Brooks's cane snapped, but he continued to batter Sumner with the length that remained. No one interfered; the attack was sudden, and most senators had departed for a midday meal. After tearing the desk free of the bolts that fastened it to the floor, Sumner finally managed to get to his feet. But almost as soon as he stood, dazed and concussed, he sank to the floor in the main aisle. Only then did Brooks step back, having delivered, by his own count, thirty blows to Sumner's head and shoulders. "I desisted," he told a congressional investigation a few days later, "simply because I had punished him to my satisfaction." Sumner lay unconscious, his head still bleeding freely, his waistcoat, shirt, and coat covered with blood.

Brooks's premeditated attack was over in a minute or two, but its impact would be lasting. News of the assault traveled quickly, with telegraph lines carrying word far and wide. In the North, people were angry at the "act of an assassin."[17] In the South, people were proud, proclaiming Brooks a hero, with the *Richmond Enquirer* announcing that "the press of the South applaud the conduct of Mr. Brooks, without condition or limitation."[18] In Kansas, however, a man named John Brown got a different message. He heard a call for retribution.

A passionate abolitionist caught up in the fight for Kansas's soul, Brown had been outraged to learn that, on May 21, 1856, pro-slavery forces from Missouri had ransacked Lawrence, Kansas, a town founded by Massachusetts abolitionists. He reached his flashpoint when, a few hours later, he got wind of the attack on Sumner in the Capitol. Brown reacted: As men on both sides had done, Brown had created an informal militia company under his leadership, its ranks including four sons and a son-in-law. Brown and his men then took what he called "radical retaliatory measures."[19] Marching into the Kansas night, they kidnapped five men associated with local pro-slavery violence and slashed them to death with broadswords.

Such carnage was exactly what Lincoln had feared—and predicted. News of what was called the Pottawatomie Massacre, named for a nearby creek, flashed across the country, a terrible counterblow to the beating Sumner had taken. Brown and company were the slaveholders' worst

John Brown, pictured here in the 1850s, fathered twenty children, failed as a businessman, but committed himself absolutely to the cause of abolition.

nightmare come to life. Abolitionism, previously a peaceable movement for change, had experienced a real-life transformation like Robert Louis Stevenson's Dr. Jekyll; this wing of the abolitionist movement was as savage and vengeful as the murderous Mr. Hyde of Stevenson's novel.

Overnight John Brown became the intimidating face of abolitionism and a man the entire South could hate. Even to Northerners, Brown seemed a dangerous man and a lawbreaker. A few years later, he would ignite an even larger firestorm.

SIX

A SUBTERRANEAN PASSWAY

[John Brown] is of the stuff of which martyrs are made.

—Samuel Gridley Howe, February 5, 1859

The neighbors knew the tall, angular man in the storm-beaten hat as Isaac Smith. A dense beard worthy of an Old Testament prophet hid the sharp planes of his face, and the fishing rod he carried suggested he was about to try to hook his dinner in a nearby stream. But the disguise didn't fool Frederick Douglass, who recognized Captain John Brown in an instant. He was unmistakable, Douglass later recalled, "lean, strong, and sinewy, of the best New England mould, built for times of trouble, fitted to grapple with life's flintiest hardships."[1]

At Brown's request, the two were about to have a clandestine meeting in an abandoned quarry near Chambersburg, Pennsylvania. The month was August 1859, but their friendship went back a dozen years. They met when Douglass, traveling the lecture circuit to promote his just-launched newspaper, stopped in Springfield, Massachusetts. The orator had been welcomed by Brown's large family—Brown would father twenty children—and partook of a simple meal of beef soup, cabbage, and potatoes.

That evening years earlier Brown had confided his master plan for emancipation. Showing Douglass a map of the United States, Brown drew his finger along the Allegheny Mountains, the rugged ridge that extended from Pennsylvania through Virginia and into Kentucky. He would start small with just twenty-five handpicked men, Brown explained. From hiding places high in the Alleghenies, his volunteers would infiltrate nearby Virginia farms and find bondsmen eager to run away. Some fugitives would be sent north on the Underground Railroad, but the strongest and bravest would join Brown's little army in the mountains, expanding the effort to emancipate as many of the enslaved as possible.[2] And it would give Blacks a larger role in shaping their destiny, as they became soldiers fighting for their own freedom.

The two men had debated the bold plan for a "Subterranean Pass-Way" until three in the morning, their faces lit by candlelight. Brown impressed Douglass, who wrote in *The North Star* that his host was "deeply interested in our cause, as though his own soul had been pierced with the iron of slavery."[3] Brown had hated slavery since the age twelve, when, having himself been treated with kindness and consideration by a slaveholder his father knew, he was horrified at the man's mistreatment of a "*negro* boy . . . who was fully if not more than [my] equal." The enslaved child was badly clothed and fed. After watching the man beat the other boy mercilessly, Brown swore an "*eternal war*" on slavery.[4]

In the years since their first meeting, Brown and his family had devoted themselves to abolitionism. As homesteaders in Gerrit Smith's Adirondack experiment, Timbuctoo, Brown welcomed Black settlers to dine at his table. He spoke to them using their surnames—as in *Mr. Jefferson* and *Mrs. Wait*—which was an unheard-of politeness from a White man. One Black neighbor child remembered Brown would "walk up to our house . . . and come in and play with us children and talk to father. Many's the time I've sat on John Brown's knee."[5] In a way that few Whites of the time did, Brown treated the Black man as his equal.

Douglass and Brown were friends, meeting over the years in Rochester and at abolitionist meetings. They found different approaches, Douglass with his lectures, Brown with action in Kansas. Douglass looked to work inside the law, looking to rewrite it; Brown followed a path that sometimes veered outside of it. But both had been stunned, in 1857, when the United States Supreme Court handed down the Dred Scott decision. It made clear that the whole simmering stew of American politics was about to bubble over.

The ruling in *Dred Scott v. John F. A. Sandford* found that the plaintiff, Dred Scott, a man born into slavery, remained the legal property of John Sandford, despite having spent years living in free states. The broad strokes of Chief Justice Roger Taney's opinion further angered

Dred Scott in an image published several decades after his death. Scott was the subject of what has been called the Supreme Court's worst-ever decision.

By the late 1850s, John Brown was notorious; he grew a beard as a disguise. In this lithograph published after his capture at Harper's Ferry, the imprisoned Brown holds a copy of the New-York Tribune. *The newspaper's editor, Horace Greeley, observed, "[Brown] seems to have infected the citizens of Virginia with a delusion as great as his own . . . that a grand army of abolitionists were about to invade them from the North."*

abolitionists because it asserted that the "negro African race" was "of an inferior and subordinate class." Blacks, wrote Taney, "had no rights which the white man was bound to respect."[6]

In the eyes of the most powerful judge in the land, people of African descent *were not* and *could not* be citizens.

ON AUGUST 20, 1859, Douglass entered the abandoned stone quarry with caution; he knew that Brown, a wanted man, would be well armed. But Brown greeted him with his usual intense gaze, his blue-gray eyes "full of light and fire."

He explained why he had summoned Douglass. The ever-larger stakes, Brown believed, required bigger action. They had talked often over the years about fomenting a slave uprising in Virginia, but the plan now consisted of much more than the creation of a subterranean passway.

Brown aimed at a new and specific target, the federal arsenal at Harper's Ferry, Virginia.*

They sat amid the rocky debris. On hearing the new scheme, Douglass immediately opposed it. His thinking had shifted from the pacifism of his early Garrison days—violence at places like Pendleton, Indiana, ended that—but for many hours that day and the next Brown and Douglass again debated the cases for and against. The entire nation would see taking the arsenal as an attack on the government, Douglass said. The country needed to be shocked, countered Brown. The assault would make the original goal of opening a conduit for runaways to the North nearly impossible, Douglass argued. He also warned the older man that he would be "going into a perfect steel-trap, and that once in he would never get out alive."[7] In short, attacking Harper's Ferry was a suicide mission. Brown remained adamant.

At last, as Douglass prepared to depart, having refused to join Brown's troops, the White man embraced his Black friend and made one last argument. "Come with me, Douglass," Brown said urgently, "I want you for a special purpose. When I strike the bees will begin to swarm, and I shall want you to help hive them."[8] He understood that Douglass could rally the Black man like no one else.

Whether out of discretion or cowardice—he could never say—Douglass refused to participate. Had he agreed to go, his life would have been very much shorter.

---

* Douglass, *Life and Times* (1881), p. 279. West Virginia did not become a separate state until 1863; in time, the name of the town would lose its possessive, and is now officially known as Harpers Ferry.

INSURRECTION!

On a moonless night eight Sundays later, John Brown held the reins of a four-horse team that pulled a wagon loaded with tools and weapons. Eighteen soldiers of his "provisional army" followed, thirteen of them White men, five Black; three others remained behind at a log schoolhouse with more supplies and ammunition. Shouldering their rifles, the raiding party marched toward Harper's Ferry, the industrial town located at the confluence of the Potomac and Shenandoah Rivers in western Virginia.

At first, Brown's plan went like clockwork. Two of his men slipped into the woods and cut the east-bound telegraph lines. The little company then crossed the Potomac on a long, covered bridge, capturing the unsuspecting watchman, whose primary worry was locomotive sparks that might set the wooden structure afire. On entering the sleeping town—it was after ten o'clock on the Sabbath—Brown headed directly for the U.S. Armory, an expanse of buildings that contained the rifle works and an arsenal.

An elderly night watchman refused them entry, but one of Brown's men pried open the padlocked gate with a crowbar, and the guard, armed with only a sword, was taken prisoner. By midnight, the invading party secured a second bridge, this one stretching across the Shenandoah River. Without firing a shot, Brown was in possession of a hundred thousand guns.

Brown was poised to embark on the fulfillment of his life's mission. "I want to free all the Negroes in this state," he told his prisoners in the armory, ". . . and if the citizens interfere with me, I must only burn the town and have blood."

After midnight the tide shifted. The first blood flowed when a

Baltimore & Ohio express train chugged into town. A Black baggage-master was shot and mortally wounded. Brown permitted the train to proceed, but its departure meant that, at 7:05 A.M., a telegram sent by its conductor arrived at the B&O main offices in Baltimore. "[Train] stopped this morning at Harper's Ferry by armed abolitionists," the train conductor reported. "They say they have come to free the slaves and intend to do it at all hazards."[9]

Brown's identity remained a secret, since the telegram referred to him only as "the Captain." But the occupation of the U.S. Armory by this dangerous man would soon become national news. "SERVILE INSURRECTION," read the front-page headline on the next day's *New-York Times.* "The Federal Arsenal at Harper's Ferry in Possession of Insurgents. GENERAL STAMPEDE OF SLAVES."[10]

The shocking reports of the uprising and of Whites and Blacks in league with one another echoed near and far. But contrary to Brown's hopes, local enslaved persons did not fly to his aid. Instead, farmers and militiamen began to arrive to fight him, alerted by two riders who, Paul Revere–like, galloped into nearby towns hollering *Insurrection! Insurrection!* As tolling church bells alerted people to an emergency, rumors circulated. Some claimed the occupying force consisted of 150 men, others that the raiding party included hundreds of Blacks and a total force of more 750 men.

By late Monday morning, the town rallied in defense, and Brown's force at the U.S. Armory faced almost constant gunfire. Just before noon a contingent of militiamen from nearby Charles Town charged across the Potomac Bridge. All but one of Brown's sentinels fled to safety, but a freedman named Dangerfield Newby became the first of Brown's men to die. After a bullet tore through his neck, Newby's lifeless body was subjected to indignities by the infuriated townspeople. They sliced off his genitals, took the ears for souvenirs, and left his bloody corpse in a gutter for roaming hogs.[11]

By then Brown held a number of hostages, including a local slaveholder named Lewis Washington, a great-grandnephew of General Washington. Brown had plotted Colonel Washington's kidnapping, believing that an association with the name Washington—in the North and South alike, the first president was revered almost as a god—would send a message about the righteousness of his cause. But now Brown had to use his hostages to negotiate a ceasefire, and he sent out a team to talk. The white flag of truce they carried would mean nothing that day, and despite the rules of war, his representative was taken prisoner.

In a defensive move, Brown consolidated his remaining men and hostages in the armory's engine house, a solid, compact building in which fire engines were stored. Here he could make his stand, but "John Brown's Fort," though made of brick, would be a fulfillment of Frederick Douglass's "perfect steel-trap" prophecy.

Brown tried to parley again, but this time the white flag met with a spray of bullets. Brown's eldest son, Watson, shot in the gut, barely managed to crawl back to the engine house. Already several men down, Brown's little army lost another fighter when Oliver, the youngest Brown son, was shot as he sighted his rifle through a crack in the double door of the engine house. He fell motionless to the floor, dead within minutes.

ENDGAME

At eleven o'clock that night, October 17, 1859, the U.S. military arrived. When word of Brown's raid had reached Washington, there had been a scramble to find a nearby fighting force. Ninety marines stationed at the Washington Navy Yard had been chosen; to command them, the secretary of war summoned Colonel Robert E. Lee, a veteran of the Mexican War and former superintendent of West Point. Lee and his troops marched in a light rain from the train station into Harper's Ferry, arriving before midnight. After reconnoitering the situation, Lee concluded that Brown and his men were not the large fighting force that had been reported, and out of concern for the safety of the hostages, he chose to take no action until morning.[12]

At 7:00 A.M., the company of marines, armed with bayonets and sledgehammers, stood just out of sight of the insurgents. Lieutenant James Ewell Brown ("Jeb") Stuart, Lee's second in command, approached the engine house under a flag of truce.

Inside, John Brown waited with what was left of his "provisional army," now down to four uninjured fighters. Despite the odds, Brown was calm. As hostage Lewis Washington later remembered, "With one son dead by his side, and another shot through, he felt the pulse of his dying son with one hand and held his rifle with the other, and commanded his men with the utmost composure, encouraging them to sell their lives as dearly as they could."[13]

An audience of some two thousand townspeople watched. Lee was in their midst, dressed in civilian garb, as Brown cracked open one of the heavy oak doors. Through the narrow gap Stuart recognized the bearded face of a man he had met once before, in Kansas, while serving in the U.S. Cavalry.

"You are Ossawattomie Brown, of Kansas?" he asked.

"Well," Brown replied, confirming the rumor that had begun to circulate in the town, "they do call me that sometimes."

Stuart delivered his message: Brown and men must surrender unconditionally.

Brown refused, asking instead for safe passage to Maryland in return for the release of the hostages. To that Stuart could not agree.

Brown remained resolute. After further back-and-forth, he said, "You have the numbers on me, but we soldiers are not afraid of death. I would as leave die by a bullet as on the gallows."

"Is that your final answer, Captain?" asked Stuart.

"Yes," replied Brown.[14]

The lieutenant bowed respectfully, but as he stepped away from the engine house, he doffed his cap, waving it high in the air. His gesture—the agreed-upon signal—meant a dozen marines leapt into action. Carrying sledgehammers, they sprinted to the building and began thumping the doors. Another team of marines ran at the building wielding a heavy ladder as a battering ram. Gunfire and smoke filled the air by the time ladder's third blow opened a ragged hole.

The first man who charged through recognized Lewis Washington, who pointed at Brown, crouching and reloading his rifle by the door. "Quicker than thought," as the marine remembered later, "I brought my saber down with all my strength upon his head."[15] Gripping the sword with two hands, he opened wounds on Brown's head and neck before delivering a thrust to the chest. Brown slumped to the ground, not dead but severely injured.

A frozen moment, just before the Harper's Ferry engine house was stormed by federal troops. Colonel Lewis Washington and the other hostages stand at left, while one of John Brown's sons bleeds to death in the foreground, in a woodcut published shortly after Brown's raid.

In a matter of a few minutes, the thirty-two-hour siege at Harper's Ferry was over. Ten of Brown's men were dead, five unaccounted for, and Brown and six others in custody. Despite the fact that Brown's army was decimated and defeated, that very afternoon the man himself rose again to do battle, this time in a war of words.

"THE END OF THAT IS NOT YET"

Brown lay on the floor, covered by a grimy quilt, his head resting on a carpetbag. His face was gaunt, his beard and hair tangled and matted with blood. He looked up defiantly at a squad of interrogators, including Virginia governor Henry Wise, Robert E. Lee, Jeb Stuart, and several congressmen. They wanted answers—and Captain Brown was more than happy to talk.

As several reporters looked on, Virginia senator James Mason led the questioning. He tried first to get Brown to identify his coconspirators, but Brown told them the plan was his idea. He brushed away questions about who had helped him. "I will answer freely and faithfully about what concerns myself . . . but not about others."

When Mason asked the purpose of the raid, Brown gave him a simple and pointed reply. "We came to free the slaves, and only that."

"How do you justify your acts?" asked Mason.

This time Brown had an even sharper reply for Senator Mason, who had drafted the hated Fugitive Slave Act. "I think, my friend," said Brown, "you are guilty of a great wrong against God and humanity . . . and it would be perfectly right for any one to interfere with you so far as to free those you wilfully and wickedly hold in bondage."[16]

Later, as the interrogation came to an end, one of the reporters asked Brown if he had anything further he wished to say. He replied with a warning to the people of the South. "Prepare yourselves for a settlement of this [slavery] question, that must come up for settlement sooner than you are prepared for it. . . . You may dispose of me easily,—I am nearly disposed of now; but . . . this negro question I mean; the end of that is not yet."

A word-for-word transcript appeared later that week in the pages of the *New York Herald.* Readers found Brown was not only unrepentant, but full of accusations. There was clarity to his thinking, too, which made it difficult for the open-minded to dismiss him as a madman. His actions at Harper's Ferry might be wrong, but he was driven by a moral and religious rectitude. Yes, he had committed a violent act, but then the men who imprisoned him were slaveholders, men who owned other men.

FREDERICK DOUGLASS, FUGITIVE

News of the Harper's Ferry raid reached Frederick Douglass in Philadelphia. It hit him "with the startling effect of an earthquake[,] . . . something to make the boldest hold his breath." But he went about his business, delivering his speech as scheduled.[17]

A few hours later, however, Douglass learned he had become the most wanted man in America.

Jeb Stuart's men had found John Brown's papers, which included maps of the Alleghenies, letters in code, and correspondence that suggested a much larger conspiracy. Tucked into the stack was a letter from Douglass to John Brown. In the estimation of Virginia governor Wise, eager to take action and quell the panic in his state over the insurrection, the letter provided sufficient evidence to charge Douglass with multiple crimes, although it made no mention of Harper's Ferry.

Despite the news, that day would be a lucky one for Douglass. A Philadelphia telegraph operator sympathetic to the abolitionist cause pocketed a wire from Washington ordering Douglass's arrest and sent word to the abolitionist of the imminent danger. By the time the telegram was delivered several hours later and the Philadelphia sheriff came knocking, Douglass was gone. Again a fugitive, he had hopped the first train to New York, where friends smuggled him onto another train for Rochester. He was on the run once more, heading for Canada, but stopping at home before he made for the border.

Douglass sent a telegram of his own, ordering one of his sons to clear out his desk, just in case the authorities arrived before he did. Though he reached his adopted hometown without incident, he soon learned that New York's lieutenant governor would honor a Virginia arrest warrant. With the authorities in hot pursuit, Douglass stepped aboard a ferry hours later, crossing Lake Ontario to Canada, like so many fugitives on the Underground Railroad before him. Six hours later, federal marshals knocked on his door in Rochester.

John Brown almost got him killed—as a Black man linked to Brown's conspiracy, Douglass believed, he would have been sentenced to death by a Virginia court. Even so, writing from Canada, Douglass defended his friend. In an essay for his paper, he called him a "glorious martyr of liberty" and compared Brown to the "heroes of Lexington, Concord, and Bunker Hill."[18] Unsure of his safety even in Canada, Douglass made his way to Quebec and sailed for England, though his thoughts remained with Brown. "John Brown has not failed," he wrote. "He has dropped an *idea*, equal to a thousand bombshells."[19]

Brown's crimes at Harper's Ferry would not go unpunished. Convicted of treason, murder, and inciting a slave insurrection, he received a death sentence. On December 2, 1859, thousands of soldiers and citizens gathered in a field outside Charles Town. In a strange glimpse of moments to come, the crowd included Professor Thomas J. Jackson of the Virginia Military Institute, soon to be a Confederate general honored with the nickname "Stonewall"; Edmund Ruffin, who would fire one of the first shots at Fort Sumter sixteen months later; and the actor John Wilkes Booth, costumed in a militia uniform. As his sentence was carried out, John Brown displayed the same extraordinary dignity he had during his interrogation and, later, at his trial, where he impressed even Governor Wise. With head hooded, arms and ankles tied, John Brown was hung by the neck until dead.

Douglass arrived in Liverpool after John Brown had met his fate on the gallows, but even across the sea Douglass recognized that John Brown had become a symbol of the national divide. Brown accomplished his goal of striking terror into the hearts of pro-slavery people, giving him the status of the devil incarnate in the South, where his actions boosted secession sentiments. In the North, the reaction was more complicated. People could see Brown as both a criminal and a holy warrior; in many towns, church bells tolled in his honor on the day he died. Henry David Thoreau equated him to Jesus Christ, calling them "two ends of a chain." The philosopher Ralph Waldo Emerson called him a saint.[20]

Almost single-handedly he pushed the national slavery debate from an angry conversation into the realm of violent combat. His actions at Harper's Ferry made people of conscience ask themselves how to carry the fight against the evil of slavery. Was such violence justified? Whether one condemned or admired John Brown, he was, in Herman Melville's words, "the meteor of war," a man who set the stakes for the immensely greater violence to follow in the new decade.[21]

SEVEN

THE DIVIDED HOUSE

> We cannot be free men if this is, by our national choice, to be a land of slavery. Those who deny freedom to others, deserve it not for themselves; and, under the rule of a just God, cannot long retain it.
>
> —Abraham Lincoln, May 19, 1856

The day before John Brown attacked Harper's Ferry, Abraham Lincoln found a telegram waiting for him on his desk in Springfield. Wired from New York, it was an invitation to speak in Brooklyn on "any subject you please." The compensation would be generous, a lecture fee of two hundred dollars.[1]

For lawyer Lincoln, fresh from traveling the circuit, the invite meant much more than money. He had speechified far and wide in the West, but never before in New York. Thus, the piece of paper he held in his hand represented a priceless political opportunity, one that dared him to think he really could find a place on the national scene.

Lincoln-the-politician was well known and well liked in Illinois, but outside his home state, his reputation had risen more slowly. At least one newspaper in the East, when deigning to mention him at all, spelled his name *Abrahm* Lincoln. His single term in Congress was a decade in the past, though stump speeches on behalf of other politicians and his occasional public pronouncements had gained him enough respect that, to his complete surprise, his name had been entered in nomination for vice president at the 1856 Republican Party convention. But that had come to nothing since he lost on the second ballot, and the ticket went nowhere when a pro-slavery Democrat, Pennsylvanian James Buchanan, became the nation's fifteenth president that autumn.

More recently, in 1858, Lincoln had run for the U.S. Senate. After winning the Republican Party nomination that June, he had delivered a speech that got people talking. Attempting to reframe the slavery debate, he had borrowed Jesus's words to proclaim that "a house divided cannot stand."* "I believe this government cannot endure," he had told his audience, "permanently half *slave* and half *free*."[2]

These were strong words, and they reverberated beyond the Illinois border. Although Lincoln softened them with a disclaimer—"I do not expect the Union to be *dissolved*—I do not expect the house to *fall*"—he also made a prediction. "I *do* expect it will cease to be divided. It will become *all* one thing or *all* the other." In his view, slavery was very much more than a regional disagreement; it posed a challenge to the country's very survival.

The "House Divided" speech, as it came to be known, had also set the stage for a series of debates with Stephen Douglas, the Democratic incumbent. Though the senator was a well-established national politician, thanks in part to his controversial promotion of popular sovereignty, Douglas understood that this time his old rival Lincoln could

---

* "And Jesus knew their thoughts, and said unto them, 'Every kingdom divided against itself is brought to desolation; and *every city or house divided against itself shall not stand*.'" Matthew, 12:25. Italics added.

Lincoln, at the podium, with Douglas, at left, on the dais, in an early twentieth-century re-creation of one of their 1858 debates.

be a real threat to both his reelection to the Senate and to Douglas's hope of running for president in 1860. "I shall have my hands full," he told one reporter, ". . . and if I beat [Lincoln], my victory will be hardly won."[3]

IT BEGINS

When Lincoln and Douglas faced off in Ottawa, Illinois, Lincoln was on a mission to change the national argument about slavery. As a sitting senator, Stephen Douglas wanted to crush his opponent, get back to Washington, and run for president two years later. Their debate that day was the first of what would be an epic series of seven across the state.

Taking the stage in Ottawa on August 21, 1858, the two could hardly have looked more different. The gangly Lincoln was a foot taller than Douglas, who was a portly bulldog of a man. Douglas was a natty dresser, while Lincoln's ill-fitting suits tended to look as if he had slept

Senator Stephen Douglas of Illinois in a campaign photograph taken when he was the Democratic nominee for president in 1860.

in them. Douglas was a polished debater with a booming voice, while Lincoln's listeners complained his near-falsetto sounded shrill.

Douglas began on the attack. He accused Lincoln of trying to create "an Abolition party, under the name and disguise of the Republican party."

Flustered at Douglas's description, Lincoln struggled to respond. He hurried through his speech, reading a large chunk of his 1854 Peoria speech. Afterward, worried at his performance, he called a meeting of his advisers to discuss strategy.

Douglas seemed to have the upper hand, but word of the contest spread. People flocked to the second debate to see and hear the verbal fireworks as "the Little Giant," as Douglas was known, tried to fight off upstart "Honest Abe." This time, at Freeport, Illinois, Lincoln went on offense. But he sounded lawyerly, his arguments overly technical. A better showing, but he still wasn't able to dominate Douglas.

The attention of the state was now on the candidates, and crowds of twelve, fifteen, and even eighteen thousand attended as the debates became even more heated. Douglas had damned Lincoln as a "black Republican," claiming that Lincoln was allied with "fred douglass, the negro," a man who, unlike Lincoln, was already an established national figure, and insisted that "Lincoln [was] the champion of the black man." In a bold-faced lie, Stephen Douglas put it out that Frederick Douglass was campaigning for Lincoln in Illinois and had been seen "reclin[ing] in a carriage next to the white driver's wife."

Lincoln responded that he did not believe in full racial equality. "I agree with Judge Douglas that [the negro] is not my equal in many respects—certainly not in color, perhaps not in moral or intellectual endowment." The furthest he could go speaking in the Black man's favor was his bread-for-work argument:

> In the right to put into his mouth the bread that his own hands have earned, he is the equal of every other man, white or black. In pointing out that more has been given you, you cannot be justified in taking away the little which has been given him. All I ask for the negro is, that if you do not like him, let him alone. If God gave him but little, that little let him enjoy.[4]

In Lincoln's mind, in the late 1850s, equality ended there. Though ahead of his time—it was his conviction that slavery must be dealt with as a moral wrong—he was still a man of his time.

In the course of the debates, Lincoln gained confidence. His energy and persuasiveness rose in the final three meetings, and he outperformed the incumbent; Douglass seemed tired and, in the last debate, hoarse. By then word of the spirited debates had spread well beyond Illinois. With big city and national papers reporting on the debates, some reprinting word for word what the candidates said, more and more people across the country began to recognize Lincoln's name.

Despite his good efforts, however, Lincoln's campaign fell short. Senate seats were still decided by state legislatures in 1858 (that wouldn't change for another half century, with the ratification of the Seventeenth Amendment), and since the Democrats managed to retain control of the Illinois house that November, Douglas was chosen by a 54–26 vote once again to represent the state in the U.S. Senate.

Yet Lincoln was far from vanquished. Illinois had been the perfect place for the larger national discussion to play out as the two brilliant orators wrestled over slavery. With his name now on many people's lips, Lincoln had become a politician many thought belonged in the discussion beside Stephen Douglas when the conversation turned to the next presidential race. Americans were beginning to recognize Lincoln's voice just might be the clearest in the Republican Party—and that he might very possibly be the party's future.

BECOMING A NATIONAL CHARACTER

The invitation a year later to speak in Brooklyn was a recognition of Lincoln's new role in national politics. The tens of thousands of people had attended the debates and the many more who had read the published transcripts, not only in Illinois but elsewhere, now had a sense of what the man stood for. In the final debate, in Alton, Illinois, he had been clear. "The real issue of this controversy," he insisted, was between "one class that looks upon the institution of a slavery *as a wrong*, and of another class that *does not* look upon it as wrong."[5]

His words appealed particularly to Republicans in the East, and they wanted to hear from the man himself; as for Lincoln, he saw that a lecture in New York was a door opened where there hadn't been one. He would get a chance to rub shoulders with men like Reverend Henry Ward Beecher, the internationally famous social reformer, whose sister was Mrs. Stowe, author of *Uncle Tom's Cabin.* Lincoln could meet mighty newspapermen like William Cullen Bryant, editor of the *Evening Post*, which had published full reports of his debates with Stephen Douglas, and attract the attention of Horace Greeley of the *New-York Tribune*, the country's only truly national newspaper. The audience would be full of other New Yorkers who ran the nation's most important city.

Lincoln could only say yes, and a date was agreed upon for him to lecture at the end of February 1860. That gave him time to prepare his speech and to organize a side trip to visit his eldest son, Robert, preparing for college at Phillips Academy in Exeter, New Hampshire. Lincoln even ordered a suit tailored for the occasion—the cost amounted to half his speaker's fee—and by the time he boarded a train headed east, the carpetbag he carried contained a thick stack of blue foolscap paper. It was his handwritten speech, one he had labored over for weeks.

In the months after Lincoln received the New York invitation, John Brown had changed everything. His actions resulted in a sudden shift in electoral politics, one that gave Lincoln another lift. Before the Harper's Ferry story broke, the odds-on favorite for the Republican nomination had been Senator William H. Seward, a patrician former New York governor and sitting senator. But as the nation recoiled from the bloodbath at the engine house, Seward's well-known slavery views—he had memorably asserted in his first speech on the Senate floor that "there is a higher law than the constitution"—seemed suddenly dangerous.[6] Adding fuel to the fire, the papers also reported that Seward once contributed twenty-five dollars to John Brown's efforts in Kansas. Now, after John Brown had taken the law into his own hands—and paid for it with his life—Lincoln's moderate perspective on slavery seemed less risky than Seward's.

Lincoln hoped his speech would be the means by which he could complete his transformation, making a formerly obscure western man a genuine contender for the Republican presidential nomination in 1860.

"THE PRINCE OF PHOTOGRAPHERS"[7]

On Monday, February 27, 1860, the day of his New York appearance, Lincoln walked north on Broadway from his hotel, together with four members of the Young Men's Central Republican Union, the sponsor of the evening's lecture. He might be from the provinces, but Lincoln understood that to promote himself he needed a suitable photograph—Mary disliked the ones he had—and he was worldly enough to know that Mathew Brady, the foremost American photographer of the day, was the man for the job.

When Lincoln entered the grand reception room at Brady's gallery on the corner of Broadway and Bleecker Street, he took in the elegant carpets, chandeliers, and gilt walls. There was an array of Brady's images on display, including one of Stephen Douglas.

Lincoln was soon introduced to George Bancroft, another man who had come to get his picture taken. Bancroft was a diplomat, two-time cabinet secretary, and the distinguished author of a book Lincoln admired, the multivolume *History of the United States of America.*

The contrast between Brady's two customers struck one of Lincoln's companions. "One [was] courtly and precise in every word and gesture," the young New Yorker observed, "with the air of a trans-Atlantic statesman; the other bluff and awkward."[8] Even before meeting the genteel Bancroft, however, Lincoln grasped that he faced a large challenge in the hours that followed. He needed to win over an elite audience full of sophisticated men like Bancroft.

On arriving two days earlier, Lincoln had checked into the busy

Astor House, one of the premier hotels in the city. The watershed in his party's politics had meant his room immediately became a reception area for visitors, some of whom he knew, though many others came to meet and to pay their respects to the stranger emerging as a plausible candidate for the Republican nomination for president. A few party people had endorsed Lincoln, and the *Chicago Press and Tribune* had announced its support back in January. While Seward remained the favorite, the dark horse from Illinois was gaining unexpected momentum.

When his turn came to go before Brady's camera, Lincoln climbed the stairs to the "operating room" to meet the man who would take his picture.[9] Brady, his face half-hidden by a flowing mustache and spade-like goatee, had spent the preceding two decades taking pictures of anybody who was anybody, making trademark images of aging historic figures like Andrew Jackson, Henry Clay, and Dolley Madison, as well as current headline-makers such as Chief Justice Roger B. Taney and President Buchanan.

Brady studied his subject, who that morning looked notably "haggard and careworn."[10] Even Lincoln's friends agreed he was an odd-looking man, his ears, Adam's apple, and blunt nose too big, his chin too sharp, with a mole that marred his right cheek.[11] Brady had his work cut out for him.

The photographer's usual preference was to make bust images, portraits that that framed the head and shoulders. But looking at Lincoln, Brady decided to make a standing portrait to call attention to his subject's imposing stature rather than to his craggy face. A table was positioned so Lincoln could rest his left hand on a small stack of books, suggesting erudition. An assistant straightened Lincoln's ribbon tie, but Brady, standing by his immense box camera and examining his subject, still wasn't satisfied. The problem, he decided, was Lincoln's collar. He asked him to pull it up a little.

The Mathew Brady likeness shot on the same day as Lincoln's Cooper Union speech. "Honest Abe" himself would later observe that the picture, which introduced him to countless Americans, put him in the White House.

"Ah," said Lincoln, a man always quick to make a joke at his own expense, "I see you want to shorten my neck."

"That's just it!" crowed Brady. The two men, one short, one tall, laughed as one.[12]

THE COOPER UNION

Just after 8:00 P.M. that evening, William Cullen Bryant introduced "an eminent citizen of the West, hitherto known to you only by reputation, . . . Abraham Lincoln of Illinois."[13] To prolonged applause, the evening's main speaker stood and made his way the podium. People were struck by his great height, rumpled black suit, and unruly black hair as he waited for the room to quiet.

When he began to speak, Lincoln looked out upon a blur of faces.

Audience members had paid twenty-five cents for admission to the Cooper Union for the Advancement of Science and Art in Manhattan, and despite snow falling and slushy streets, the great hall was nearly full. "His manner was to a New York audience a very strange one," thought one observer. But Lincoln set about presenting his methodically constructed constitutional argument.[14]

His debates with Douglas a year and a half earlier had been superb practice, the dialectic of their back-and-forth sharpening his thinking. But in the four months since receiving the invitation to speak here in New York, Lincoln had spent uncounted hours in the Illinois State Library where, guided by his training as a lawyer, he drafted what was in effect a legal brief. This evening he would reargue the Dred Scott case and refute Chief Justice Taney's insistence that the Framers of the Constitution approved of slavery in the territories.

Lincoln went back in time to count heads. As he told the roughly 1,500 people in the room, he found that "a clear majority" of the signers, twenty-one of thirty-nine, had opposed the extension of slavery into the territories.[15] Thus, said Lincoln, *Dred Scott v. Sandford* had been wrongly decided.

His argument was detailed, but he won the rapt attention of his audience. Moving from the subject of Dred Scott, he took issue with the Southerners who tried to demonize him and his party. He defended the Republicans from the accusation that they were "revolutionary." Not so, he argued. "You charge that we stir up insurrections among your slaves. We deny it; and what is your proof? Harper's Ferry!" Lincoln insisted that "John Brown was no Republican; and you have failed to implicate a single Republican in his Harper's Ferry enterprise." In fact, Lincoln said, "John Brown's effort was peculiar. It was not a slave insurrection. It was an attempt by white men to get up a revolt among slaves, in which the slaves refused to participate. In fact, it was so absurd that the slaves, with all their ignorance, saw plainly enough it could not succeed."[16] Although Lincoln held a quiet admiration for the man, he carefully distanced himself from John Brown.

The long speech required an hour and a half to deliver, with frequent interruptions for cheers and loud applause from his listeners. But his central argument was clear: Slavery could not, must not, be extended to the territories. It wasn't a matter of abolishing slavery; in fact, the Fugitive Slave Act, hateful though it might be, was the law of the land and, Lincoln said, had to be enforced. "Wrong as we think slavery is," he said in conclusion, "we can yet afford it to leave it alone where it is." But in his final words, he sent a clarion call to all good men. "Let us have faith that right makes might, and in that faith, let us, to the end, dare do our duty as we understand it."

Lincoln had chosen not to deliver a stump speech. Instead, he offered an argument, a statement of principles, a defense, and a call to arms. He held his audience spellbound as he persuaded them of his positions. He won their admiration for him as a man and—all importantly—as a candidate. As one partisan reported, "the house broke out in wild and prolonged enthusiasm. I think I never saw an audience more carried away by an orator."[17] People tossed handkerchiefs in the air and gave the speaker an ovation that lasted ten minutes.

The evening was a triumph. Lincoln himself was surprised at the wave of congratulations and hearty handshakes that came his way. After the full text of his speech appeared, as previously arranged, in five New York newspapers the following day, he received multiple invitations inviting him to speak. His trip to visit his son in southern New Hampshire rapidly became a campaign tour, with nine addresses in Connecticut, Rhode Island, and New Hampshire. He would turn down invitations from Philadelphia and other eastern cities in order to return to Illinois, but the Cooper Union speech elevated him. If Senator Seward remained the front-runner, he now had a worthy challenger in the contest for the Republication nomination for president, which would be decided in Chicago, in the month of May, at the party convention.

Whatever private doubts Lincoln had about running were banished, too. When Illinois senator Lyman Trumbull inquired about the coming fight for office, Lincoln wrote back, with wry understatement, "I will be entirely frank. The taste *is* in my mouth a little."[18]

DEATH OF A DAUGHTER

In exile on the other side of the Atlantic, Frederick Douglass sheltered with friends in Yorkshire, England. In the early months of 1860, he used their home in the mill town of Halifax as a base from which he traveled to speaking engagements in England and Scotland, making stops in Newcastle, Edinburgh, and other large cities. His expanding gifts as a writer were on full display in the lectures he gave. One in particular he would deliver dozens of times over a period of years; he called it "The Trials and Triumphs of Self-Made Men." He espoused a theory: Success, he believed, was to be credited to the individual, "ascribed to brave, honest, earnest, ceaseless heart and soul industry. By this simple means—open and free to all men—whatever may be said of chances, circumstances, and natural endowments—the simple man may become wise, the wise man wiser."[19]

The separation was difficult for Douglass's wife and children back in Rochester, but ten-year-old Annie took it the hardest. The baby of the family, she was born in 1849 after her father's return from his long first exile to the British Isles. Knowing only the stability of their Rochester lives, she was a lively and happy child, much cherished by her family. But the departure of her father weighed on her.

The trial and execution of John Brown was an added burden for the young girl. In early 1858, Annie had become a nearly constant companion of Brown when the old man spent several weeks as a boarder in the Douglass household, insisting upon paying his host three dollars a week. Quietly planning his passway escape plan, he employed Annie's older brother Lewis to courier his letters to and from the post office. Like a man playing with toy soldiers, he engaged the Douglass children with model fortifications he made of boards and showed them drawings of his mountain installations. The young Douglasses found it all more intriguing than their father did, and to Annie their eccentric boarder became "dear old John Brown, upon whose knee she . . . often sat."[20]

Early in 1860, Douglass received a letter from Annie. "I am proceeding in my grammar very well for my teacher says so," she wrote to her distant father. Well aware of his passions, she transcribed a poem she had memorized for school: "He is not the man for me / Who buys or sells a slave / Not he who will not set him free / But send him to his grave." More ominously, the innocent, ordinarily cheerful girl acknowledged her sense of loss. "Poor Mr. Brown is dead," she told her father. "[T]hey took him in an open field and . . . hung him."[21]

Frederick Douglass and his high-spirited youngest child, Annie, when she was about five years old. Her unexpected death at age eleven brought her father back from his exile in England in early 1860.

On March 13, 1860, Frederick Douglass was rocked by the news that his beloved daughter, "the light and life of my house," was dead.[22] He was shocked and surprised; it was worst possible piece of news he could have received and the most unexpected. The circumstances were inexplicable; after having written to him, in December, she had become withdrawn and lost the ability to speak and to hear. No doctor could explain why, beyond guesswork about a possible fever of the brain. Douglass himself would attribute her retreat from life to "over-anxiety for the safety of her father and deep sorrow for the death of dear old John Brown."[23]

Brokenhearted at the loss, Douglass's fear for his own safety seemed suddenly unimportant. The report of her death reached him in Glasgow and, after collecting his things in Halifax, he hurried to Liverpool and found passage on the first available ship. He made landfall in Portland, Maine, then traveled across Canada to Rochester, taking care not to call attention to himself. A funeral cortege of thirty-five carriages had long since delivered Annie's remains to her grave, but Douglass did his best to console his family. Friends would report that the father would mourn the loss of his little daughter for years to come.[24]

During the spring of 1860, Douglass lay low, wary that he might still be a wanted man. But the feverish pursuit of Brown's alleged co-conspirators was rapidly cooling. A congressional committee formed to investigate had found no evidence of a larger conspiracy involving Douglass or anyone else beyond Brown and his provisional army.* Even before the June publication of the Senate Select Committee report,

* "There is no evidence that any other citizens than those there with Brown were accessory to this outbreak." "Select Committee Report on the Invasion and Seizure of the Armory and Arsenal at Harper's Ferry, June 15, 1860." Senator Mason, the man who had interrogated the bloody Brown hours after his capture, chaired the committee. Senator Jefferson Davis, who would soon resign from the U.S. Senate and become president of the Confederate States of America, was also a member.

Douglass resumed his public life, traveling and speaking without fear of arrest. The nation's central focus had become the future. Electoral politics—in particular, the race for the presidency—filled the air.

THE RAIL SPLITTER

In the month of May, Lincoln made his first and last major appearance of the 1860 presidential campaign.

His nomination was far from a certainty even in Illinois, but when he arrived in Decatur, the site of the state's Republican convention, he did have a national strategy in mind. If he could win the unanimous support of the Illinois delegation, together with a few delegates from other states, he might prevent Seward from winning a majority on the first ballot. That would give his supporters a chance to harvest delegates previously committed to others for subsequent tallies. Going in, however, everyone in Lincoln's camp agreed: He needed a boost to win big among his in-state neighbors and to enhance his appeal beyond his region.

Richard Oglesby, Lincoln's Illinois campaign manager and, later, a governor of Illinois, cogitated on how to generate enthusiasm for his rawboned candidate. He wondered whether they might devise a down-to-earth nickname, something like Andrew Jackson's "Old Hickory." Or a catchy slogan, such as "Tippecanoe and Tyler, Too," the memorable phrase that had helped power William Henry Harrison, victor at the Battle of Tippecanoe, along with his running mate, John Tyler, to victory in the election of 1840. Oglesby wanted to capitalize on Lincoln's folksy and likable manner, maybe with a moniker that conveyed not that he was a fine lawyer but that he was a down-to-earth westerner who could spin a yarn and speak for the common man.

A conversation with Lincoln's cousin John Hanks provided the answer.

A dozen miles down the road from Decatur lay the fields that Lincoln helped clear as a last act of family duty before leaving his father's homestead. Lincoln and Hanks had enclosed the acres with split-rail fences.

"John," Oglesby asked Hanks, "did you split rails down there with old Abe?"

"Yes, every day," was the reply.

"Do you suppose you could find any of them now?"

Hanks thought he could.

"Then," said Oglesby, "come around and get in my buggy, and we will drive down there."[25] A plan was taking shape in Oglesby's mind.

With no building large enough to house the convention, the citizens of Decatur constructed an oversized tent of canvas and lumber, which they called a wigwam, complete with stands for the delegates, a platform, and a podium. When the convention opened, on May 9, the wigwam was packed with an overflow crowd estimated at more than four thousand. But the undoubted high point of the two-day meeting came when Oglesby announced from the dais that John Hanks "desired to make a contribution to the Convention." Right on cue, Hanks and an assistant marched down the center aisle carrying two well-worn fence rails of split walnut. They were decorated with flags, streamers, and a banner that read, "Abraham Lincoln, The Rail Candidate for President in 1860."

According to one newspaper account, "the effect was electrical." The spontaneous applause was deafening, and Lincoln, seated in the tent, rose to speak. He hadn't split a rail in years, but with his long lean frame and weathered features, he looked like a man who had. He told the crowd he could not be certain that these ones in particular were his handiwork, but that, yes, certainly he "had mauled many and many better ones" as he had "grown to manhood."[26] After he finished speaking, he was met with another long ovation. When he left Decatur the following day, he did so with the unanimous support of the Illinois delegation.

For purposes of the campaign, he became "the Rail Splitter." A week later, the national Republican convention convened in a Chicago auditorium, its entrance framed with a pair of fence rails. Observing tradition, Lincoln himself did not attend, but waited in Springfield. A local newspaperman found Lincoln in his law office with encouraging news: On the first ballot, Seward outpolled him only 173½ to 102, with the other 40 percent of the votes almost evenly divided among three other candidates.

Lincoln made his way to the telegraph office, where he learned the second ballot was closer, the count Seward 185½, Lincoln 181. Lincoln, as one congressman put it, became "the second choice of everybody" because people thought he might be more electable than Seward, despite his relative lack of experience.[27] At the candidate's next stop, the offices of the *Illinois State Journal*, Lincoln opened a telegram addressed to him: The nomination was his. He shook hands all around before heading for home to Mary and his boys. "Well, Gentlemen," he said in parting, "there is a little woman at our house who is probably more interested in this dispatch than I am."[28]

EIGHT

THE ELECTION OF 1860

I want every man to have the chance—and I believe a black man is entitled to it—in which he can better his condition.

—Abraham Lincoln, speech in New Haven, Connecticut, March 6, 1860

During the summer and fall, Lincoln did no campaigning. At the insistence of his advisers, the Republican nominee remained in Springfield, since whistle-stop tours and other city-to-city campaigning would not become a standard part of American presidential politics for another two decades. But Lincoln found other ways to keep himself in the conversation.

For one, he had compiled transcripts of his debates with Stephen Douglas, pasting them in a scrapbook. He arranged with an Ohio publisher to set them into type and publish them as a book, thinking they might be useful as a campaign document. He was right: Just as his ideas had resonated with the large crowds from the debate stage in 1858, the published debates fueled curiosity about the rising political star in 1860. The book became an unexpected bestseller, selling more than thirty thousand copies and finding its way into parlors and taverns across the North.

Then there was the picture. Immediately after the convention, Mathew Brady released his photograph of Lincoln, as recorded on the day of the Cooper Union speech, and within a week, the covers of two major national publications, *Harper's Weekly* and *Frank Leslie's Illustrated Newspaper*, featured woodcuts based on the picture.* Brady had touched up the image in the darkroom, softening the lines and shadows on Lincoln's face. To those inclined to think so, the man looked positively presidential—somber and serious, tall and confident, unafraid

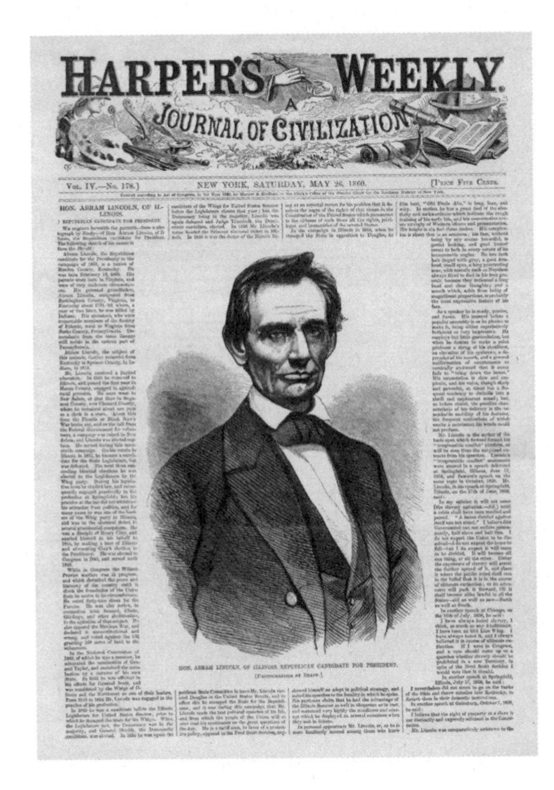

HARPER'S WEEKLY. A JOURNAL OF CIVILIZATION

Vol. IV.—No. 178.] NEW YORK, SATURDAY, MAY 26, 1860. [Price Five Cents.

HON. ABRAM LINCOLN, OF ILLINOIS.

HON. ABRAM LINCOLN, OF ILLINOIS, REPUBLICAN CANDIDATE FOR PRESIDENT.

Brady's first photograph of Lincoln, cropped to his head and upper torso and reworked as a woodcut for newspaper publication.

* *Harper's Weekly*, May 26, 1860, which rendered the candidate's name *Abram*; and *Frank Leslie's Illustrated Newspaper*, May 20, 1860. Twenty and thirty years, respectively, would elapse before the technology permitted photographs to appear in newspapers and magazines.

to engage the viewer's eye. Lincoln's Cooper Union speech, already widely reprinted in newspapers, appeared in numerous new editions.

Lincoln set up shop in the Illinois State House in offices lent to him by the governor, but on issues of policy, he made a point of keeping his thoughts to himself. At the Republican national convention he had put out the word he would make no deals—"I authorize no bargain and will be bound by none"—and he steadfastly refused to associate himself with the liberal, anti-slavery wing of the party. He wanted to appeal to easterners and westerners, to radicals and conservatives, to former Whigs and wavering Democrats. The strategy seemed to work, as the common ground he shared with most Republicans helped consolidate the party behind him.

He hired a new secretary, a bright young journalist name John Nicolay. A reporter from the *Chicago Press and Tribune* arrived, and with Lincoln's cooperation, John Locke Scripps published a thirty-two-page biography. A nation hungry to learn more about Illinois's little-known man bought more than a million copies of the brief life. Hundreds, perhaps even thousands, of biographies and profiles appeared in print, and lithographers Currier & Ives soon found a market for frameable prints. Painters and sculptors arrived to take Lincoln's likeness. Countless images of Lincoln-as-rail-splitter proliferated.[1] The symbolism reverberated just as Oglesby hoped. As one letter to the editor of the *Illinois State Journal* put it, "[Lincoln's] rails, like his political record, are straight, sound and out of good timber."[2]

In June his candidacy got a large boost when the opposition Democrats split. His old nemesis Douglas won the support of the Northern wing of the party with his promise of popular sovereignty. The South chose a Kentuckian named John C. Breckenridge who favored imposing a strict slave code on the territories. A fourth major candidate, Tennessean John Bell, would also be on the ballot, running under the banner of the anti-secessionist Constitutional Union Party, further dividing the

The political mythology of rail-splitter Lincoln has produced countless artistic renditions of Lincoln in his homesteading days.

electorate. But a dark cloud had rolled in, shading all the candidates and parties, Lincoln especially.

The threat of disunion hung over the election. Lincoln's outspoken opposition to the spread of slavery—and the growing chance that he might actually win—meant that leaders in some Southern states had begun to threaten they would withdraw from the Union if he did. That would be a calamity, everyone agreed, but as the preelection months slowly passed, Lincoln continued to insist the slave states would never carry out such a threat.

ELECTION DAY 1860

In Springfield, Lincoln, though growing cautiously optimistic at his chances, was "bored—*bored badly.*"[3] Since he wasn't campaigning, he had little to do but wait and read news reports—until 2:00 A.M. on

Wednesday, November 7, when a telegram arrived carrying the news that New York's voters had put him over the top. "I then felt as I never had before," Lincoln remembered, "the responsibility was upon me."[4]

He won only 40 percent of the popular vote, but his 180 Electoral College votes, out of 303, were more than enough for a solid victory. Breckenridge led the rest of the field with 72. Douglas finished last with a mere 12 electoral votes.

After decades in which slaveholders—the "slave power," the "slavocracy"—controlled national politics, a president who wanted to steer his country away from slavery won the White House. However, because of his anti-slavery stance, Lincoln collected zero votes—literally none—in most of the South. In the nine states south of Virginia his name did not even appear on the ballot.

THE TREMONT TEMPLE RIOT

Three days after the presidential election of 1860, South Carolina began the process of secession. The move to disunion was driven by the prospect that Lincoln would interfere with slavery, which was the very foundation of Southern culture and prosperity. There was much anger in other Southern states, too, but Lincoln still insisted he wasn't particularly worried, assuring a Kentucky friend in a confidential letter that "the good people of the South [will] . . . find no cause to complain of me."[5] As president-elect, he waited quietly in Springfield for his March 4 inauguration day; until then he could do little but instruct his supporters to "Stand firm."[6]

In contrast, Frederick Douglass boarded a train in Rochester, his destination Boston. As engaged in the battle for freedom as ever, he was booked to deliver a lecture in honor of John Brown, who had been executed exactly one year earlier. To Douglass's surprise, he found that the anger evident in the South had infected even staid old Boston.

For most of 1860, Douglass observed the election from a distance. After the Republican convention, he told subscribers of *Douglass' Monthly* that "Mr. Lincoln is a man of unblemished private character . . . one of the most frank, honest men in political life."[7] But for Douglass, that was not enough. Lincoln neither favored giving the vote to Black citizens nor thought them the White man's equal. He wasn't promising abolition in slave states, since his anti-slavery policy ended with the territories, where he swore to prevent slavery's spread. That left Douglass with no choice. "I cannot support Lincoln," he had told Gerrit Smith in July.[8]

Despite being disgruntled that Lincoln didn't take a radical abolitionist position, Douglass recognized what a Lincoln win could mean to his cause. It would be a victory for "anti-slavery sentiment," he wrote in the September issue of his newspaper. It would be a major blow to the slavocracy if the "government [was] divorced from the active support" of slavery.[9] After decades of controlling Washington politics, "The slaveholders know that the day of their power is over when a Republican President is Elected."[10] Now, with Lincoln set to take up residence on Pennsylvania Avenue, the slave states began their exodus.

To Douglass, it smacked of revolutionary change. He did not want war, but he wasn't afraid of that; he worried about compromises that might be made to avoid it. There was talk in Washington and many Northern states of walking back protections for the Black man in order to placate angry Southerners. Douglass felt certain that concessions granted to the South could only undermine his quest for freedom.

The site of the John Brown commemoration was the Baptist Tremont Temple, a converted theater a block from Boston Common. At first, the arriving crowd consisted of mostly Black faces, since the church was racially integrated, as were the local schools (Boston had been the first city in the United States to integrate its classrooms five years before). But as the hall filled, the audience grew more varied. From wealthy Beacon Hill came well-dressed gentlemen, many of whom were merchants and millowners dependent upon cotton from the plantation South. Dock-workers who unloaded cotton bales in the port of Boston founds seats. So did laborers from the city's rough-and-tumble North End, men who competed for jobs with free Blacks. The balance of the crowd had shifted from African Americans and abolitionists to a "mob of gentlemen," some of whom were simply ruffians, hired by the merchants, to cause trouble.[11] They came to condemn—not to celebrate—"John Brown and his aiders and abettors."[12]

The unruly crowd greeted the call to order with loud hissing and catcalls. When they refused to quiet, the moderator stepped from the stage, walked down the aisle, and took a man who had been heckling him by the collar. Before he could drag him to the door, the man's companions separated the two men.

A Black minister called for order, but the crowd shouted him down. Abolitionists responded by offering three cheers for their featured speaker, Frederick Douglass, who was seated on the stage. The disrupters shouted, "Put him out." With the room filling with shouting, mad laughter, and scuffles, the police arrived to quell the growing chaos.

Douglass rose to his feet, asking to be heard. The chair refused to recognize him, and this time it was the abolitionists who screamed. When Douglass tried to yell over the din, someone in the audience screamed, "Throttle him!" In Boston, the nation's intellectual center, a genteel place and a longtime stronghold of abolitionism, the scene was set for a riot.

A resolution was offered condemning Brown and the "fanatics of the Northern States" who had supported him and "even now [are] attempting to subvert the Constitution and the Union."[13] Douglass demanded to be heard, and the police struggled to keep order, breaking up fights in

The normally peaceful Tremont Temple in Boston was the scene of a violent riot when Douglass was invited to speak on December 3, 1860, the one-year anniversary of John Brown's execution.

the crowd. This time the chair permitted Douglass to speak, but warned him to be brief.

He bellowed over the jeers and shouts, quoting the Bible. "The freedom of all mankind was written on the law by the finger of God!" He condemned the proceedings, yelling, "You are serving the slaveholders!"

"Put a rope around his neck," someone yelled back.

When a large man in the front threw fresh insults at him, Douglass fired back. "If I was a slave driver, and had hold of that man for five minutes, I would let more daylight through his skin than ever got there before." When the outraged crowd began throwing chairs at the stage, the police, who stood between Douglass and the angry crowd, tried to seize him; when he slipped from their grasp his supporters cheered. "We will not yield our place on the platform! No, by God," he swore, still fighting off policemen intent upon subduing him. One grabbed him by the hair; several dragged him to the edge of the stage and threw him down the stairs. He jumped to his feet and ran, pursued by some of the rabble.

Douglass was finally gone. But before a vote could be taken on any resolutions, he was back. His hair disheveled and clothes torn, he walked with dignity down the aisle. This time no one put a hand on him. He mounted the stage, with every eye fixed on him. The incandescently angry man resumed his seat without saying a word. On orders from the police, the chairman declared the meeting over.[14]

With passions running high from Boston to Baton Rouge, Mr. Lincoln's war was about to begin. It would be a war that jeopardized the basic beliefs of the nation as stated in the Constitution, and the president would be at the center of the conflict, always seeking strategies to bind the nation together. As an outsider, as an agent for change, Frederick Douglass would play a very different—and yet essential—part in determining the outcome.

NINE

MR. LINCOLN'S WAR

I have no purpose, directly or indirectly, to interfere with the institution of slavery in the States where it exists.

—Abraham Lincoln, First Inaugural Address, March 4, 1861

Chilled by the cold drizzle, the crowd of more than a thousand admirers waited at Springfield's Great Western Railroad depot. Just before 8:00 A.M. on the morning of February 11, 1861, the throng of friends and neighbors got what they were waiting for: an opportunity to bid Illinois's favorite son farewell before he traveled east to be sworn in as president of the United States.

When President-Elect Abraham Lincoln emerged from the station, the crowd parted as he walked toward the train. He shook many hands and offered greetings to familiar faces. He climbed aboard the last car and, standing on its rear platform, turned to face his well-wishers, solemnly removing his tall beaver hat. At first, he said nothing, honoring the silence. His bodyguard, the six-foot-four, barrel-chested Ward Hill

Lamon, remembered that the seconds that ticked by "were as full of melancholy eloquence as any words he could have uttered."[1]

Lincoln had prepared no words for this moment, but memory and emotion provided them.

> My Friends,
>
> No one not in my situation can appreciate my feeling of sadness at this parting. To this place, and the kindness of these people, I owe everything. Here I have lived a quarter of a century, and have passed from a young to an old man. Here my children have been born, and one is buried. I now leave, not knowing when or whether ever I may return, with a task before me greater than that which rested upon Washington. Without the assistance of that Divine Being who ever attended him I cannot succeed. With that assistance I cannot fail. Trusting in Him, who can go with me and remain with you, and be everywhere for good, let us confidently hope that all will yet be well. To His care commending you, as I hope in your prayers you will commend me, I bid you an affectionate farewell.[2]

As scheduled, at eight o'clock sharp, the train departed for his journey to Washington, D.C.

In the preceding weeks, Lincoln had endured what some called the "secession winter." As South Carolina's convention debated whether to secede, he considered how he could respond if states actually left the Union. He read a book about the 1832 nullification crisis, during which President Jackson angrily warned South Carolina he could at any time order an army into the Palmetto State to enforce the federal law the state threatened to defy. Finding no easy answer in Jackson's threats, the new president also focused on his cabinet appointments. As a relatively inexperienced politician, he chose to nominate previous political rivals for their experience. He made a key choice in December, selecting William Seward to become his secretary of state. However, by the time the New Yorker accepted Lincoln's offer shortly after Christmas, South Carolina had withdrawn from the Union. There was no denying that the nation was coming apart.

Yet, until his inauguration, scheduled for March 4, 1861, Lincoln lacked legal standing to act. He could and did attempt to exercise a quiet persuasion, writing to Alexander H. Stephens, an influential voice in the South and a friend from his time in Congress. Stephens went on the record after Lincoln's election as opposing secession, and the president wrote to ask the diminutive Georgian to pass the word that he had no plans to interfere with the internal workings of the Southern states. In a letter he marked as "for your eye only," Lincoln wrote, "Do the people of the South really entertain fears that a Republican administration would, *directly*, or *indirectly*, interfere with their slaves?" Not so, he promised Stephens. "I wish to assure you, as once a friend, and still, I hope not an enemy, that there is no cause for such fears."[3] Unfortunately for Lincoln, his assurances had little or no impact, and in the new year six states—Mississippi, Florida, Alabama, Louisiana, Texas, and Georgia—followed South Carolina's lead, seceding in a period of four weeks. Adding insult to injury, on the very day Lincoln departed Springfield, his erstwhile friend Stephens took the oath of office as provisional vice president of the newly established Confederate States of America.

Above all else, Lincoln wished to preserve the Union. Many years before, as a self-tutored youth looking for an anchor in his life, he embraced the law and in particular the principles laid out in the Constitution. More recently, in the wake of the wrongheaded interpretation offered by Roger Taney and others, Lincoln retrenched, placing his faith in the simpler fabric of the Declaration of Independence with its promise of liberty for all. That premise—*liberty for all*—became his benchmark, his ultimate point of reference in the months and years to come.

GLIMPSES OF LINCOLN

Frederick Douglass and Abraham Lincoln had yet to meet. Nor would they come face-to-face on February 18, 1861, although Douglass did lay distant eyes on Lincoln early that Monday morning, when the Presidential Special hissed to a halt in Rochester, New York.

The train ran late so Lincoln's planned speech from the balcony of the Waverly Hotel had to be scrapped. Instead, he again stood on the rear platform at the end of the four-car train. His family rode in a saloon car, which was accompanied by a passenger car for dignitaries and the press, and one for baggage.

"The train only stopped a few minutes," Douglass reported in the next issue of *Douglass' Monthly*, "but Mr. Lincoln had to make a speech to the assembled thousands who had come to greet him." Douglass found what Lincoln said disappointing. "His short speech here," he told his readers, "did not touch on the great question of the day."[4]

For the most part, Lincoln's frequent addresses on the twelve-day trip east would not focus on policy; it was planned to be a getting-to-know-you tour. He would glad-hand political supporters at important stops while, in lesser towns, he showed himself briefly to give the curious the opportunity, as he put it in typically self-deprecating terms, "of observing my very interesting countenance." One youthful Ohio politician who came for a look, future president Rutherford B. Hayes, thought that Lincoln's bow to his public was oddly mechanical. "His chin rises," Hayes wrote to a friend, "his body breaks in two at the hips—there is a bend of the knees at a queer angle."[5]

Lincoln revealed little about his plan for a rapidly dividing nation. He told a Cincinnati audience that "I deem it my duty . . . [to] wait until the last moment, for a development of the present national difficulties, before I express myself decidedly what course I shall pursue."[6] He reassured the citizens of Cleveland that "the crisis, as it is called, is altogether an artificial crisis."[7] But his calm words were contradicted by actions elsewhere when the Confederate States of America declared its existence in its new capital of Montgomery, Alabama. In the eyes of Southerners, one nation officially became two with the inauguration of Jefferson Davis on February 18. Even if the CSA had yet to gain formal recognition as anything more than an illegal rebel conclave, the dilemma Lincoln faced grew by the day.

A light note was struck at what became the best-remembered stop on the trip. In the weeks before the election, Lincoln had received a letter from a girl named Grace Bedell. "If you let your whiskers grow," the eleven-year-old advised, ". . . you would look a great deal better for your face is so thin. All the ladies like whiskers and they would tease their husbands to vote for you and then you would be president." Lincoln wrote back, offering no promises but musing that people might think a beard "a silly affectation."[8] Yet by the time his train paused four months later in Westfield, New York, Miss Bedell's hometown, a Quaker's beard lined Lincoln's chin. To the delight of onlookers, he invited young Miss Bedell to join him on the platform. He planted an appreciative kiss on the young girl's head, and many of the newspapermen accompanying him had a field day. "WHISKERS WIN WINSOME MISS" read one headline; another, "OLD ABE KISSES PRETTY GIRL."[9]

PINKERTON'S PLOT

A few days later, a man named Allan Pinkerton brought much darker news. Lincoln had been getting hate mail for months and, more recently, there had been nonstop rumors that he was in physical danger. He refused to take any of them seriously, but this new warning came from a well-regarded Chicago detective agency that specialized in railroad security. At a meeting behind closed doors in his Philadelphia hotel, said Lincoln, "Pinkerton informed me, that a plan had been laid for my assassination."[10]

Pinkerton explained that in Baltimore, a pro-South city in a slave state, secessionist sentiments were at a boil. The presidential entourage was supposed to arrive in the Maryland city, change trains, and then proceed to Washington. With Lincoln's travel schedule public knowledge, a group of conspirators, according to Pinkerton, intended "to create a mob of the most excitable elements of society" at Baltimore's Camden Street Station. In the confusion, hired assassins would "rush forward, shoot or stab the President elect" and then "fly back to the shelter of the rioters."

Pinkerton urged Lincoln to proceed immediately—and secretly—to Washington. Still unwilling to believe the murder plot was real, Lincoln refused Pinkerton's plea to cancel the rest of his appearances. But the warning shook him, and his speech on Washington's birthday, at the very place where the Declaration of Independence had been signed, was almost a confession.

"I have never had a feeling politically that did not spring from the sentiments embodied in the Declaration," Lincoln said, and the crowd at Independence Hall erupted in cheers. He promised to try to save the country to protect liberty, the principle he honored above all others. Then, perhaps for the first time, he admitted to a shadow of doubt. "But, if this country cannot be saved without giving up that principle—I was about to say I would rather be assassinated on this spot than to surrender it." His strong words met with applause, though few knew what was really on Lincoln's mind.[11]

A second warning of the danger to Lincoln arrived when Frederick Seward, son of the soon-to-be secretary of state, delivered a report of the plot to kill the president; it came from Winfield Scott, the commanding general of the U.S. Army. This time, Lincoln agreed to cut short his tour, and Pinkerton quickly put in place a scheme for Lincoln's safe passage.

After delivering a speech that evening in Harrisburg, Lincoln ate a quick supper. Leaving all but the burly Lamon behind, the guest of honor then slipped out, exiting through the side door of his hotel, an overcoat draped over his shoulders to disguise his narrow physique. In place of his widely recognized stovepipe hat, Lincoln drew a soft woolen hat over his head. He and Lamon then stepped into a waiting carriage, which carried them to a special Pennsylvania Railroad train.

The party became three with the addition of Allan Pinkerton. By midnight the train for Baltimore—this time with Lincoln hidden behind the curtains of a sleeping berth that one of Pinkerton's agents had reserved purportedly for an invalid—was racing southward. At 3:30 A.M. it arrived in slumbering Baltimore, where the sleeper car was pulled by a team of horses to Camden Street Station, from which southbound trains departed. By sunrise, the unfinished U.S. Capitol, its dome half-obscured by scaffolding and cranes, came into view.

Lincoln arrived in Washington sleepless, but safe, in time for breakfast at the Willard Hotel. He was joined in the afternoon by Mary and their

This photograph of the unfinished Capitol—Lincoln made the completion of the dome an objective during his first term—was taken the day of the sixteenth president's inauguration.

Opposition Democratic newspapers satirized Lincoln's furtive trip to Washington after he was warned of a possible assassination plot. In this cartoon he is portrayed running fearfully for his life.

sons, who, still aboard the Presidential Special, had passed through Baltimore without incident.

When word got out, Lincoln's enemies were quick to mock his midnight escape. A prominent national magazine ran a scathing political cartoon, lampooning him as a coward, running for his life, with the heading "The Flight of Abraham."[12] But Frederick Douglass understood why Lincoln acted as he had. As he saw it, Lincoln had run a dangerous gauntlet, very much like "the poor, hunted fugitive slave [who] reaches the North, in disguise, . . . crawling and dodging under the sable wing of night.[13] Douglass could not help but look back to 1838 when, as Fred Bailey, his younger self made his own dangerous escape.

INAUGURAL BLUES

A week later, soldiers were everywhere. On inauguration day, March 4, 1861, sharpshooters scanned the crowd from the rooftops. Infantrymen patrolled Pennsylvania Avenue. Companies of cavalry blocked the side streets, and there was even a battery of howitzers on a nearby hill, just in case. The air was alive with fears not only for Lincoln's safety but of impending war.

An open carriage and military escort arrived at noon to ferry Lincoln from his hotel to the Capitol. Once there he made his way to a large platform constructed on the east portico. He removed his new silk hat, preparing to address the crowd, but after a moment of confusion—*where would he put it?*—Stephen Douglas, one of the several hundred dignitaries arrayed behind him, stepped forward. "Permit me, sir," he said, taking the hat. Lincoln put on his steel-rimmed spectacles and began to read. Twenty-five thousand listeners were spread before him on the Capitol grounds.

His address attempted to heal the nation's wounds, but it fell largely on deaf ears. In faraway Rochester, New York, Frederick Douglass in particular saw Lincoln's words as nothing short of an abject betrayal. He had wanted from Lincoln an explicit damnation of slavery, but when Douglass read the speech, the new president's promises fell far short of that.

Instead of denouncing secessionists, Lincoln was cool and reasonable. "I have no purpose," he explained, "directly or indirectly, to interfere with the institution of slavery in the States where it exists." He appealed to Unionists in the South, promising to uphold existing laws, including the Fugitive Slave Act. He did take a hard line regarding secession, calling it unlawful, but added, "there needs to be no bloodshed or violence; and there shall be none, unless it be forced upon the national authority." Above all, he saw defending and maintaining the Union as his duty, but he would not order the first gun fired. He wanted "harmony," not "anarchy."

The tightly crafted ending to his speech was a poetic plea:

> We must not be enemies. Though passion may have strained, it must not break our bonds of affection. The mystic chords of

> memory, stretching from every battle-field, and patriot grave, to every living heart and hearthstone, all over this broad land, will yet swell the chorus of the Union, when again touched, as surely they will be, by the better angels of our nature.[14]

With the speech at an end, Chief Justice Roger Taney, withered, bent, and nearly as tall as Lincoln, administered the oath of office to the sixteenth president. Afterward the crowd cheered and cannons were fired in tribute. But in the days to come, none of the seven states that seceded accepted Lincoln's olive branch. John Bell, one of Lincoln's opponents in the election, thought Lincoln's speech amounted to a declaration of war. President Buchanan was uncertain what to think. "I cannot understand the secret meaning of the document," he told one reporter. Some observers suspected they knew why Buchanan seemed confused; he appeared to them to have fallen asleep.

FREDERICK'S FRUSTRATION

After Frederick Douglass read the published version of Lincoln's address, he penned a screed that was one and a half times its length. The speech was pure hypocrisy, he concluded, a "double-tongued document." The angry and disappointed Douglass condemned Lincoln: His willingness to enforce the Fugitive Slave Act made him nothing less than "an excellent slave hound," said Douglass, adding in disgust that Lincoln thought "the poor bondman should [be] returned to the hell of slavery."[15] Douglass was furious that Lincoln had said he would sign a proposed amendment to the Constitution, then under consideration in Congress, which would guarantee slavery in the Southern states *forever.* In Douglass's eyes, Lincoln's "better angels," so oblivious to the freedom of the Black man, were poor angels indeed. Lincoln's words shattered Douglass's hopes that this man would end slavery.

Douglass's disappointment was so great he considered leaving his country. No fan of colonization, he was on record as strongly opposed to sending his people back to Africa. But abolitionist friends had begun advocating for African American emigration to the Republic of Haiti, the largely Black island nation that had gained independence from France. In the next issue of his newspaper, he reported having booked tickets to travel there, departing New Haven, together with his eldest child, daughter Rosetta. The Haitian consulate invited him, and he told his readers that, as a journalist, he wanted to "do justice to Haiti, to paint her as she is."[16]

The first guns of war, fired on April 12, would change his—and Mr. Lincoln's—plans.

THE FALL OF FORT SUMTER

In his first hours as president, Abraham Lincoln found a ticking time bomb on his desk. It took the form of a dispatch from the commander of Fort Sumter. U.S. Army major Robert Anderson reported that food and other supplies were running low. Unless provisions arrived soon, his small garrison would have no choice but to surrender to Confederate forces. In the time it took Lincoln to read Anderson's words, the immense national challenge of the seceding South shrank to the fate of a tiny, two-acre island in Charleston Harbor.

Lincoln asked the advice of the men he'd selected for his cabinet. At first, most of them, including Seward, saw no alternative to withdrawal from the fort. The head military man, General Winfield Scott, concurred. But Lincoln refused to give up Fort Sumter so easily, and he dispatched his friend Ward Lamon to Charleston to talk with South Carolina governor Francis Pickens and Major Anderson. As days became weeks, the president heard voices pro and con from Congress—*You must do something!* said some, while others warned, *Anything you do could start war!*—but he ordered preliminary preparations be made to ship supplies to Anderson. Finally, on April 4, he issued the order that humanitarian aid of food and medicine be sent to Fort Sumter. It was to be done "peaceably," Lincoln ordered, if at all possible.

A lifelong insomniac, Lincoln stayed awake nights worrying about much more than Sumter. Holding the fort was a point of principle, since he had sworn in his inaugural speech to "hold, occupy, and possess" federal property, but he devoted a great deal of thought to the much larger canvas of slave country. His hope had been that Unionists in the South would persuade hotheaded "fire-eaters," the angry men who wanted war, to rethink secession and voluntarily rejoin the Union. That now seemed laughably unlikely—and giving up Sumter would amount to both capitulation and recognition of the Confederacy.

Some newspapers began to question the new president's leadership: "HAVE WE A GOVERNMENT?" screamed one headline.[17] But Lincoln knew one thing for certain: He wanted to avoid firing the first shot. He did not want to be the aggressor, which might provoke more states to leave the Union. In particular, he worried about Virginia, just across the Potomac, where the state's politicians waited and watched.

On Friday morning, April 12, 1861, Confederate president Jefferson Davis brought the stalemate to a crashing conclusion. On Davis's order, Confederate cannons boomed in Charleston at 4:30 A.M., firing the first of some four thousand shots and shells from what seemed like all directions. With first light, Fort Sumter's artillerists began to return fire, but Major Anderson ordered his gunners to fire sparingly to conserve limited supplies of powder and cartridges.

Confederate cannonballs preheated in a furnace ignited fires in

Before becoming president of the Confederate States of America, Jefferson Davis served honorably as a soldier, secretary of war, and, at the time this picture was taken just before the Civil War, U.S. senator from Mississippi.

Fort Sumter's wooden barracks. The projectiles traced soaring arcs in the sky, remaining in the air for thirty seconds and more as spotters shouted warnings of incoming shot, shells, or mortars.[18] Remarkably, no one was killed during the artillery exchange, but fighters on both sides knew what the result of the assault would be.

The Union men managed to hold on until the next day, when Anderson negotiated terms. His men marched out of the fort with heads high and saluted the Stars and Stripes as the flag was lowered. They would not be prisoners of war, instead boarding a civilian ship bound for New York. But Fort Sumter was lost, surrendered to an enemy in the first battle of what could now only be called civil war.

In Washington, Lincoln was far from shocked; in a way, he had engineered this fight. The surprise was that the nation immediately rose from the sleepy limbo in which it had been suspended for months. When Lincoln acted quickly to issue a proclamation, calling upon the remaining

Union states to muster seventy-five thousand militiamen to put down what he called the "insurrection," men raced to volunteer. The Northern states unified in a way that had seemed impossible days before. Stephen Douglas, addressing a huge rally in Chicago, defined the new stakes. "There are only two sides to the question. Every man must be for the United States or against it. There can be no neutrals in this, *only patriots—or traitors.*"[19] No one in the North seemed to disagree.

The South reacted differently: They believed Union troops were being assembled to suppress their rebellion. The response of North Carolina's chief executive was typical. "I can be no party to this wicked violation of the laws of the country," said Governor John Ellis, "and to this war upon the liberties of free people."[20] In a matter of weeks, his state, along with Virginia, Tennessee, and Arkansas, joined the Confederacy. Mr. Lincoln's play for peace had failed spectacularly. He truly had a war on his hands.

REACTION IN ROCHESTER

"God be Praised!" wrote Frederick Douglass in response to the news. "As a friend of freedom, . . . we have no tears to shed, no lamentations to make over the fall of Fort Sumter." His worry that Lincoln would compromise on slavery to keep the peace went up in the billowing gun smoke over Charleston. "The slaveholders themselves," Douglass continued, "have saved our cause from ruin!"[21]

He canceled his trip to Haiti. "The last ten days have made a tremendous revolution in all things pertaining to the possible future of colored people in the United States," he told readers of *Douglass' Monthly.* "This is not the time to leave the country."[22]

Douglass saw with his own eyes his countrymen rising to defend the Union. On May 3, eight companies of militia volunteers boarded trains in Rochester. "[Their] departure was a thrilling spectacle," he noted in his paper, yet also a sad one. He saw the tears of wives wishing their husbands farewell and heard the "mournful sobs of mothers, as they part from their sons."[23] The safe return of their men was far from certain.

As one of the twenty thousand who crowded State Street to cheer the departing soldiers-to-be, Douglass was seized by another powerful emotion. No person of color—not a single one—departed Rochester that day to fight for liberty and for the freedom of his people in particular. This was not out of fear or a reluctance to join the ranks; dark-skinned men, even in the North, were simply not welcome to go to war shoulder to shoulder with Whites.

Douglass never doubted he and his brethren could fight. He claimed that "One black regiment alone would be . . . the full equal of two white ones." When asked what his people were prepared to do "in the present solemn crisis," Douglass could only express his frustration. "Would to God you would let us do something!"[24]

Douglass himself would do everything in his power to persuade the government to let freedmen and fugitives fight to liberate their brothers, an unwinnable fight in the spring of 1861. Though he failed initially, his words would prove prophetic. "If this conflict shall expand to the grand dimensions which events seem to indicate, the iron arm of the black man may be called into service."[25]

CONTRABANDS OF WAR

A month after the fall of Fort Sumter, three men in a boat approached another Union fort. They were fugitive bondsmen seeking sanctuary, and after rowing across Norfolk harbor on the night of May 23, 1861, the unarmed Black men presented themselves to the sentries at the five-sided federal stronghold called Fortress Monroe.

A crude rendering on a Civil War–era envelope that pictured General Benjamin Butler, on horseback, meeting up with Black sappers and miners volunteering to work for the Union.

They were taken to the fort's commanding officer, Major General Benjamin Butler, a large and pugnacious man with a walrus mustache. He had only just arrived to take charge of the strategically placed citadel, which overlooked the channel linking the James River and the Chesapeake Bay, but General Butler already knew he needed all the help he could get to improve the dilapidated installation.

While the Confederacy had bondsmen to dig ditches, build fortifications, and perform a hundred other support duties, the Union did not. So when Frank Baker, Shepard Mallory, and James Townsend explained they had been working on enemy artillery earthworks but escaped across Confederate lines out of fear of being shipped south—which, to them, might mean permanent separation from their families—Butler put them to work right then and right there. Unexpectedly, his snap decision also put him in the middle of a new and heated debate about enslaved people.

Before the day was out, a messenger arrived under a flag of truce. Acting as agent for the local Confederate commander, the officer claimed that Butler, given his "constitutional obligations," was compelled to return Baker, Mallory, and Townsend as human property under the terms of the Fugitive Slave Act. Having been a lawyer in Boston before Lincoln commissioned him a general, Butler considered the facts of the case. Even in the absence of a clear precedent, Butler saw his duty. He was quick to reject the man's demands.

"Virginia passed an ordinance of secession and claims to be a foreign country," he reminded his visitor. Thus, the Union general explained, "I am under no constitutional obligations to a foreign country, which Virginia now claims to be."

Butler's argument went further. The two sides were, after all, at war, and the men were "being used against the government by its foes." To return them to their owner would aid and abet the enemy. The three runaways were, under the terms of international law as Butler saw it, military goods and could be lawfully seized by a combatant in wartime.

"I shall hold them," he said firmly, "as *contraband of war.*"[26]

That ended the two men's conversation—but opened a much broader one. The term "contraband of war" came immediately into general use after other enslaved persons heard about the "freedom fort" and presented themselves at the gate of Fortress Monroe. Within two months, the contraband count of men, women, and children neared nine hundred at Fortress Monroe alone; eventually there were many thousands at other Union outposts. Contraband camps and contraband schools were created after Congress memorialized the legality of Butler's position with the Confiscation Act, which stated that a slaveholder forfeited his bondsmen if he permitted them to be used in service to the Confederacy.

But the new law had another immediate and important effect, too. It made the old Fugitive Slave Act unenforceable.

Lincoln signed the Confiscation Act into law, though not without misgivings. He found himself making a hard choice. He recognized he must do everything he could to appease the border states of Kentucky, Missouri, Maryland, and Delaware, since he could not afford to lose any more states to the Confederacy. None of the four in question had seceded yet, but each of the states did permit slavery. That meant Lincoln, like a tightrope walker, had to find just the right balance in his slavery policy. He could tip only so far in either direction as he tried to satisfy the free states' anti-slavery demands and to appease the slaveholders in the border states to keep them in the fold. If he had to choose between freeing enslaved people and saving the country, he felt he had no choice but to choose unity.

The strategy was a delicate one, only made more so by the contraband controversy, since the Confiscation Act represented a new stage in putting the powers of the Constitution to use regarding what was clearly the war's central issue. As Frederick Douglass saw it, the whole conversation was just "another compromise, [with] the old virus left to heal over."[27] He was interested less in a cautious balance than in radical change.

The president had other problems to confront, the biggest one the reality of a war about to unfold. In military terms, the bombardment in Charleston Harbor had been a small matter and a total mismatch. The outcome was never really in doubt, with forty-three Confederate artillery pieces firing upon just eighty-seven ill-fed Union soldiers. Now, however, both sides were raising large new armies to confront each other.

The North had certain large advantages. Its population numbered more than twenty million, compared to nine million in the South, where more than a third were enslaved people. The Union's industrial capacity was vastly greater, too, with roughly 90 percent of the nation's factories and double the railroad lines. As a result of this calculus, many a Union man thought the war would be short, perhaps a simple matter of capturing the new Confederate capital at Richmond, Virginia, a hundred miles into Confederate territory. Lincoln's first request for seventy-five thousand volunteers reflected this expectation, with a term of enlistment of just ninety days.

On the other side, Southerners, who believed they were fighting to preserve their honor and their way of life, chose not to see the war in stark David-and-Goliath terms. They had the distinct advantage of fighting a defensive war, and they would be fighting it with some of the best military minds that West Point had produced. Many in the South looked forward to the first land battle of war, confident they would prevail, while many Northerners told one another their army would easily overpower the undermanned South. July 21, 1861, would be the first test.

THE BATTLE BY BULL RUN

On that Sunday, three months after Sumter's fall, on July 21, 1861, the armies faced off in Manassas, Virginia. Goaded by such newspaper headlines as the *Tribune*'s "FORWARD TO RICHMOND! FORWARD TO RICHMOND!," an impatient Lincoln had ordered General Irvin McDowell's army to move south. Although headed for the capital of the Confederacy, the Army of the Potomac would first have to deal with the enemy's line of defense at Manassas Junction. Just twenty-five miles from Washington, it was a railroad hub alongside a tributary to the Occoquan River called Bull Run.

Slowed by a shortage of horses and mules, McDowell's army of thirty thousand men marched into nearby Centreville on July 18. His opponent, General Pierre Gustave Toutant-Beauregard, the man who had been in charge at Fort Sumter, camped nearby with some twenty thousand soldiers. McDowell chose not attack immediately, and Confederate president Jefferson Davis took advantage of the pause, ordering reinforcements. General Joe Johnston's army of ten thousand men, then in the Shenandoah Mountains, clambered into train cars for the long ride to Manassas.

By the time McDowell took the offensive three days later, everyone knew a battle was about to unfold. Half a dozen senators and their wives, newspaper correspondents, and other spectators rode out from Washington in their carriages, arriving with picnic baskets and opera glasses, anticipating a day's entertainment as the Union army won a great victory. Lincoln chose to remain in Washington, attending church that Sunday morning, but on his return to the President's House, he got his first reports of the battle.

Under a clear blue sky, a large Union force moved on the left flank of the Confederate line. As one Virginia soldier reported, "The enemy were thick in the field, and the long lines of blue could not be counted."[28] The spectators cheered the puffs of artillery smoke. The woods lining Bull Run rang with gunfire and the shouts of combatants. Early on, the Rebels were forced to retreat, and Lincoln was informed late in the afternoon that the battle looked like a clear Union victory. A *New York Herald* reporter telegraphed his editor, "I am en route to Washington with details of a great battle. We have carried the day."[29]

Such optimistic reports could not have been more wrong, as Lincoln learned when a despondent Secretary of State Seward arrived at 6:00 P.M. The same man who had repeatedly predicted the war would end in thirty days brought devastating news: McDowell's army was in full retreat. The momentum of the battle had shifted when a Confederate general, rallying his men in the face of Union attack, gestured with his sword at a nearby hilltop. He called to his troops, "Look, men, there is Jackson standing like a stone wall! Let us determine to die here, and we will conquer! Follow me!" The man on the hill—who at that moment gained the nickname "Stonewall"—was Thomas J. Jackson. He and his

Lincoln's Republican rival for the presidency and later his secretary of state, William Henry Seward. As one contemporary said of him, he "had a head like a wise macaw."

men bought the Rebels time, holding their position until late afternoon. By then the trains carrying Joe Johnston's fresh troops arrived, and their fierce counterattack shifted the momentum of battle.

Lincoln lay awake all night at the White House, prone on a couch in the Cabinet Room, listening to a series of firsthand accounts of the battle. The Union army had cut and run, he was told. The scene had been utter chaos as soldiers and civilians retreated in disorder and humiliation to Washington, terrified the enemy might be in hot pursuit.

The follow-up attack never came, but at dawn Lincoln looked down on Pennsylvania Avenue. From his window he saw stragglers stumbling back from the battle in a driving rain, some of them collapsing and being tended to by medical staff. The Union had lost its first big battle, and Lincoln suspected perhaps for the first time that he faced a long, hard-fought war.

TEN

WAR IN THE WEST

> Fire must be met with water, darkness with light, and war for the destruction of liberty must be met with war for the destruction of slavery.
>
> —Frederick Douglass, *Douglass' Monthly*, May 1861

Eight hundred miles west, the Civil War was playing out in miniature. More than half of Missouri's White citizens were pro-Union but, as a slave state settled by Southerners, it was also home to powerful and militant secessionists. Its pro-slavery governor confronted a legislature dominated by Unionists. Bloody guerrilla fighting broke out across the state, and a worried President Lincoln dispatched Major General John C. Frémont to keep the peace. He wanted the famous and charismatic general to keep Missouri in the Union—but he underestimated the size of Frémont's ego.

Frémont was famous, known as "the Pathfinder" after his heroic expeditions with the United States Topographical Corps. People admired him as if, in making his maps, he had lassoed the great beauty of the boundless Far West. He had been acting governor and one of California's first senators before becoming the Republican nominee in the three-man 1856 presidential election. Having resumed his army service after Fort Sumter, he was rich, well connected, and a man John Hay, one of Lincoln's secretaries, extolled as "upright, brave, generous, enterprising, learned and eminently practical."[1] He also held strong anti-slavery convictions.

On the very day of the Manassas debacle, Frémont boarded a train headed for St. Louis to take charge of the war's western theater. There, far from Washington, his job would be to create an army out of almost nothing, with little help from the distant federal government, and to figure out how to keep the Mississippi River open to trade. Lincoln personally told Frémont he had "carte blanche," trusting him to "use your own judgement."[2] The general soon made one visionary call when he put a former army captain of dubious reputation in charge of Missouri's Southeastern District, and Ulysses S. Grant thus began his rapid rise. But another Frémont decision seriously undermined Lincoln's border-state strategy.

In St. Louis, a town with strong Southern sympathies, Frémont was widely despised as a Yankee whose troops more or less occupied their city. It didn't help that his wife, Jessie Benton Frémont, daughter of Missouri's late senator Thomas Hart Benton, chafed at the social limitations imposed on women; "she had a man's power, a man's education, and she did a man's work in the world," remarked one friend.[3]

As instructed, Frémont attempted to consolidate Union control in Missouri. He took over railroads, converted ferry boats to gun boats, and trained recruits. But Confederates had swept into southern Missouri. Insurgents from Arkansas and Tennessee stole horses, burned bridges, cut telegraph lines, and attacked Unionist settlements. They recruited some forty thousand Missourians, too, some of whom, after taking part in guerrilla activities, would quietly return to their farms. Missouri appeared to be on the verge of falling into Confederate hands.

General John C. Frémont—famous as the "The Pathfinder" for his mapping expeditions in California and the West—and his formidable wife, Jessie Benton Frémont, daughter of Senator Thomas Hart Benton.

On August 30, 1861, Frémont declared martial law across the state. Set into type, his proclamation was nailed to posts and trees. Its harsh terms ordered that guerrilla fighters arrested behind Union lines be tried by court-martial; if found guilty, they would be executed. But Frémont didn't stop there. The property of Confederate sympathizers would "be confiscated," stated the order, "and their slaves, if any they have, are hereby declared free men."[4]

In addressing slavery, an institution he despised, Frémont went a long step further than General Benjamin Butler. In an unprecedented move, one he made without consulting Lincoln, the Pathfinder had issued an emancipation proclamation.

A NOBLE DEED

Many Republicans back east enthusiastically embraced Frémont's decree. Gerrit Smith called it "a noble deed of a noble man." There were torchlight parades in Northern cities in celebration. Newspapers endorsed the proclamation, among them the *New York Evening Post*, which told its readers, "Mr. Frémont has done what the Government ought to have done from the beginning."[5]

To Harriet Beecher Stowe, Frémont's gesture had the aura of revelation: "The hour has come," she wrote, "and [so has] the man."[6] As far as she was concerned, the hero Frémont had earned another star to pin on his chest.

Frederick Douglass was more practical. He wanted to believe Frémont's order was "the hinge, the pivot upon which the war was to turn." But having been disappointed before, he also understood that the news from St. Louis might be too good to be true. He admitted to readers of his newspaper that he experienced "the deepest anxiety" as he waited and wondered "whether that remarkable and startling document was the utterance of the Major-General, or that of the Cabinet at Washington [and] whether . . . the President would approve it or condemn it."[7]

He did not have to wait long.

TIMING IS EVERYTHING

Like so many others, Lincoln learned of Frémont's martial law declaration from a newspaper. He could hardly believe his eyes: His carefully calibrated border-state strategy, based on the promise not to interfere with slavery where it legally existed, was suddenly worth less than the paper he held in his hand. Sent west to put out a fire, Frémont had fueled it.

Lincoln rarely revealed his emotions in public and, though deeply angry, he kept his temper. Writing confidentially in what he called "a spirit of caution and not of censure," he told Frémont that the promise to liberate the enslaved would "alarm our southern Union friends"—and *that* he did not wish to do. Frémont's policy, Lincoln pointed out, could lead not only to Missouri's secession but might also "ruin our rather fair prospect" for keeping Kentucky in the Union. The president asked Frémont—but did not order him—to amend his proclamation to align with the terms of the recently passed Confiscation Act. In short, Lincoln wanted Frémont's promise of emancipation to go away.[8]

Despite the reasonable tone of Lincoln's letter, Frémont took offense. As a man steeped in the ways of the military, he read it as a reprimand. He sat down and wrote a firm response, in which the self-satisfied general assured his commander in chief that he remained quite certain of his own judgment. His proclamation, he assured Lincoln, was "right and necessary." He had decided to buck the chain of command. He wasn't going to admit he was wrong. Nor would he comply with Lincoln's soft order to amend the proclamation. Unless the president publicly directed him to modify his proclamation, Frémont would make no "correction."[9]

This second surprise from Frémont's pen came to Lincoln's attention on September 10, 1861, when Jessie Benton Frémont arrived at the White House. At 10:00 P.M., after spending more than fifty hours on a train from St. Louis, Mrs. Frémont waited for Lincoln in the Red Parlor. Her dress was smudged with coal soot, but, tired as she was, she felt at home, having known nine previous occupants of the White House, extending back to Andrew Jackson. Decades before, Old Hickory had played with her hair as he talked politics with her father, Senator Thomas Hart Benton, for decades one of the most powerful politicians in Washington.

To Mrs. Frémont, the wait seemed like forever. Then, when Lincoln finally did appear, he took the sealed letter but did not meet her eye.

This was not the welcome Mrs. Frémont expected. Nor did Lincoln offer her a seat as he stood beneath the chandelier reading her husband's pointed rejection of his thinking. He folded and pocketed the letter before he spoke.

"I have written to the General and he knows what I want done," he said coldly.

His unhappiness with the situation was painfully obvious. Not only had Frémont refused to do as he was told, he had sent his wife to deliver the message. This meeting was virtually unprecedented: Women in mid-nineteenth-century America played no direct role in government, and they certainly didn't seek meetings with presidents. But Jessie Frémont was not a conventional woman. She was her father's daughter, confident and almost fearless, and, like her husband, she possessed a cocky certainty that he was right.

Despite Lincoln's obvious displeasure, she launched into a detailed justification of her husband's action, explaining to the president the workings of international politics. The nation needed England, France, and Spain, she argued, and in order to win their support, emancipation was necessary. Her foreign policy knowledge was considerable, her presentation clear and impassioned.

But Lincoln wasn't going to be lectured to. "I had to exercise all the awkward tact I have to avoid quarrelling with her," he told John Hay. Before she could complete her case, the frowning Lincoln, his patience at an end, interrupted.[10]

"You are quite a female politician," he said dismissively.

Although his words signaled their conversation was at an end, Lincoln did offer Jessie Frémont a few parting words. Her husband should have sought consultation before issuing his edict, he scolded. Furthermore, he said, "The General should never have dragged the Negro into the war. It is a war for a great national object and the Negro has nothing to do with it." For once, Lincoln put the matter in bold terms: Whatever his heart told him about the evils of slavery, his brain, ever pragmatic, put saving the country first.

The next day a letter bearing Lincoln's signature went west, ordering General Frémont to rescind the emancipation provision in his proclamation. As Frémont demanded, Lincoln took responsibility, releasing a copy of his letter to the newspapers. The notion of an emancipation proclamation was off the table, but the subject was far from closed.

DISCOURAGED, NOT DEFEATED

For Frederick Douglass and his allies, the joy they shared at Frémont's declaration soured after Lincoln made public his order gutting the proclamation. Douglass saw the retraction as the work of a "crafty lawyer." He wrote that he still wanted to believe that Lincoln's reputation for being "honest and humane" was deserved, but what he actually saw was a willingness to pacify slaveholders. He did not for a moment accept the border-state argument. Missouri, Kentucky, Delaware, and Maryland were the "mill-stone about the neck of the government," he wrote, "and their so-called loyalty has been the very best shield to the treason of the cotton States."[11] Lincoln's rejection of Frémont's policy was appeasement, plain and simple.

Although Douglass wanted to appeal to the president directly, he had no channel to the White House; his best option was his *Douglass' Monthly.* In an essay he titled "Fighting Rebels with Only One Hand," he asked, "Why does the Government reject the Negro? Is he not a man? Can he not wield a sword, fire a gun, march and countermarch, and obey orders like any other?" Douglass was convinced that not only could a Black man fight as well as a White one, but, as he wrote that September, "in a war such as this, [one Black regiment] . . . would be worth to the government more than two of any other." The refusal to arm Blacks was foolish. "Men in earnest don't fight with one hand, when they might fight with two, and a man drowning would not refuse to be saved even by a colored hand."[12]

Douglass's words may have reached Lincoln, but his harsh criticisms of Lincoln's border-state policy certainly angered a great many people. When he arrived that November to lecture in Syracuse, New York, he saw advertisements promoting his speeches—but side by side were handbills that called him "Thief!," "Rescal," "Traitor!!!"—and worse. It invited all comers to "give him a warm reception at this time for his insolence, as he deserves."[13]

Yet his appearance on Thursday evening, November 14, would not be a repeat of the Tremont Temple riot the previous year. As Douglass told the story, "even the rattlesnake sounds an alarm . . . and in this instance the Mayor of the city, the police authorities and citizens took the alarm and provided amply for the attack."

The owner of the theater rejected calls from his neighbors to the close the doors to his hall. When he was reminded that "Frederick Douglass was a Negro," he replied that his "principles of freedom applied to humanity not to color." The mayor reached into the city coffers and came up with three hundred dollars to pay for regular police and a special fifty-man security force.

Additional recruits were summoned. They marched into Syracuse, where they guarded the theater entrance, bayonets fixed on their rifles. When Douglass arrived, the mayor locked arms with Douglass, endangering his own safety to shield his guest from mob violence.

Once at the podium, Douglass looked out over a packed house, and spoke freely of Lincoln, Frémont, the war, and emancipation. His reward was a standing ovation and a sense on leaving Syracuse that he was far from alone. As the tumultuous year 1861 came to an end, Douglass could only hope that his voice would carry to faraway Washington.

ELEVEN

TO PROCLAIM OR NOT TO PROCLAIM

I may walk slowly but I don't walk backward.

—Abraham Lincoln

In the early months of 1862, reports of real military progress reached Washington. Ulysses S. Grant captured Forts Henry and Donelson in Tennessee. Newspapers dubbed him "Unconditional Surrender Grant," playfully adapting the hero's initials, and President Lincoln promoted him to major general. Unfortunately, while Grant demonstrated he was a man willing to fight, George B. McClellan, the commanding general of the entire army, seemed to lack that desire. To his frustration, the president saw much planning and preparation but little action in the eastern theater. Adding to his burden was a family tragedy.

On February 20, 1862, eleven-year-old Willie, the son said to be most like his father, lost a battle with typhoid. "My boy is gone—he is actually gone," Lincoln told one of his young secretaries before burying his face in his hands. "He was too good for this earth. God has called him home."[1]

The grieving president then nursed and consoled his youngest boy, Tad, age eight, still ill with the disease, and coped with his wife's devastating grief. Mary took to her bed, unable even to attend her son's funeral, which left Washington tongues wagging about her mental stability. Dressed in widow's weeds, she consulted with spiritualists in hopes of communing with Willie.

Despite the other pressures, Lincoln remained attentive to the ebb and flow of public opinion regarding slavery. He had long foreseen an eventual end to enslavement, but he sensed growing pressure in the country to do something sooner rather than later. An organization called the Emancipation League had been formed the previous November by Boston abolitionists. And in late February, at New York's Cooper Union, Frederick Douglass called for the destruction of the slave system. His well-publicized and persuasive address to the league met with "most hearty and enthusiastic applause."[2]

Mary Lincoln during her husband's first term, all dressed up as if for a ball.

Hoping to make at least a little progress, Lincoln looked for a legal framework. He understood that most slaveholders had more invested in their bondsmen than they did in their land. That meant, he reasoned, they needed to be compensated if they were to give up their bondsmen willingly. In a close rereading of the Constitution, he focused on the founding document's promise of "just compensation" for private property appropriated by the federal government.

Thinking the principle might apply to slavery, he chose Delaware as a test case. It was a slave state but one where less than 1 percent of Whites were slaveholders. When he proposed to several influential Delaware men that the federal government might pay four hundred dollars per person in return for emancipation, they expressed a willingness to move the idea forward. Lincoln drafted sample legislation for their consideration, but when once the discussion became public, vocal opposition immediately arose. The biggest objection was that emancipation could lead to equality—and few Whites, north or south, were ready to acknowledge racial equality in 1862. The bill was never formally introduced.

Lincoln also began talking, once again, about colonization. He mentioned it in his annual message, the precursor of today's State of the Union address, urging Congress to appropriate funds for establishing contraband colonies. He called for recognition of Haiti and Liberia, possible destinations for American freedmen. But these ideas, too, met with more opposition than enthusiasm, since even his own Republican Party remained deeply divided on almost every matter concerning the Black man.

On the other hand, the radical wing of the party wasn't about to let the matter drop. When pressed in one meeting, Lincoln shifted the conversation as he often did by telling a story. He told his guests that he first heard the yarn as a boy.

One spring, a group of Methodist preachers made their way across

Illinois. Two of the parsons discussed how, with its waters running high, they could get across a fast-moving river some distance ahead. The two talked and talked, and the more they talked, the more heated their words became. When their argument turned into a quarrel, a third member of their company stepped in.

"Brethren," said the older man, "this here talk ain't no use. I never cross a river until I come to it."[3]

Lincoln the raconteur left his visitors laughing, releasing the tension. Yet the moment also cleverly conveyed to them his attitude toward emancipation: *Not yet, gentlemen. That decision lies in our future.*

SLAVERY IN THE CAPITAL

From a distance, Frederick Douglass observed the president. Some of what he saw, he liked; just as often, Lincoln's words and actions made him angry. But in April 1862, Lincoln gave him reason to rejoice.

Early in the month, Grant and his forces in the west prevailed at the bloody two-day Battle of Shiloh, fought on April 6–7. The cost was staggering, with the combined Union and Confederate casualties numbering 23,746, losses greater than those in all previous American wars combined. In the east, the Union continued to pin its hopes on General McClellan's army, which, after months of delays, again had the Confederate capital of Richmond in its sights. But the outcome of McClellan's so-called Peninsula Campaign could not be known for some time, since his strategy involved moving the Army of the Potomac by ship to a Virginia peninsula between the James and York Rivers and then marching fifty miles to the Confederate capital. Lincoln had no choice but to be patient.

In Washington, however, the issue of slavery emerged as a matter of intense local interest. A dozen years earlier, during Lincoln's one term as congressman, his bill to end the slave trade in the District of Columbia went nowhere. Now, as president, he watched as new legislation to end slavery altogether in the nation's capital worked its way through the legislative process. Introduced in January, the two-pronged bill called for the immediate emancipation of the enslaved in Washington and three hundred dollars in compensation to owners for each man freed.

One of its most outspoken supporters was Charles Sumner. After his brutal beating at the hands of Preston Brooks in 1856, the Massachusetts senator had not returned to the Senate for three years. His flesh wounds repaired quickly enough, but, concussed and traumatized, Sumner consulted a series of more than a dozen doctors, even going to Europe to seek care. During his long absence from Washington, his chair remained vacant, a perpetual reminder of the violence perpetrated on his person by South Carolina congressman Brooks. When he had finally returned to the Capitol, he demonstrated that his hatred of slavery and his contempt for slaveholders remained unshaken. In June 1860, he proclaimed in a major speech to the Senate that "barbarous in origin; barbarous in law; . . . barbarous in spirit; barbarous wherever it shows itself, Slavery must breed Barbarians."[4] Sumner was a man the slavocracy loved to hate. He reciprocated the feeling with a hard passion, which he displayed on March 30, 1862, when he took to the floor of Congress to speak in favor of ending slavery in Washington.

"It is the first instalment of the great debt," he told his fellow senators, "which we all owe to an enslaved race, and will be recognized as one of the victories of humanity." There was an ongoing argument about whether people raised in bondage and deprived of education could live and thrive independently; the dispute continued, too, on where they should go. Yet the Senate approved the measure within a week of Sumner's impassioned pitch, and eight days later the House passed the measure. On April 16, 1862, Lincoln signed into law the District of Columbia Compensated Emancipation Act.

Never before in the nation's history had a federal statute given freedom to the enslaved, and Frederick Douglass felt a joy he had not since the war began. He wrote directly to Sumner. "I trust I am not dreaming but the events taking place seem like a dream."[5] On Douglass's good days, he imagined Lincoln might do more to right racial injustice. In light of this new law, Douglass felt a cautious optimism. "A blind man can see where the president's heart is," he told an audience in Rochester. "He is tall and strong but he is not done growing, he grows as the nation grows."[6]

THE GO-BETWEEN

Mary Todd grew up with enslaved servants in her household, people she remembered lovingly. Influenced as a girl by family friend Henry Clay and, later, by her husband, Mrs. Lincoln believed that compensated emancipation was the best route to ending slavery. But after the arrival of Charles Sumner in the Lincolns' lives, in early 1861, Mary's thinking shifted. As she herself admitted, she became an "ardent abolitionist."*

Sumner had hoped the national attitude toward emancipation would shift with Lincoln's election, and when the president-elect first reached Washington, in February 1861, Sumner had gone to visit the stranger who would lead the country. The senator left their meeting at Willard's Hotel with mixed feelings. He thought the man lacked dignity and social poise: Lincoln, on noting they were roughly the same height, suggested in his lighthearted way that they "measure backs" by standing back-to-back to determine who was taller. The solemn Sumner refused, observing that this was "the time for uniting our fronts against

* "Mr. Sumner says he wishes my husband was as ardent an abolitionist as I am." Randall, *Mary Lincoln* (1953), p. 355.

the enemy and not our backs." To his credit, however, Sumner also recognized "flashes of thought and bursts of illuminating expression" in Lincoln's conversation.[7] As for the savvy Lincoln, he saw clearly what sort of man Sumner was. "I have never had much to do with bishops where I live," he told a journalist, "but, do you know, Sumner is my idea of a bishop."[8]

Yet the humorless and pompous Sumner had gained easy access to the White House after Mary took a liking to him; she and the senator began to have "delightful conversations & often later in the evening."[9] She thought the aging and cultured bachelor was good company, and he devotedly came to their open evenings at the White House. He went with the Lincolns to the theater and escorted Mary to the opera when Mr. Lincoln was too busy. Mary and Sumner corresponded, too, covering topics from poetry to politics, from military news to "the great cause of abolition."[10]

In part through Sumner's influence, Mary became a convert to the cause. And Sumner emerged as the man in the middle, a link between the worlds of Frederick Douglass—a Sumner confidant—and Mr. and Mrs. Lincoln.

JULY FOURTH

On Independence Day 1862, Charles Sumner visited the White House. His mission was to persuade Lincoln to make emancipation his cause, too.

This was a regular topic for the two men. On the very day that the fighting began at Fort Sumter sixteen months before, the senator tried to convince the president that emancipation would not only be right but legal. As Sumner recalled that April 1861 conversation, "I . . . told . . . him that under the war power the right had come to him to emancipate the slaves."[11] Lincoln had listened respectfully, then ignored the advice, since it flew in the face of his larger promises to the country and his border-state strategy.

Despite mixed feelings about the bossy Bostonian, Lincoln had learned to work with Sumner. As chairman of the Senate Committee on Foreign Relations, Sumner was an ally as the president navigated the unfamiliar waters of international affairs, and his counsel proved particularly valuable when Lincoln and Secretary of State Seward did not see eye to eye. Mary described evenings when Sumner visited Pennsylvania Avenue, and he and "my darling husband would laugh together, like *two* school boys."[12]

On July 4, 1862, Seward arrived dressed to the nines, wearing a maroon vest and lavender trousers. Lincoln's attitude toward ending slavery continued to evolve; after much brooding, he now agreed in private that slavery must forcefully be abolished.[13] He said as much the previous December to Sumner, observing that "the only difference between you and me on this subject of emancipation is a month or six weeks in time."[14] Slavery must not be a permanent institution, Lincoln now believed, although he continued to resist a broad emancipation order, calling it a "thunderbolt" that he would use only as a last resort. But Sumner was insistent: *This is the time, Mr. President, to unleash that thunderbolt.*

The timing was significant. For more than a decade, abolitionists had held July Fourth celebrations. Their festivals were patriotic, and often included a reading of the Declaration of Independence. But the commemoration by Black and White abolitionists had also been critiques of the oppression and cruelty of slavery.

In 1852, Frederick Douglass had given an in-your-face speech at an Independence Day celebration, asking his audience of mostly White abolitionists, "What to the American slave, is your Fourth of July?"[15] He answered his own pregnant question: "To him, your celebration is a sham; your boasted liberty, an unholy license; your national greatness, swelling vanity; your sounds of rejoicing are empty and heartless."

Two years later at another July Fourth celebration, this one in Massachusetts, William Lloyd Garrison went further, burning a copy of the U.S. Constitution, along with the Fugitive Slave Act, warning that "the only remedy" to slavery was a dissolution of the union."[16]

When he had walked back Frémont's declaration the previous fall, Lincoln did so in part because he believed Frémont's decree illegal. According to Lincoln's interpretation, seizing property under the guise of military necessity was legitimate only "as long as the necessity lasts." To take an enemy's "property, real and personal" forever, he confided in a friend, was "*purely political,* and not within the range of *military* law." It wasn't the proper province of a general, he felt, and as president he could not permit Frémont to make political decisions.[17]

But in July 1862, when Sumner renewed his argument, he argued that, as the nation's chief executive, Lincoln *could* legally issue an emancipation proclamation. And on this day of all days, Sumner told Lincoln, an emancipation decree would be a "reconsecration" of independence and freedom.

Lincoln listened, but still he could not agree. In his homespun way he said no, telling Sumner that across-the-board emancipation would be "too big a lick."

Even in the face of a counterargument—namely, that "big licks" were exactly what the present situation called for—Lincoln stayed firm, and Sumner departed unsatisfied. Never one to give up the high ground, however, he was undeterred and returned two hours later. This time he told Lincoln, "You need more men, not only at the North, but at the South, in the rear of the Rebels: *you need the slaves.*"[18] This was a different argument: *Arm the freedmen. They can help you win the war.*

On this day, this argument also failed to persuade, and Senator Sumner left the White House twice disappointed at the president's refusal to act.

THE PROPOSAL

Less than three weeks later, on July 22, 1862, Lincoln called a cabinet meeting to order. Everyone in the room knew the war effort was going badly. With McClellan in retreat from Richmond, the Peninsula Campaign was a failure. Lincoln saw few good military options, and as he later told the story, he believed "that we had about played our last card, and must change our tactics or lose the game."[19]

With these dark thoughts in mind, he had set to work on an executive order to alter the nation's war strategy. With the draft before him, he told his advisers that he'd summoned them "[not] to ask their advice

In this political cartoon, English artist John Tenniel portrayed a desperate but wily Lincoln. The caption reads "Lincoln's last card," a metaphor for the Emancipation Proclamation, and an accompanying poem reads, in part, "From the Slaves of Southern rebels / Thus I strike the chain."

but to lay the subject matter of a proclamation before them." After they heard him out, he wanted their "suggestions" as to how to proceed.

Then he read aloud the Preliminary Emancipation Proclamation. It contained the extraordinary promise that that, as of January 1, 1863, "all persons held as slaves" within the rebellious states "henceforth shall be free." The proclamation would not declare all the enslaved free, since it specifically excluded those in the border states that remained in the Union.

The response to Lincoln's recitation was a mix of doubts, hopes, and surprise. The secretary of the treasury worried that financial instability could result. The postmaster general expressed concern its release might seriously reduce the Republican Party's prospects in the November elections. The attorney general thought Lincoln should publish the proclamation, but Secretary of State Seward offered the shrewdest advice.

"Mr. President, I approve of the proclamation, but I question the expediency of its issue at this juncture." With the recent military setbacks in mind, he proposed a delay. "I suggest, sir, that you postpone its issue until you can give it to the country supported by military success." Seward worried it would be seen as "the last measure of an exhausted government, a cry for help . . . on the retreat."

Lincoln saw the wisdom of the secretary of state's suggestion. As he later remembered, "I put the draft of the proclamation aside, . . . waiting for a victory." The war would go relentlessly on, and the document that could—and would—change American history remained a closely held secret as Lincoln waited for the tide to turn in the Union's favor.

PROPOSAL NOT ACCEPTED

Though Frederick Douglass knew nothing about Lincoln's July 22 meeting, he had plenty to say about a small, invitation-only gathering at the

White House on August 14, 1862. Of the five Black guests, none was a well-known abolitionist or a man of national stature; four had been enslaved and not all could read and write. Even so, the meeting was unprecedented. No president had ever summoned a "Deputation of Negroes," as Lincoln called it, to talk about a matter of national interest.

Though he did not utter the word, *emancipation* was the great weight that burdened Lincoln's thoughts. If he were to free the enslaved, would his war strategy collapse like a house of cards? He understood that, as one senator told his brethren in the Capitol, "There is a very great aversion . . . [to] having free negroes come among us."[20] Northern citizens who feared that freedmen would take their jobs and marry their daughters might reject such a change. Could emancipation lead to another secession? The departure of even one border state might just tip the balance in favor of the Confederacy.

After he shook hands with his Black guests, Lincoln began to talk about the relocation of people of African descent to other countries. This wasn't going to be an exchange of ideas about colonization—"I do not propose to discuss this," he told his guests—but he wished to present a plan. And he wanted their cooperation.

As he rarely did, Lincoln talked at and down to his guests. "You and we are different races. . . . Whether it is right or wrong I need not discuss, but this physical difference is a great disadvantage to us both, as I think your race suffer very greatly, many of them by living among us, while ours suffers from your presence. In a word, we suffer on each side. If this is admitted, it affords a reason why we should be separated. . . .

"The institution of Slavery," he continued, has "evil effects on the white race. See our present condition—the country engaged in war!—our white men cutting one another's throats. . . . But for your race among us there would not be war."

Lincoln offered a proposal. He wanted a vanguard of African American volunteers, prominent men and their families, to pack up and leave.

He hoped their departure would launch an exodus. "I want you to let me know whether this can be done." It would be, he concluded, "for the good of mankind."

His words were met by a stunned silence.

As the men rose to leave, one of them, a minister, promised to "hold a consultation and in a short time give an answer" to Lincoln's proposal.

"Take your full time," Lincoln replied, "no hurry at all."[21]

The audience ended, but a reporter from the *New-York Tribune* in attendance produced a full transcript of the meeting, and as Lincoln had hoped, many newspapers published in full what he had said. His words were a message intended to mollify a very much larger audience of White men.

When Frederick Douglass read Lincoln's verbatim words, he was enraged. He fired back angrily in *Douglass' Monthly*, ripping into the fallacy of Lincoln's arguments. Lincoln was wrong to suggest that both the war and slavery could be blamed on the enslaved. To Douglass, that notion was laughable, and he turned Lincoln's fondness for the folksy back at him. "No, Mr. President," Douglass wrote sarcastically, "it is not the innocent horse that makes the horse thief, not the traveler's purse that makes the highway robber, it is not the presence of the negro that causes this foul and unnatural war, but [the fault] of those who wish to possess horses, money and negroes by means of theft, robbery, and rebellion." No, Douglass insisted, blame for the Civil War does not belong to the Black man.

In his bitter response, Douglass also called Lincoln "a genuine representative of American prejudice." The president appeared "silly and ridiculous," said Douglass, and the facts affirmed it. For anyone who could do basic arithmetic, the idea of colonization was preposterous: In 1860, there had been more than four million enslaved persons and another half million free Blacks. That made the president's request for volunteers—he asked his five visitors to find him "twenty-five able-bodied men, with a mixture of women and children"—ridiculous in the extreme. Sending millions of Douglass's brothers and sisters abroad or to Africa? That wasn't remotely feasible.

To Douglass, the meeting demonstrated Lincoln's "hypocrisy." After the joy at the emancipation in the capital, Douglass's estimation of Lincoln crashed to a new low. He feared above all that this supposedly antislavery president had just demonstrated "his pride of race and blood [and] his contempt for negroes."[22]

When a member of Lincoln's cabinet, Postmaster General Montgomery Blair, reached out to try to enlist Douglass's help in the colonization scheme, Douglass responded as he no doubt would have if he had been in the room when Lincoln delivered his colonization lecture. "We have readily adapted ourselves to your civilization," Douglass wrote to Blair. "We are American by birth and education, and have a preference for American institutions. . . . [W]hy, oh why! may not men of different races inhabit in peace and happiness this vast and wealthy country?" Having rejected colonization, Douglass offered a better idea. "Instead of sending any of the loyal people out of the country, it seems to me that at this time our great nation should hail with joy every loyal man, who has an arm and a heart to fight as a kinsman and clansman, to be marshalled to the defence and protection of a common country."[23] It was an argument he had made before and would again: *Let us Black men be soldiers!*

The voice from Rochester was far from the only one to mock Lincoln's proposal. The idea of exporting Blacks was rejected outright by many, and signatures were collected for a petition that advocated the deportation not of formerly enslaved people but of slaveholders.

IN AUGUST, LINCOLN made a camouflaged pronouncement in his quiet campaign. Horace Greeley, the widely admired editor of the *New-York*

Tribune, provided an opening to Lincoln that he used to prepare the public for the emancipation document locked in his desk drawer.

In the August 20, 1862, issue of his paper, Greeley's editorial addressed President Lincoln directly. Titled "The Prayer for Twenty Millions," the editor's passionate plea called for emancipation; Greeley argued that eradicating slavery was necessary for the Union to prevail. Two days later, Lincoln responded by taking sides—all of them—without settling upon one.

"My paramount object in this struggle," Lincoln wrote, "*is* to save the Union, and is *not* either to save or to destroy slavery. If I could save the Union without freeing *any* slave I would do it, and if I could save it by freeing *all* the slaves I would do it; and if I could save it by freeing some and leaving others alone I would also do that."[24]

Lincoln's political dance was delicate: He offered assurance to nervous Northerners that he did not wish to turn this war into a crusade for abolition. At the same time, he alerted anti-slavery advocates that abolition also featured in his thinking. Although Douglass was pleased to see Lincoln's restatement that it was his "personal wish that men everywhere could be free," he also wrote to his friend Gerrit Smith, "I think the nation was never more completely in the hands of the slave power."[25]

Lincoln continued to bide his time as the debate whirled around him. He still needed the U.S. Army to deliver a victory before he could make his move.

THE BATTLE OF ANTIETAM

August brought little good news. As the month ended, Lincoln absorbed the shock of a second military defeat at Bull Run, a failure that meant Confederate troops now camped within easy striking distance of Union territory.

Emboldened by his smashing win, Confederate general Robert E. Lee took the offensive, marching his fifty-five thousand men into Maryland. This played into Lincoln's fears, because the citizens of Maryland, a border and slaveholding state, could still switch sides. A Southern victory on Northern soil might also sway foreign opinion, winning the CSA valuable recognition in Europe. Lincoln could envision a chilling scenario in which Lee was about to set the stage to win the war.

The situation was shocking enough that Lincoln, though not a deeply religious man, "made a solemn vow before God." The loss of son Willie may have disposed him to see worldly matters in a more spiritual light, but whatever the cause, Lincoln resolved that if his army managed to drive Lee back across the Potomac, "I would send the Proclamation after him."[26] Freeing the enslaved became a matter of waiting for divine intervention.

When it came to earthly practicalities, Lincoln put his faith in General McClellan one last time, and on September 15, he got good news. Waiting in the telegraph office near the White House, Lincoln was among the first to hear that the Army of the Potomac had driven back the Rebels at the Battle of South Mountain, Maryland. He wired McClellan back, "God bless you, and all with you. Destroy the rebel army, if possible. A. Lincoln."[27]

George McClellan possessed a talent for painstaking preparation but no genius for strategy. By the time McClellan was satisfied his troops were ready to reengage with the enemy, Lee's army had advanced another ten miles and settled into a defensive position on a low ridge near Antietam Creek, in Sharpsburg, Maryland. On the other side of the stream, McClellan's eighty-thousand-man Army of the Potomac was ready on September 17, 1861.

The fight began with an early-morning engagement in a cornfield near a plain but picturesque whitewashed church. For hours, combatants on the ground fought hand to hand with bayonets and rifle butts, while cannoneers on both sides turned the formerly peaceful farmland

Photographer Alexander Gardner's image of the aftermath at the Battle of Antietam. When this image of the Dunker Church and other detailed images of the battlefield were put on display in Mathew Brady's New York gallery, the New-York Times *said of the pictures that they "[brought] home to us the terrible reality and earnestness of War."*

into "artillery hell." The thirty-acre stand of tall, browning stalks changed hands more than a dozen times in the chaotic combat.

By late morning, the fighting shifted south to a sunken road, which would be remembered as Bloody Lane. The carnage would leave some five thousand men in both blue and gray uniforms lying in a river of blood. By midafternoon the center of action moved again, this time to the vicinity of a stone bridge. Strategic blunders by Union general Ambrose Burnside—his army was stymied by a bottleneck of his own making as it attempted to cross "Burnside's Bridge"—gave the outnumbered Confederates precious time, and reinforcements arrived to join the fight.

Lincoln scrutinized the battle reports as they arrived in Washington, but neither side seemed on its way to a resounding victory; still, if only because of its sheer numbers, the Union army did manage to hold the upper hand when the day ended. On Thursday both sides remained in place, stacking their dead and treating the wounded—and the casualty numbers were appalling. More than ten thousand men fell on each side, making that smoke-filled Wednesday the bloodiest day not only of the war but in all of American history, before and since. At last, on Friday, reports arrived that Lee and his army had begun to retreat.

Lincoln retired for the weekend to his cottage at the Soldiers' Home, a sprawling national cemetery on a hilltop four miles from the White House. There, on Saturday, he finally learned that Lee had gone back across the Potomac to Virginia soil. Antietam was the success that Seward prescribed, even if the battle was, in strictly military terms, little better than a draw. The next day when the president returned to Pennsylvania Avenue, honoring his holy promise, he locked himself in his office to work on his executive order.

Before he issued the Emancipation Proclamation on Monday, September 22, 1862, he remarked to a friend, "It is my last card, and I will play it and may win the trick."[28]

"THIS RIGHTEOUS DECREE"

Four hundred miles separate the White House from Rochester, New York, but after Frederick Douglass read the newspaper on Tuesday, September 23, 1862, the distance seemed much shorter. "We shout for joy that we live to record this righteous decree," he wrote on behalf of himself, his readership, and the four million in bondage.[29] He believed—he hoped—deliverance was at hand.

Forever free were the magic words. As of January 1, 1863, Lincoln promised, "all persons held as slaves"—albeit those in the designated states and territories—"shall be then, thenceforward, and forever free." And the promise was delivered with an iron fist: Lincoln commanded that the U.S. Army and Navy were to perform "no act or acts" that would interfere with freedmen exercising "their actual freedom." Had Douglass drafted it, the language might have been different; Lincoln had written it not in the manner of an abolitionist tract, making no mention of the Almighty or even righteousness. Despite the lawyer talk, which put the order in military terms as a necessity of war, Douglass understood the Emancipation Proclamation could be "the turning-point in the conflict between freedom and slavery."[30]

Lincoln did not free all the enslaved: As in the preliminary draft, the proclamation excluded those in the border states and in the occupied areas of Virginia, Louisiana, and Tennessee. In practice, it would have little impact in Confederate territory, over which the United States no longer had jurisdiction, but it did enunciate a new policy. It shifted the relationship of the government to slavery. In boldface, it meant Lincoln had extended the idea of liberty to the Black man.

Yet in autumn 1862, elated as he was, Douglass worried whether or not Lincoln would make the preliminary permanent. He wanted to believe that Lincoln would issue the proclamation in one hundred days, as promised, making it a legal instrument with the force of law, but until he did, it was merely a contingency. Douglass could only wait and hope that nothing transpired to dampen Lincoln's enthusiasm.

In November, such concerns did arise after Lincoln's party suffered serious setbacks in the midterm elections, even in his home state of Illinois. Some voter frustration could no doubt be blamed on war weariness, but no one doubted that emancipation had cost the party, too.

A concerned Harriet Beecher Stowe traveled to the White House for tea on December 2, hoping to get confirmation. When she entered the room, the long tall Lincoln rose from a seat before the fire to greet the famous author. He took Mrs. Stowe's tiny hand in his—the small-boned woman weighed less than a hundred pounds—and offered a welcoming witticism. "So this is the little lady who made this big war," he said to the author of *Uncle Tom's Cabin.* Warm as their meeting was, Lincoln did not confide his intentions, though their mutual friend Charles Sumner soon wrote to Stowe that "the Presdt. has repeatedly assured me of his purpose to stand by his proclamation."[31]

On January 1, 1863, the president and first lady took their place in the Blue Room at the White House at 11:00 A.M. for the traditional New Year's Day presidential reception. They greeted hundreds of diplomats, military officers, members of Congress, judges, and other worthies. They shook the hands of a great mass of average people, too, who had waited in line to wish the president well. When the doors finally closed in the early afternoon, Lincoln retired upstairs to his office.

In the preceding weeks and days, he had repeatedly reworked and revised the proclamation. He amended some wording and added a solemn closing line that made reference to the law, military necessity, the principle of justice, and even the Lord. "Upon this act," he wrote, "sincerely believed to be an act of justice, warranted by the Constitution, upon military necessity, I invoke the considerate judgement of mankind, and the gracious favor of Almighty God."

With a freshly inked copy of the final version before him, he raised his wooden pen, dipped its steel nib into the inkwell, and moved to affix his signature. But his large hand trembled badly as "a superstitious feeling came over me." When he recalled he had just spent three hours having his hand pumped by hundreds of people, the sensation passed quickly. "I never in my life felt more certain that I was doing right than I do in signing this paper," he said to Secretary of State Seward as he slowly, carefully, signed his full name.

With that he looked up, smiled, and said quietly, "That will do."[32] Lincoln had traveled many miles to reach this juncture, but his thinking about the Black man was still evolving.

Francis Carpenter's 1864 painting, The First Reading of the Emancipation Proclamation before the Cabinet, *which memorialized the first hearing of Lincoln's landmark decree two years earlier. Edwin M. Stanton, secretary of war, is at left, and Secretary of State Seward sits across the table from Lincoln.*

THE PRELIMINARY EMANCIPATION PROCLAMATION announced the previous September had been a promise; the document itself stated that Lincoln's presidential order would not take legal effect until he issued the final version, on January 1, 1863. That moment was awaited with great anticipation, not least in Boston.

In the Massachusetts capital, Henry Wadsworth Longfellow, the nation's most famous poet, organized a jubilee concert, held on New Year's Day, at the Music Hall, an event much ballyhooed in the papers. The White audience came to listen to Ralph Waldo Emerson, Beethoven's Fifth Symphony, and the cheers that greeted Mrs. Stowe, who took a bow from the balcony rail with tears in her eyes. Less than two blocks away, a largely Black audience of some three thousand people attended a day-long vigil at the Tremont Temple, as ministers delivered prayers and abolitionists spoke. With expectation in the air, everyone—whether of African or European ancestry—waited for the final word from Washington.

There was still no news when Frederick Douglass stepped to the Tremont Temple podium that evening; the absence of news was worrisome, and he sensed "a visible shadow . . . falling on the expecting throng." But the good news did arrive when a messenger burst into the hall, shouting, "It is coming! It is on the wires!" The audience leapt to its feet. People uttered prayers, clapped, unleashed shouts of joy, and tossed their hats in the air. Lincoln had honored his word.

Front and center, Douglass broke into song, his deep baritone leading the ecstatic crowd in singing "Blow Ye the Trumpet, Blow," a favorite of his friend John Brown. "*The year of jubilee is come,*" they sang. "*The year of jubilee is come! / Return, ye ransomed sinners, home.*" The songs and celebrations, the embraces and tears, lasted well into the night.

Before dawn, however, Douglass exited, leaving the warmth of the hall for the cold January night. Snowflakes gently fell as he made his way to the train station. He would remember this night as the happiest of his life. Yet the war wasn't over; the Confederacy hadn't been vanquished. Joyful as the moment was, Douglass knew his work was not yet complete.[33]

TWELVE

TURNING POINT AT GETTYSBURG

We are not to be saved by the captain, at this time, but by the crew. We are not to be saved by Abraham Lincoln, but by that power behind the throne.

—Frederick Douglass, "Our Work Is Not Done," December 1863

Frederick Douglass answered a knock at his door. On February 23, 1863, he greeted a small man with a great bushy beard that reached his chest and clothes that might have belonged to a bigger brother. Recognizable behind the whiskers were the familiar features of Douglass's old abolitionist comrade George Luther Stearns.

Stearns had accumulated a large fortune as an industrialist and merchant, but, like Gerrit Smith and Charles Sumner, he adopted antislavery as the great public passion in his life. His estate outside Boston had been an important station on the Underground Railroad, and

Stearns helped underwrite John Brown's little army at Harper's Ferry several years before.

He and Douglass exchanged greetings before Stearns, a shy man who spoke with a pronounced stutter, explained he was freshly off the train from Boston. His trip was prompted partly by the Emancipation Proclamation, since Lincoln had added to its final version a new section concerning Black enlistments. It stated that Blacks "will be received into the armed service of the United States to garrison forts, positions, stations, and other places, and to man vessels of all sorts."[1] The Black man could now go officially to the front lines to fight for his freedom.

Stearns needed Douglass's help in implementation. The governor of Massachusetts had obtained President Lincoln's permission to organize a Black regiment, but since the Bay State had fewer than two thousand Black male inhabitants of military age, the Massachusetts Fifty-Fourth Regiment would have to reach beyond the commonwealth's borders to recruit soldiers. Who better than Frederick Douglass to help create what Stearns called "a true John Brown corps"?[2]

Emancipation thus represented a transformation, and not only for Black Americans. With Lincoln's shift in policy, the U.S. Army could now fairly be called an army of liberation. The stubborn refusal of the Rebels to listen meant Lincoln had been forced to abandon his hopes for reconciliation. That also meant Douglass and Lincoln now shared the same objective: Nothing, absolutely nothing else, was more important to either man than winning this war. Lincoln's presidency and the survival of the Union depended upon defeating the Rebels—but so did emancipation. Both men understood the proclamation was a war measure: If the South won independence, emancipation in the region would almost certainly be proclaimed at an end, disappearing into thin air. Only victory—and a constitutional amendment—could assure permanent freedom for Black Americans.

The idea of arming Black men was obviously not new. Almost two

Massachusetts abolitionist George Luther Stearns. Commissioned a major by Secretary of War Stanton, he recruited some 13,000 Black troops to fight for the Union.

years before, just after Fort Sumter fell, Douglass himself wrote an essay titled "How to End the War" for his newspaper. Typeset in all caps, he pleaded, "LET THE SLAVES AND FREE COLORED PEOPLE BE CALLED INTO SERVICE, AND FORMED INTO A LIBERATING ARMY."[3] He made the case again after the first disaster at Bull Run. But his, Sumner's, and other men's calls to arms had met with stony resistance.

Many in the upper ranks of the military doubted the bravery and intelligence of African Americans and worried that White soldiers would refuse to serve with them. Early in the war Lincoln had feared that putting formerly enslaved men in uniform would undo his administration's argument that this was a war about Union rather than abolition. If he enlisted Blacks, he told Sumner in mid-1862, "Half the army would lay down their arms and three other States would join the rebellion."[4]

During that year, however, cracks appeared in the no-Black-soldiers policy when several generals quietly raised Black units. The Kansas

Colored Volunteers came into being. The War Department authorized recruitment of up to five thousand former bondsmen in South Carolina, although the secretary of war also warned that his order "must never see daylight, because it is so much in advance of public opinion."[5] Attitudes had been slowly shifting. Speaking of his role in the First Regiment Louisiana Native Guards, also known as the Corps d'Afrique, one officer saw both sides of the argument. "I felt a little repugnance at having anything to do with negroes, but having got fairly over that, am in the work. They are just as good tools to crush rebellion with as any that can be got."[6]

In the weeks before Stearns knocked at the door, Douglass had given a series of speeches. He brought new fervor to audiences from Massachusetts to Michigan. At each engagement, his three-hour lecture was interrupted by applause and cheers, and sometimes his words provoked laughter, too. But never was he more serious than when he spoke of an army of Black soldiers.

"I want to assure you," he told audiences in Philadelphia, Chicago, and New York's Cooper Union, "that the colored man only waits for honorable admission into the service of the country."[7]

Stearns's invitation to join the effort delighted Douglass, who leapt at the chance to find men for the Massachusetts Fifty-Fourth. Within days, he published a broadside titled "Men of Color, to Arms!" He urged other Black men to prove themselves, to disprove the doubters. "This is our golden opportunity. Let us accept it, and forever wipe out the dark reproaches unsparingly hurled against us by our enemies. Let us win for ourselves the gratitude of our country."[8]

The Emancipation Proclamation also produced a valuable shift in attitudes abroad. Great Britain, France, and other European countries previously wavered on whether to recognize the Confederate States of America as a new nation. But Lincoln's proclamation cleared the air. As one American ambassador put it, "Everyone can understand the significance of a war where emancipation is written on one banner and slavery on the other."[9] One Londoner who grasped the larger significance was Karl Marx, the German scholar who watched events in America carefully from his seat in the reading room at the British Museum. To Marx, the Emancipation Proclamation was "the most important document of American history since the founding of the Union."[10]

Opening the door to Black enlistment was also crucial to the Union's military fortunes, and as reports drifted in about the effectiveness of the units, Lincoln overcame his earlier doubts as he saw a double opportunity: He needed more men, since recruitment was down, but now he also embraced a psychological advantage. When he wrote to Tennessee governor Andrew Johnson to try to persuade him to form a Black fighting force, he made the case. "The bare sight of fifty thousand armed, and drilled black soldiers upon the banks of the Mississippi, would end the rebellion at once."[11] Frederick Douglass couldn't have agreed more. Yet these troops faced a double danger. As if marching into battle weren't enough, men of color faced the added risk of mistreatment in the event they were captured. Jefferson Davis and his generals made clear the rules of war regarding the humane treatment of prisoners of war did not extend to the Union's new recruits.

By the time George Stearns departed, just hours after he arrived, the Douglass family had made a deeply personal contribution to the new "sable arm" of the military. Among the first volunteers recruited for the Massachusetts Fifty-Fourth were Frederick Douglass's sons Charles and Lewis. In a matter of weeks, Frederick Douglass himself would travel to Boston, where, on May 28, 1863, the ladies of Boston society waved their handkerchiefs in farewell as Black soldiers marched up Beacon Street before entering Boston Common. A crowd of more than a thousand people cheered the regiment as it drilled. From there the troops marched on to Battery Wharf, where they were serenaded by throngs of well-wishers singing "John Brown's Body."

At the docks, Frederick Douglass went aboard the steamer *De Molay*

The eldest of Frederick Douglass's three sons, Lewis Henry Douglass joined the Massachusetts 54th in 1863, fought at Fort Wagner, and gained the rank of sergeant major. Prior to one battle, he wrote home to his future wife, "Should I fall in the next fight killed or wounded I hope to fall with my face to the foe. If I survive I shall write you a long letter."

to wish the soldiers well. Among them were Sergeant Major Lewis Douglass and another hundred men that the elder Douglass had personally recruited. The Fifty-Fourth was a cross section of African American society, with freemen and former bondsmen from twenty-four states, including five that had seceded. They were off to war, bound for South Carolina.

"WHAT WILL THE COUNTRY SAY!"

The Civil War ground cruelly on. Despite new commanders—at last, Lincoln gave in to his frustrations and replaced George McClellan—no one seemed able to best General "Bobby" Lee. At Fredericksburg, Virginia, in December, Lee's forces had routed the Army of the Potomac, which sustained nearly three times the casualties of Lee's troops. As Lincoln watched thousands of wounded and defeated men arrive in Washington, he lamented, "If there is a worse place than hell, I am in it."[12]

With the arrival of spring, he tried to take an optimistic view. After four days spent reviewing the army in northern Virginia, he thought his generals might be in a position to engineer a major victory. On May 6, 1863, at the White House, he awaited word of just such a hoped-for triumph at Chancellorsville, a crossroads town some sixty miles south of the capital. "I expect the best, but I am prepared for the worst," Lincoln admitted to a reporter from the *Chicago Tribune*.[13]

Two friends who arrived that afternoon waited in an anteroom for the president. At three o'clock he appeared, but both Noah Brooks, a California journalist, and Dr. Anson Henry, whom Lincoln knew from his days as a young lawyer in Springfield, Illinois, were shocked at his appearance. As ashen as the gray wallpaper, he held out a telegram.

"Read it. News from the army." His voice trembled with emotion.

In a celebratory image painted more than three decades after the war, General Robert E. Lee is being cheered by his troops after the resounding victory at Chancellorsville, Virginia.

The report from the front was bad—worse than bad. Union commander "Fighting Joe" Hooker had been manhandled by Lee, his army forced to retreat back across the Rappahannock River. The battle at Chancellorsville was over, a catastrophic loss for the Union.

Brooks kept his composure. Dr. Henry's eyes filled with tears; he was so swamped with emotion he could not speak.

Lincoln paced, his hands clasped behind his back. Brooks thought he never looked "so broken, so dispirited, and so ghostlike." Finally, the president broke the silence.

"My God! my God! What will the country say! What will the country say!"[14]

Before the question could be answered, a carriage pulled up below and Lincoln hurried out of the room, headed for the army encampment to meet with his defeated generals. Trying to absorb another battle lost, despite the fact that the Union fielded a much larger force, Lincoln was asking himself, *What does this mean?* He was afraid the answer was, *The Union could lose this war.*

THE INVASION OF PENNSYLVANIA

A few scraps of better news arrived. On May 27, official reports lauded the bravery of two African American regiments that marched repeatedly into deadly Confederate fire at Port Hudson, Louisiana. The performance of the Black troops defied widely held racist views. "It is no longer possible," acknowledged their commanding general, "to doubt the bravery and steadiness of the colored race."[15] At a base near Vicksburg, Mississippi, on June 7, Black troops fought off a Confederate bayonet charge, leading a War Department official to observe that "the bravery of the blacks in the battle of Milliken's Bend completely revolutionized the sentiment of the army with regard to the employment of negro troops."[16] The racist prejudice against Black soldiers showed signs of fading into history.

Closer to the capital, however, the news remained bad. On the heels of his big win at Chancellorsville, Robert E. Lee persuaded Confederate president Jefferson Davis to take the fight to Northern soil. By mid-June, reports reached Lincoln of the Army of Northern Virginia crossing the Potomac into Union territory, and by June 27, the invaders marched over the Mason-Dixon Line into Pennsylvania. Although Lee had lost his most trusted general, "Stonewall" Jackson, killed at Chancellorsville, his army was headed for Harrisburg, the state capital, and perhaps Baltimore, Philadelphia, or even Washington. No one could be sure.

Lincoln saw a larger danger. If Lee's campaign succeeded, he might cleave the country in two and thereby force peace talks. So far, Lincoln refused to speak the name of the Confederate States of America; there was no such country in his mind, just an illegal brotherhood of traitors with no legal standing. However, if he were forced to accept a negotiated end to the war, the CSA might become a separate nation, permanently dividing the Union and ending the war on slavery.

Nervous and agitated, Lincoln changed generals again, giving command in the east to Pennsylvanian George Meade. Meanwhile, the Confederates continued their campaign, taking the Pennsylvania cities of Carlisle and York. A major military confrontation became inevitable when Meade's army, strategically positioned to shield Washington, reached the Maryland–Pennsylvania line on June 30, and Lee concentrated his forces near a little college town named Gettysburg.

On July 1, a Confederate division seeking supplies in open farmland north of the town encountered two Union cavalry brigades. After skirmishing with the much larger enemy force, the Federals beat a hurried retreat through the streets of Gettysburg, halting just south of the village. By daybreak on day two of the Gettysburg fight, four Union corps arrived and established defensive positions on Cemetery and Culp's Hills.

The pastoral setting—a fertile valley surrounded by low ridges, with a mix of wooded hills and babbling brooks—became an immense campground for two armies, with some seventy thousand men in gray facing off against roughly one hundred thousand dressed in Union blue.

That day, the Rebels directed heavy assaults on a number of Union positions, including those remembered today as Little Round Top, the Devil's Den, and the Wheatfield. At three o'clock on the afternoon of July 2, Meade telegraphed the War Department. "If I find . . . the enemy is endeavoring to move to my rear and interpose between me and Washington, I shall fall back."[17]

The possibility of retreat was not at all what Lincoln wanted to hear, but, as if he didn't have enough to concern him, a messenger arrived with word that Mary Lincoln had been thrown from her carriage, striking her head violently on the road. He arranged for a nurse to care for her, but he parked himself at the telegraph office for hour after hour, monitoring the faraway battle, waiting for news. He simply could not draw his attention away from Gettysburg.

Day three at Gettysburg would be decisive. Lee directed his attack at the center of the Union line at Cemetery Ridge, delivering one of the largest artillery bombardments of the war to soften up Union resistance; then he ordered his infantry to attack. At 2:00 P.M. nine brigades, commanded by Major General George Pickett and two other officers, numbering more than twelve thousand men, began to march deliberately across open fields. As they covered the thousand-yard distance, shell and solid shot from Union artillery tore holes in the oncoming formation. Once the Rebels were within four hundred yards, the Federals' canister, rifle, and then musket fire proved equally murderous. In barely an hour, the assault was over. In one of the best-remembered and most analyzed military moments of all time, Pickett's charge had failed. A decimated Confederate force retreated, leaving dead and wounded strewn across the landscape. The battle over, Meade's army held the field, though the casualty total for the

Some military historians believe that artist Peter Rothermel was trying to capture the carnage of Pickett's Charge in this 1872 image. Although the war's outcome remained uncertain in July 1863, the Battle of Gettysburg, in retrospect, was clearly a turning point.

two armies exceeded fifty-one thousand dead, wounded, and missing. Gettysburg set a terrible new record for the deadliest battle of the war.

The next day, Lincoln went to visit General Daniel Sickles, a New Yorker who had been wounded at Gettysburg. Just back in Washington, his right leg amputated, he lay on a blood-stained stretcher in a house on F Street. When Lincoln arrived with son Tad, aged ten, at his side, he took a chair next to Sickles. The two men "discussed the great battle and its probable consequences." Lincoln wanted to know everything about the victory.

When Sickles completed his account of the battle, he asked, "Well, Mr. President, I beg pardon, but what did you think about Gettysburg?"

"I suppose some of us were a bit 'rattled,'" Lincoln admitted. "[But] we did right handsomely."[18] He kept his feelings, his worries, his hand-wringing to himself.

In the coming days, Lincoln faced good news and bad. Mary took a turn for the worse, her head wound infected, before recovering. Lee managed to slip back into Confederate territory—Lincoln was deeply frustrated at the Confederates' escape—but, still, Lee's northern offensive was at an end. Then news from the west arrived. After weeks of resisting a Union siege, Confederate forces at Vicksburg, Mississippi, had surrendered to Ulysses S. Grant on July 4. Vicksburg's strategic location meant the Union gained control of the Mississippi, the key supply line to the heart of the country.

Although the momentum of the war shifted, the Confederacy was far from defeated. The Union still needed more soldiers and, with the Mississippi now controlled by Union ships, Lincoln wanted to broaden his recruitment campaign among Blacks. "I believe it is a resource," he wrote to Grant, "which, if vigorously applied now, will soon close the contest."[19]

FORT WAGNER

For Frederick Douglass, July was an unhappy month. His high spirits tumbled after a recruitment speech, delivered in Philadelphia on July 6, to some five thousand people. The applause was thunderous, but only a few volunteers stepped forward. Arguments that had worked for months—hundreds upon hundreds of men enlisted after hearing his pitch—rang hollow when everyone in the room knew about the recent slaughter in Gettysburg.

On his way home to Rochester, Douglass narrowly escaped the Draft Riots that rocked New York for three days and three nights. What began as a protest of recent draft laws degenerated into a race riot, with Whites lynching Black men on light poles, murdering Black women, and setting fire to their homes, businesses, and even the Colored Orphan Asylum;

"the cry of the crowd," Douglass wrote, was "beat, shoot, hang, stab, kill, burn and destroy the Negro."[20] Passing through the city, he escaped the mob's wrath only because the "high carnival of crime and reign of terror" was unfolding well uptown from the Chambers Street rail station where he changed trains.[21]

In Rochester, the news got suddenly personal with reports of the Fifty-Fourth's assault on Fort Wagner. The Massachusetts regiment exhibited great bravery in attacking the fortification at the mouth of Charleston Harbor, on July 16, but enemy artillery took a terrible toll on Sergeant Major Lewis Douglass and his men. Their commanding officer, a young White Bostonian named Robert Gould Shaw, was killed. Half of his six hundred men became casualties, including Lewis, and many of the officers. "Men fell all around me," the wounded Lewis

In this 1890 lithograph, the commander of the Massachusetts 54th, Colonel Robert Gould Shaw, brandishes his sword over his head as the soldiers in his regiment engage in hand-to-hand combat with the enemy. Shaw was killed at the Battle of Fort Wagner, along with 272 of his men.

Douglass wrote home. "A shell would explode and clear a space of twenty feet, our men would close up again. . . . How I got out of that fight alive I cannot tell."[22]

His father, though relieved his son survived, was deeply upset about bureaucratic issues. For one, White soldiers were paid thirteen dollars a month, while the new Black soldiers, 80 percent of whom had been in bondage, received the wages of laborers, seven dollars a month plus a clothing allowance. Even worse, Douglass felt a new sense of betrayal at the government's failure to stand up for the soldiers he recruited. According to reports from the front, many captured Black troops were executed by their Confederate captors, and others returned to slavery and forced labor.

Douglass stewed for a few days before writing to George Stearns, who had been commissioned a major in the army to aid in his recruitment efforts. He directed his anger at the president's inaction: "No word comes from Mr. Lincoln or from the War Department, sternly assuring the Rebel Chief that inquisitions shall yet be made . . . when a black man is slain by a rebel in cold blood . . . or caught and sold into slavery."[23] Douglass had decided he was finished with recruiting "until the President shall give the same protection [to Black] . . . as to white soldiers."

Major Stearns, who could ill afford to lose his best recruiter, wrote back promptly. He suggested Douglass take the matter into his own hands and pay "a flying visit to Washington."[24] Douglass took Stearns's urging seriously, and in a matter of days he departed on the two-day rail journey to the nation's capital.

At last, the time had come for the president and the freedom fighter to meet.

THIRTEEN

A BLACK VISITOR TO THE WHITE HOUSE

You say you will not fight to free negroes. Some of them seem willing to fight for you.

—Abraham Lincoln to James Conkling, August 26, 1863

Many years earlier, another kind of railroad—the Underground Railroad, that mix of the real and the metaphorical—carried a runaway named Fred Bailey to freedom. By August 1863, things had changed. A free man for many years, the former fugitive bore the surname Douglass. A world-famous writer and lecturer, he rode in a first-class sleeper, befitting his elevated station, on the Baltimore & Ohio Railroad. But for the first time in twenty-five years, he was again in slaveholder territory, with slavery still legal in border state Maryland.

At his destination, the most important residence in the land, the White House, the weary traveler hoped to meet with the president of the United States.

When Frederick Douglass emerged from the rail station in Washington, his body humming from two long days of continuous rail travel, he stood just a few blocks from the Capitol with its still unfinished dome. Even on a Sunday night, the city around him was a teeming military depot, its streets alive with the rumbling of army wagons and the clatter of galloping cavalry. Markets, saloons, and slaughterhouses lined Pennsylvania Avenue. Almost two dozen mansions, schools, churches, and hotels converted to makeshift hospitals were crowded with wounded. Peddlers occupied street corners, and many a storefront offered the services of an undertaker. Crippled veterans maneuvered on crutches, some of them missing limbs. The smell of sewage, death, and garbage hung in the air.

In the fading evening light, Douglass could see he was far from the only visitor. Strangers looking to find loved ones wounded in the war flooded the city. Washington's new citizens also included many formerly enslaved people hoping to establish new lives as free men and women in a place where bondage was now forbidden.

The well-dressed Douglass stood out amid the hustle-bustle. Wherever he went people noticed this proud forty-four-year-old. The long black coat he habitually wore could not disguise his muscular build, and his large head of salt-and-pepper hair was combed back to frame his handsome face. "[His] appearance," as one friend observed, looked "stamped by past storms and struggles, [and] bespeaks great energy and will power . . . in the face of all odds."[1]

He would need that confidence the next day when he went about the business of persuading President Lincoln to come to the aid of the Black soldier.

ON AUGUST 10, 1863, Frederick Douglass took a place in line. The White House reception room was crowded with people waiting, since

President Lincoln made it a practice to meet with idea men, office seekers, the curious, and even foreign visitors who came to his door. Douglass handed his calling card to an assistant, and as "the only dark spot" among a roomful of White visitors, he readied himself to wait for hours or even days.[2] Rumor had it that some people waited a week.

The White House was his second stop that sultry Monday morning. The thermometer was well on its way to the hundred-degree mark when Douglass had arrived at the War Department, accompanied by Samuel Pomeroy, a Massachusetts-born reformer who went to the territories during the days of Bleeding Kansas. Now a senator from his adopted state, Pomeroy added authority to Douglass's mission, and the pair were ushered into the office of Secretary of War Edwin Stanton.

They were allotted a half hour. Stanton, famous for his brusque and impatient manner, listened intently as Douglass articulated his argument that all soldiers be paid the same. Stanton hedged—equal pay, he said, was a hard sell in the face of prejudice in the ranks—but he said he believed equal treatment could come in time.

With his large scraggly beard and small wire-framed glasses, Ohioan Edwin Stanton became the secretary of war in 1862, the year this photo was taken. He was an able man doing a demanding job and uninterested in small talk. Douglass reported after their half-hour meeting that Stanton's "manner was cold and business like throughout but earnest."

Much more surprising, the secretary countered with a proposition. *Would Douglass join recruiting efforts in the Mississippi Valley?* Now, barely an hour later, Douglass waited at the White House, his head spinning. Stanton had offered him the post of assistant adjutant. He had asked him how soon he could be ready; Douglass had told him two weeks. After meeting with the president, he would go home, where, Stanton promised, he would soon receive an official legal document making him an officer. Douglass felt deeply honored at being granted an officer's commission, perhaps the first ever to a Black man in the U.S. Army. Yet he could not help but worry about the great personal danger of working in the deep South. Many a Black man had been murdered or lynched for much less, and his fame would surely put a target on his back.

His thoughts were interrupted.

"Mr. Douglass?"

Barely two minutes had elapsed since he'd submitted his card, but one of the president's aides invited Douglass and Senator Pomeroy to follow him. As the trio shouldered their way through the throng, Douglass heard one voice complain, "Yes, dammit, I knew they would let the n—r through." He left the room to the crackle of bitter laughter.

Moments from meeting the president, Douglass did not know what to expect. For years, he had been writing and speaking publicly about Lincoln's politics and positions, and more often than not, his remarks were highly critical. His personal feelings about the man fluctuated between hopeful and impatient, exhilarated and angry. Lincoln wielded immense power and at last had done right with the Emancipation Proclamation, yet at other times he fell far short of Douglass's expectations. Just days before, the president rose in Douglass's estimation after he declared a one-for-one policy: For every Union soldier executed by the Rebels, a Rebel soldier would be put to death. Douglass was grateful, but he still held many reservations about Mr. Lincoln.

Douglass took in Lincoln's office, a moderate-sized room and a hub of activity. A desk with an overflowing pigeonhole organizer rested against one wall, a marble fireplace on another. A small mountain of documents obscured an oak table. There were two sofas, chairs arranged in no apparent pattern, and several young men, secretaries and clerks, came and went, their conversation a low buzz. Inevitably, Douglass's gaze fell upon the tall man sitting between two west-facing windows, surrounded by maps and books and stacks of paperwork. Lincoln's long legs stretched out in front of him.

The stresses of his job had robbed Lincoln of twenty pounds. The well-muscled torso of his younger years had narrowed, his face grown more angular. Even on first look, Douglass noticed the "long lines of care . . . deeply written on Mr. Lincoln's brow." But the president's expression opened and brightened with the announcement of his visitors.

Lincoln rose laboriously from the low chair. On reaching his full height, he extended his hand in cordial welcome. As they shook, Douglass again was struck by Lincoln's face. This time, as he wrote to Stearns the next day, "[I] saw at a glance the justice of the popular estimate of his qualities expressed in the prefix '*Honest*' to the name of Abraham Lincoln. I have never seen a more transparent countenance."

Seizing the moment, Douglass launched into an explanation of his reason for calling. But Lincoln stopped him.

"Mr. Douglass," said the president kindly, "I know you; I have read about you, and Mr. Seward has told me about you. Sit down. I am glad to see you."

A candid conversation followed. Douglass thanked Lincoln for extending protections to Black soldiers, but added that, in his estimation, the president had been "somewhat slow" in doing so.

In great earnestness, Lincoln acknowledged that, yes, he had been accused—by others, as well as Douglass—of a reluctance to confront the problems of the Black man. As Stanton had done an hour before, he cited the need to overcome popular prejudices. The best means of meeting the objections to treating the races equally, Lincoln said, would be the performance of Black soldiers on the battlefield. As fighters and patriots, their "bravery and general good conduct," said Lincoln, would set the stage for him to protect the proclamation.

But Lincoln wasn't finished. He clearly took Douglass seriously, wanting his guest to understand his larger approach to such a matter. "I have been charged with being tardy," he acknowledged. "[B]ut, Mr. Douglass, I do not think that charge can be sustained." He was firm. "I think it can be shown that when I take a position, I think no man can say I retreat from it."

Douglass had underestimated Lincoln; as had so many others, he found himself captivated by the manner in which Lincoln listened patiently to his every word, then responded in measured and thoughtful ways; he spoke, Douglass thought, "with an earnestness and fluency of which I had not suspected him." Lincoln would ask for particulars; Douglass replied in detail. Their exchange was truly a conversation; they did not speak at cross-purposes as Lincoln looked to find common ground. If the justifications the president offered for actions or inaction did not always wholly satisfy Douglass—and they did not—the Black man nonetheless recognized he was in the presence of a "humane spirit."

As the meeting drew to a close and Douglass prepared to leave, he told Lincoln of Stanton's invitation to be a recruiter in Mississippi. He showed him the pass he was issued, which bore Stanton's name. Lincoln examined it then put his own pen to it.

"I concur," he wrote. "A. Lincoln. Aug. 10, 1863."

When Douglass departed, Lincoln was left with a sense that his guest deserved his reputation as a tough-minded, articulate, and passionate man who could hold his own in a discussion with any man. The Black man arrived as both adversary and ally; perhaps their exchange could tip the balance toward cooperation? With his parting words, Lincoln tried to leave that door open.

Douglass in a photograph taken in January 1863 when he was in Michigan on a speaking tour.

"Douglass," he said in farewell, "never come to Washington without calling upon me." In fact, they might have met again just a few days later, when Douglass was invited to tea, but had previously agreed to deliver a lecture that afternoon. The carriage Lincoln had dispatched to bring Douglass back for the afternoon returned without a passenger. Both men anticipated there would be other days and more conversations.

DOUGLASS'S MEETING WITH LINCOLN left him deeply impressed by "the gravity of his character," and for the first time, he felt that he could take Lincoln at his word. He wrote to Stearns saying he believed the president would "stand firm," that he felt confident "slavery would not survive the war." For Douglass, that was a very large leap of faith. He accepted both that Lincoln's thinking was evolving and that the pragmatic president might be right in thinking that if he moved too quickly with regard to the rights of African Americans, the move toward equality could blow up in their faces.

Douglass's trip to Washington was a watershed moment for him personally, too, with Stanton's offer of a job and a commission. The opportunity prompted him to change his life: He would immerse himself in the work of recruiting, reinventing the life he had lived for decades. He decided to complete one final issue of *Douglass' Monthly*, publishing it as a valedictory in which he bid goodbye to his devoted readers. He would move on from his publishing life to this new and important work.

The visit to the White House changed him in another way, too. "I tell you I felt big there," he told a Philadelphia audience a few months later. Lincoln received him, a man once enslaved, "precisely as one gentleman would be received by another." The president shook his hand, offered him a seat, listened to his concerns, and responded with directness and honesty. The reassurance stayed with him, and his respect for Lincoln grew. He believed that if the country survived, the name "Honest Abraham" would be written "side by side with that of Washington."

Although Lincoln left no record of their conversation—Douglass wrote several—their meeting sent a loud message to the country.* The president of the United States had opened the doors of his office—and opened his mind—to the most essential representative of a largely voiceless population of millions. The terms were set for an unprecedented partnership of two self-made men concerned most of all with what Lincoln would soon call "a new birth of freedom."

---

* John Hay, one of Lincoln's secretaries, noted in his diary on August 11 that "Fred Douglass in company with Sen. Pomeroy visited the president yesterday." Hay, *Inside Lincoln's White House* (1997), p. 309.

A NEW CEMETERY IN GETTYSBURG

In the autumn, Abraham Lincoln received an invitation: *Would he deliver "a few appropriate remarks" at the dedication of a cemetery in Gettysburg?*[3] Although the busy president turned down most requests to speak, this one he accepted, agreeing to travel to Pennsylvania for the solemn event. The purpose of the war had shifted in his mind, and the time to say so publicly just might be November 19, 1863, in front of acres of mourners at a new rural cemetery.

After the July battle, teams of soldiers, Confederate prisoners, and civilians had raced to bury the eight thousand dead. With the stench of decaying flesh poisoning their nostrils, the gravediggers choked on smoke as nearby fires consumed immense pyres of dead horses and mules. Many graves were so shallow that the fresh soil could not disguise the human silhouettes beneath. The gravediggers scrawled the names of some Union soldiers on crude wooden markers, but the Confederate dead were buried anonymously in trenches.

Since the fallen were interred where they died, unquiet graves dotted Gettysburg's cornfields, orchards, gardens, and woodlands. But survivors of the dead soon began to arrive. They opened some of the fresh graves, looking for the remains of beloved husbands, fathers, and sons to take them home for burial; the strangers they uncovered were hastily reburied. According to one local civic leader, the result was that "in many instances arms and legs and sometimes heads protrude and my attention has been directed to several places where the hogs were actually rooting out the bodies and devouring them."[4] The whole town had become a graveyard, with fetid air and fouled streams, and the dead dishonored. Something had to be done.

An interstate commission was established to create a proper cemetery. Acreage was purchased. A landscape designer laid out the burial ground, arranging the graves not in rank and file but in curves that fit the contours of the site and gave the dead of no one state a higher status than another's. The federal government agreed to provide pine coffins, and over a period of weeks and months, the Union dead would be methodically exhumed, identified, and reinterred. No Rebel corpses would be welcome on this hallowed ground.

In preparing his remarks for Gettysburg, Lincoln could not follow his usual writing routine. With only two weeks' notice prior to the dedication of the new Soldiers' National Cemetery, he simply didn't have time for his usual extensive library research and writing. But the president already had a template in his head.

Back in July, just after the victories at Gettysburg and Vicksburg, he spoke to a small crowd outside the White House. The good war news put him in a philosophical mood, and after a band serenaded him, he gave a short speech. He mused aloud about the Declaration of Independence. How long ago was it, he asked his audience—perhaps "eighty odd years"?—since that memorable July Fourth? He cited the "self-evident truth" at the core of that founding document, namely, "that all men had been created equal." Rather than develop the thought, however, he stopped short and thanked the band for the music.

"Gentlemen, this is a glorious theme, and the occasion for a speech," he said, "but I am not prepared to make one worthy of the occasion." Now, a few months later, as he prepared to fulfill his duty in Pennsylvania, he decided to complete those thoughts.[5]

His would not be a long oration, he told his friend Noah Brooks, but "short, short, short."[6] That suited the organizers, who had invited Edward Everett, a former senator and secretary of state and a renowned orator, to give the main address. As his last-minute invitation indicated, Lincoln's inclusion was almost an afterthought.

When Lincoln began the six-hour trip to Gettysburg the day before the event, he carried a sheet of paper with a few lines jotted on it, amounting to perhaps half the speech. Working on a piece of cardboard during the train journey, he added to his text, then in his room that night and early the following morning, he worked on the draft. By the time he left his room to join Seward on a carriage tour of the battlefield, he carried a completed manuscript copied out in his steady, even handwriting, a few last edits squiggled between the lines.

With the dedication scheduled for two o'clock, every road into town was jammed with carriages, buggies, wagons, and carts filled with out-of-towners. Special excursion trains brought visitors, too, and Gettysburg's population of three thousand would be at least five or six times that with the visitors who made their way to the new cemetery's hilltop a half mile south of town. The fighting had taken place on every side, with Culp's Hill just northeast and Seminary Hill to the north and west. Well south lay Little Roundtop, Big Roundtop, and Devil's Den.

The speakers and other notables gathered on a platform lined with chairs. The program began with the invocation, a long and emotional prayer, followed by a hymn played by a band. Edward Everett came next. As Everett delivered a two-hour address, Lincoln listened intently, and Secretary of State Seward sat nearby with his arms folded, his hat drawn down over his eyes to shield them from the bright sun. When Everett finished, he bowed to the audience, and they responded with polite applause. A hymn followed.

Lincoln's turn came next. Rising from his chair, he stepped forward, carrying his address in his left hand. He adjusted his spectacles but delivered the speech from memory. He spoke slowly, clearly, so his tenor voice would carry his words to listeners at the greatest distance, to whom the speakers resembled doll-like figures on a toy wagon.

272 WORDS

In the brief speech destined to be remembered as the Gettysburg Address, Lincoln told no stories. In speaking to the thousands gathered to mourn the death of many good men on the field of battle, he offered no parables. But he did compress layers of meaning into just ten sentences.

> Four score and seven years ago our fathers brought forth on this continent, a new nation, conceived in Liberty, and dedicated to the proposition that all men are created equal.
>
> Now we are engaged in a great civil war, testing whether that nation, or any nation so conceived and so dedicated, can long endure. We are met on a great battle-field of that war. We have come to dedicate a portion of that field, as a final resting place for those who here gave their lives that that nation might live. It is altogether fitting and proper that we should do this.
>
> But, in a larger sense, we can not dedicate—we can not consecrate—we can not hallow—this ground. The brave men, living and dead, who struggled here, have consecrated it, far above our poor power to add or detract. The world will little note, nor long remember what we say here, but it can never forget what they did here. It is for us the living, rather, to be dedicated here to the unfinished work which they who fought here have thus far so nobly advanced. It is rather for us to be here dedicated to the great task remaining before us—that from these honored dead we take increased devotion to that cause for which they gave the last full measure of devotion—that we here highly resolve that these dead shall not have died in vain—that this nation, under God, shall have a new birth of

> freedom—and that government of the people, by the people, for the people, shall not perish from the earth.[7]

As he had done in his off-the-cuff speech on July 7, Lincoln reminded his listeners of the founding proposition that "all men are created equal." This time he got the year exactly right—the Declaration of Independence, he noted, had been issued "four score and seven years ago"—but then went straight to the purpose of the day. They assembled to dedicate "a final resting place for those who here gave their lives."

Lincoln rooted his words in the earth ("this ground," he called it) and "the brave men" who fought upon it. But his message was not only a matter of mourning. His called to his listeners, to his countrymen, "for us the living," to dedicate themselves to "the unfinished work."

The last sentence of his artfully crafted speech laid out his challenge clearly.

Without saying so directly, the president acknowledged the passing not only of the Gettysburg dead but of the old republic, wounded and divided by war. In its place, he proclaimed a reborn nation, one in which freedom would be redefined, not solely as freedom from tyranny but by adopting the even higher standard of equality. Lincoln had already extended the pledge of freedom to the enslaved; here, in Gettysburg, Lincoln was asking his countrymen to join him in carrying out a transformation of the nation's society, extending freedom to all, Black and White.

The great underlying truth of Lincoln's 272 words was that the president no longer regarded the war as a battle to restore the Union to what it had been. His goal was to reconstitute the nation based upon human liberty, making explicit what was implied in the Declaration of Independence. To those listening with care, it was both a shock to hear and shockingly obvious.

Frederick Douglass had not traveled to Gettysburg that November; thus he wasn't physically present to hear Lincoln speak. But he did not

The war took a toll on the president; as it wore on he began to look older and more gaunt.

need to be there nor to read between the lines of Lincoln's concisely constructed three paragraphs to grasp the profound message. To him, it was distinctly familiar: Lincoln made the same constitutional argument that Douglass had been putting forth for years.

The year 1863 thus saw a major upturn in the Union's fortunes. On a much smaller scale, it was also a time when a new relationship began to emerge, as Lincoln and Douglass became men of shared purpose.

FOURTEEN

THE MISSION OF THE WAR

Every Slave who Escapes from the Rebel States is a loss to the Rebellion and a gain to the Loyal Cause.

—Frederick Douglass to Abraham Lincoln, August 29, 1864

Frederick Douglass never received his officer commission. He got no explanation, though he guessed that Secretary of War Stanton had had second thoughts over the "radical and aggressive" idea of making a Black man an officer. Angry and disappointed at the broken promise, Douglass decided not to report to Vicksburg, Mississippi. As he explained later, "I knew too much of camp life and the value of shoulder straps in the army to go into the service without some visible mark of my rank."[1]

With the fight slowed to a halt by winter weather, Douglass crisscrossed the North in late 1863 and early 1864, delivering lectures. Freed of his newspaper commitments, he spoke in such distant places as Peoria, Illinois, and Portland, Maine. "I am, this winter," he acknowledged in a letter, "doing more with my voice than with my pen."[2] He gave variations of the same speech in Rochester, Philadelphia, New York, Boston, Baltimore, and Washington. He titled the talk "The Mission of the War."

That speech and Lincoln's Gettysburg Address grew from the same central idea: To Douglass, the Gettysburg address was another "declaration that thereafter the war was to be conducted on a new principle."[3] But in the two hours he spent educating, entertaining, and challenging his audiences, he carried the argument beyond the groundwork he'd shared with Lincoln.

He spoke first of the war, which he described as "a rebellion which even at this unfinished stage of it counts the number of its slain not by thousands nor by tens of thousands, but by hundreds of thousands." After counting out the terrible price the nation was paying, he asked why: "For what all this desolation, ruin, shame suffering and sorrow?" Then he answered his own question: "We all know it is *slavery*."[4]

He picked up Lincoln's theme of "a broken Constitution and a dead Union" and described the alternative. "What we now want is a country—a free country—a country not saddened by the footprints of a single slave—and nowhere cursed by the presence of a slaveholder." Lincoln had spoken of a "new birth of freedom" in his Gettysburg Address; Douglass wanted a "national regeneration" that would lead to a "new order." Theirs was an overlapping vision.

Among his stops on the tour was New York's Cooper Union, still the nation's most prestigious lecture hall. His audience was the Women's Loyal League, a group that advocated a constitutional amendment to end slavery. The crowd cheered, applauded, and laughed; once he was interrupted when the women, along with the handful of men present, "[sprang] to their feet, swinging their hats, and shouting 'Hear, hear.'"[5]

Douglass admitted to the ladies that his earlier reservations about President Lincoln had yet to be erased. He still did not entirely trust the president and reminded his listeners that "Mr. Lincoln wants Union." But he made abundantly clear what Frederick Douglass wanted.

"I end where I began—no war but an Abolition war; no peace but an Abolition peace; liberty for all, chains for none; the black man a soldier in war, a laborer in peace; a voter at the South as well as at the North; America his permanent home, and all Americans his fellow countrymen." He wanted to gain for his brothers the full rights of other Americans: If they were good enough to fight, should they not have the right to vote? The Women's Loyal League agreed, greeting his call for Black suffrage with "Great Applause."[6]

Once again, the outspoken Douglass was out ahead of Lincoln. If real equality could be accomplished, he concluded, "our glory as a nation will be complete, our peace will flow like a river, and our foundation will be the everlasting rocks."*

This cartoon—labeled "Long Abraham Lincoln a Little Longer"—was published in Harper's Weekly *shortly after Lincoln won a second term in office.*

A VISIT TO THE FRONT

One June day in 1864, Abraham Lincoln boarded a river steamer headed for City Point, Virginia. He felt like he carried a great albatross on his

---

* In his thoughtful analysis of this speech and, in particular, its conclusion, biographer David Blight points out that Douglass anticipated Martin Luther King Jr.'s "I Have a Dream" speech ninety-nine years later. Blight, *Frederick Douglass* (2018), p. 421.

shoulders, the weight of a war that, at the start, many believed would last no more than ninety days. Now, after more than three years of fighting, no end was in sight.

In March, Lincoln had ordered his most successful general, Ulysses S. Grant, to come east. Grant arrived at the White House for their first meeting dressed in his rumpled traveling uniform, but Lincoln took an immediate liking to the man. Grant was all business, unpretentious, with a leadership style that won his soldiers' admiration. The two westerners shared similar origins, and the commander in chief respected both the general's willingness to speak his mind and his tenacious fighting style. Lincoln put him in charge, promoting Grant to the military's highest rank, general-in-chief of the United States.

Three months later, on Tuesday, June 21, 1864, Lincoln wondered at the wisdom of his choice. As the steamboat carrying the president approached City Point, Grant's campaign to take Richmond had fallen short, just as his predecessors' attempts had. Grant's army sustained major setbacks fighting its way toward the Confederate capital, first at the Battle of the Wilderness, then at Spotsylvania Court House. More recently, the outcome at the Battle of Cold Harbor ended badly. With no victories and almost fifty thousand casualties in three battles, Lincoln decided he needed to talk with his head general.

More than the war burdened Lincoln. A presidential election loomed less than five months away. No president since Andrew Jackson had served a second term, but Lincoln felt his work was far from complete. After a tough political fight, Lincoln had managed to gain the Republican Party nomination a few days before, so his name would appear on the November ballot, along with a new vice presidential candidate. Andrew Johnson, a former senator from Tennessee, was a compromise choice as a Democrat and a Southerner; he earned Lincoln's respect after sticking with the Union when his state seceded, and Lincoln appointed him Tennessee's military governor. Many Republicans thought that as a border man Johnson would strengthen the ticket, but with so many Americans weary of the war, the outcome of the election was far from a sure thing.

From the upper deck of the steamer, General Grant's headquarters came into view. The encampment consisted of tents for the general and his officers on a high bluff overlooking the James River. A long wooden staircase descended the steep stope to the wharves, warehouses, and hospitals, where Lincoln could see Grant and his aides making their way to the water's edge to greet him. Once the hawsers held the ship fast, the military men came aboard. Lincoln reached out "his long angular arm [and] wrung General Grant's hand vigorously" and greeted each of his officers.[7] "There was a kindliness in his tone and a hearty manner

This pencil sketch of Lincoln, son "Tad," and General Grant was drawn by Winslow Homer, then working as an illustrator. After the war, Homer became one of America's most admired painters.

of expression," one of them remembered, "which went far to captivate all who met him."

Lincoln came by choice, looking to escape the pressures of Washington, but also to consult with his army brain trust; as one general observed, "I am of the opinion that he considered himself a good judge of the time when operations should commence."[8] The president and generals talked awhile in one of the ship's cabins then went ashore. Lincoln mounted a horse, a large bay belonging to Grant named Cincinnati, and accompanied by Grant, he rode out to visit the army of more than one hundred thousand troops. He made an odd sight in his tall silk hat and frock coat, an outfit better suited to Pennsylvania Avenue than a dusty army encampment. "But the soldiers loudly cheered Uncle Abe," his unusual figure unmistakable. "As he had no straps," reported a Grant aide-de-camp, "his trousers gradually worked up above his ankles, and gave him the appearance of a country farmer riding into town wearing his Sunday clothes."

In particular, Lincoln wished to see the Black units of the Eighteenth Corps, and they welcomed him as their liberator. "They cheered, laughed, cried, sang hymns of praise, and shouted . . . "God bless Master Lincoln!' and 'Lord save Father Abraham.'" Lincoln himself was moved to tears, according to one of Grant's adjutants. "The scene was affecting in the extreme, and no one could have witnessed it unmoved."

The president remained with his army for two days, sleeping on the ship and spending his daylight hours reviewing the troops and studying the operations of the army. The visit cemented his working relationship with Grant. Lincoln found his general-in-chief's certainty that he would capture Richmond reassuring, and in the opinion of Secretary of Navy Gideon Welles, who saw Lincoln on his return, "His journey has done him good, physically, and strengthened him mentally and inspired confidence in the General and army."[9]

Lincoln wanted to win the election and, more important, the war.

With his view of the war as a fight for equality, he needed to do both to deliver the freedom he promised to the enslaved and, in particular, to Black volunteers, now a crucial part of the Union army and numbering more than a hundred and thirty thousand men.

A few weeks earlier he was asked by a Kentucky newspaperman to explain his thinking about slavery and Black soldiers. "I am naturally anti-slavery," Lincoln replied. "If slavery is not wrong, nothing is wrong." He told the correspondent that after his earlier reluctance, he accepted Black fighters because he saw it as his only good choice. On the one hand, he could "surrender . . . the Union, and with it, the Constitution"—clearly an unacceptable option—or he could draw upon "the colored element."[10]

A year into the experiment, with so many Black men having proved themselves in battle, he expressed only confidence that they were a "resource which, if vigorously applied now, will soon close the contest."[11] Grant agreed, confiding in Lincoln his belief that "arming the negro . . . is the heavyest blow yet given the Confederacy." Their shared confidence served their belief, as Grant put it, "that slavery must be destroyed," to erase the "stain to the Union."[12]

THE DARKEST OF DARK DAYS

The war news in July got worse. Hoping to distract Grant from his siege at Richmond, CSA general Jubal Early marched across Maryland and brought the war within earshot of the Capitol. Lincoln rode out in mid-July, curious to see the enemy with his own eyes. Once again a conspicuous figure in his frock coat and tall hat, he climbed atop a parapet at Fort Stevens, located within the Washington city limits and a few hundred yards from the Confederate encampment. Only after a man standing nearby took a bullet in the leg from a Rebel sharpshooter did Lincoln remove himself from the line of fire. The next day Early and his men

According to legend, when Lincoln stood exposed on the rampart at Fort Stevens, in July 1864, future Supreme Court justice Oliver Wendell Holmes Jr. yelled, in what is likely an apocryphal version of the story, "Get down, you damn fool, before you get shot!"

COURTESY DC PUBLIC LIBRARY, THE PEOPLE'S ARCHIVE

retreated back toward Confederate lines, but the intimidating message was clear: Lee's army was alive and well and not far away.

As July ended, Grant's siege in central Virginia met with a spectacular failure at the Battle of the Crater. After a month of secretly tunneling beneath a Confederate fortification, Union engineers set off an immense explosion, blowing a hole in the perimeter defense at the city of Petersburg, a few miles south of Richmond. A last-minute decision not to lead the follow-up attack with a division of United States Colored Troops meant that the White soldiers on point were poorly briefed. Most of a Rebel regiment died instantly in the blast, but the defenders recovered quickly; they were soon slaughtering Union men who, slow to attack, wandered into the crater. Confederate general William Mahone joked it was "a turkey shoot." General Grant called it "the saddest affair I have seen in this war."[13]

A new strategy emerged in a conversation Lincoln had with a minister named John Eaton, who arrived in Washington on August 10, 1864. Though a chaplain in an Ohio regiment, the Dartmouth-educated Eaton was tasked with establishing camps for freed Blacks in the lower Mississippi Valley. Nearly two years into the job, he supervised more than one hundred thousand people. Many of the fugitives in his care arrived barefoot and dressed in rags, traumatized, illiterate, and with few skills. Some were sick and dying. But Eaton, an ardent abolitionist before the war, found his calling in guiding the transformation of the so-called "contrabands" to free citizens. At Davis Bend, an island plantation confiscated from Jefferson Davis's family, his "Negro Paradise" produced food and goods for the Union army. Theirs was a new community of independent farmers, one based on freedom, not bondage. According to some reports, it was now the most productive plantation in the region.

On arriving in Washington, Eaton went directly to the White House. Lincoln, who was thinking long and hard about the postwar fate of millions of Blacks formerly in bondage, knew of Eaton's Mississippi project and wanted to know more. The minister was struck by how much the president already knew and his keen understanding of what the changed circumstances meant to Black men being compensated for their labor, to Black women who felt safer than they ever had, and to their children, many of whom were attending school for the first time.

Pleased by Eaton's report, Lincoln shifted the conversation to soldiering as a "means by which the Negro could be secured in his freedom, and at the same time prove a source of strength to the Union." When Reverend Eaton told Lincoln that, on his way east just days before, he'd talked at length with Frederick Douglass after a lecture in Toledo, Ohio, an idea struck the president. As Eaton recalled the moment, "The greatest man of his time asked me, with that curious modesty characteristic of him, if I thought Mr. Douglass could be induced to come to see him."[14]

With Reverend Eaton as the connection, Lincoln invited Douglass to the White House for what would be their second meeting. He wished to discuss what he called the "grapevine telegraph."

THE GRAPEVINE TELEGRAPH

On Lincoln's orders, a carriage met Frederick Douglass at Washington's Baltimore & Ohio rail station nine days later. After arriving safely at the White House, Douglass learned that the president was already in a meeting, so he settled into a seat in the reception room, awaiting his turn.

Douglass began to read, but his mere presence was out of the ordinary; more than just another dark-skinned servant in the nation's largest house, he was a Black man waiting to meet with the president, man-to-man. Yet Douglass felt surprisingly at home—until Joseph Mills, a Wisconsin Republican, entered the room.

Unable to process what he saw, Mills froze. He couldn't take his eyes off Douglass.

Minding his own business in a corner chair, Douglass sensed the weight of the man's stare and looked up. Caught in the act, fixed by Douglass's return gaze, Mills felt obliged to address him. He inquired who he was.

"I am Frederick Douglass," was the solemn reply.[15]

Out of shock or outrage, Mills, a man to whom words like "darkey" and "Sambo" came easily, summarized his exchange with Douglass in his diary. He mentioned his waiting-room conversation to Lincoln when he later met with the president, and jokingly raised the subject of "miscegenation," a recently coined label for racial intermarriage, which was seen by racist Whites as a danger to racial purity.*

---

* The word first appeared in a pamphlet titled *Miscegenation: The Theory of the Blending of the Races, Applied to the American White Man and Negro* (1864). The work

Lincoln's reaction to his conversation with Mills was not recorded, and for Douglass, too, the brief encounter was unimportant and got no mention in his subsequent writings or correspondence. Most likely the moment simply blurred with countless others in Douglass's experience when he looked into the eyes of a White man and saw the disregard of someone who thought him less than a man.

As he waited, Douglass did not know the purpose of his summons. He was, however, acutely aware of his recent harsh criticism of President Lincoln. Again and again Douglass had told audiences that the commander in chief was still not doing enough to protect Black soldiers. The shocking aftermath of the Battle at Fort Pillow was a telling example.

Back in April, a Rebel force commanded by General Nathan Bedford Forrest overwhelmed a Union earthworks just upriver from Memphis, Tennessee. In the aftermath of the fight at Fort Pillow, the victors massacred more than two hundred Union soldiers, killing mostly Black troops but also Whites who courageously fought at their side. According to survivor accounts, Forrest's men executed wounded soldiers in cold blood, murdered Black soldiers who held their hands in the air, and shot those fleeing for their lives and freedom in the back. They even buried some wounded men alive. General Forrest, who after the war helped found the Ku Klux Klan, boasted that the bloodletting dyed the Mississippi River red for a distance of two hundred yards. Lincoln's failure to retaliate after Fort Pillow planted new doubts in Douglass's mind concerning the president's commitment to his Black soldiers.

On entering Lincoln's office a few minutes later, Douglass noticed the toll the presidency had taken in the twelve months since their last meeting. Lincoln looked pale, his eyes sunken and dark below his fur-

---

of two New York newspapermen, it was a hoax, part of a political disinformation campaign to advance the cause of those opposed to racial equality. The term would have surprising staying power both in the United States and abroad, with usage up to the present day by contemporary White supremacists.

rowed brow. This was a bewildering moment for the nation but particularly so for Lincoln, and with surprising candor, he got right to business. Douglass quickly learned that the causes of the president's "alarmed condition" were the gloomy war prospects and the November election.[16]

The most powerful newspaper editor in the country, Horace Greeley, now urged Lincoln to negotiate with the Rebels. That very morning, a Washington paper, the *Daily National Intelligencer*, featured the headline "How to Make Peace." George McClellan, formerly Lincoln's head general, was emerging as Lincoln's likely Democratic opponent in November, and the president worried that in the current climate, with the public clamoring for peace, McClellan, who wanted to negotiate with the Rebels, might win. If he did, Lincoln told Douglass ominously, a peace might be forced on the country that left slavery in place. Emancipation was at risk since a Lincoln loss might translate to no constitutional amendment ending slavery. With only about one in twenty of the enslaved in the South freed, that meant the war would have accomplished little.

Lincoln produced a letter from his desk. Faced with the public accusation that he valued abolition more than peace, he had drafted a response. He handed the handwritten sheet to Douglass. It would be another public letter of the kind that, over the course of his presidency, he published periodically to explain his positions.

As he read it, Douglass realized Lincoln proposed uncoupling emancipation and peace, promising not to make the first a precondition for the second. It was a political compromise, one that Douglass himself could never accept. But he could also see that Lincoln was taking him into his confidence. Lincoln looked to him, a Black man, for help as he shaped a matter of policy. Asking Douglass for his counsel seemed an extraordinary compliment, an unprecedented gesture of profound respect.

When Douglass looked up from the document, Lincoln asked, "Shall I send forth this letter?"

Douglass did not mince words; the question demanded a truthful answer.

"Certainly not," he replied.

Lincoln asked why. Again, Douglass spoke directly.

"It would give a broader meaning than you intend to convey; it would be taken as a complete surrender of your anti-slavery politics, and do you serious damage."

Having asked for Douglass's opinion, he got it in no uncertain terms. Lincoln shifted the discussion.

"The slaves are not coming so rapidly and so numerously to us as I had hoped."

Douglass understood both why this was so and why Lincoln did not see it. The president was a White man who, despite growing up poor and in an isolated place, never had blinders imposed on what he saw or learned. Douglass's experience differed; he knew more fugitives didn't come running to the cause simply because they did not know that they were welcome in the U.S. Navy and Army. In the Cotton Kingdom, as during Douglass's childhood in rural Maryland, slaveholders kept enslaved persons ignorant of events and circumstances beyond the boundaries of their plantations. He laid it out for Lincoln, and after a moment's consideration, Lincoln began again.

"Well," he told Douglass, "I want you to set about devising some means of . . . bringing them into our lines." The idea was to organize "a band of scouts, composed of colored men . . . to go into the rebel States, beyond the lines of our armies, and carry the news of emancipation, and urge the slaves to come within our boundaries."[17]

Lincoln asked for *his* help; as if that wasn't extraordinary enough, Douglass sensed a feeling of déjà vu in the air. Lincoln's proposal was a reworking of the "Subterranean Pass-Way" that John Brown had laid out in Douglass's Rochester parlor half a dozen years before. Back then, Douglass decided at the last minute not to go along with what had become a dangerous military action, the raid on Harper's Ferry. Now, however, the man asking Frederick Douglass to take part in a secret, pass-the-word plan was the nation's chief executive.

Douglass was more than a little surprised that Lincoln was embracing a radical idea well beyond what he expected of the usually cautious president. But he agreed to think on the matter and report back.

One of Lincoln's secretaries appeared, announcing the arrival of Connecticut's governor. William Buckingham was a strong Union supporter and personal friend of the president. Douglass immediately stood, saying, "I must not stay to prevent your interview with Governor Buckingham."

Lincoln wouldn't have it. "Tell Governor Buckingham to wait," he instructed his young aide, "for I want to have a long talk with my friend Frederick Douglass."

"Mr. Lincoln, I will retire," Douglass offered again. Once again, Lincoln insisted he stay.

Their conversation last for hours. Despite stark differences, the two men's backgrounds dovetailed. Once children of no importance, they had found vastly different routes to prominence. Men of powerful intellect, each possessed a gift for words and deep empathy for other men. A friendship, that inexplicable bonding of shared instincts, truly took shape in the White House that day.

When a secretary interrupted once more to remind Lincoln that the governor awaited, he again asked that Buckingham be told to wait. Douglass read this gesture as a huge compliment, "the first time in the history of the Republic when its chief magistrate had found an occasion or shown a disposition to exercise such an act of impartiality between persons so widely different in their positions and supposed claims upon his attention." Nor did Governor Buckingham, who was admitted as Douglass was ready to depart, take the least offense at being asked to wait. Douglass felt embraced by Lincoln's magnanimity.

. . .

IN THE DAYS and weeks that followed, Lincoln took Douglass's advice regarding his public letter: He did not publish it, in part because his new, if informal, adviser on slavery affairs, Frederick Douglass, told him not to.

As for Douglass, he left his meeting with Lincoln excited at the new task, and he promptly consulted with "several trustworthy and Patriotic Colored men" regarding a recruitment network. Ten days later, he wrote as he'd promised to do, sketching a plan for spreading the word in the South. He proposed finding two dozen agents who would identify local subagents to persuade enslaved men to join the "Loyal Cause." He suggested a rate of pay (two dollars a day) and that a general agent (presumably himself, though he did not say so) take charge of the scheme. The plan was preliminary, he said, while assuring Lincoln that "All with whom I have thus far spoke on the subject, concur in the wisdom and benevolence of the Idea."[18] Before Lincoln could read the letter from Rochester, however, the momentum of the war shifted.

In Chicago, the Democratic convention at the end of August chose George McClellan as its nominee for president. But the party's platform, which argued the war was a failure and the time had come to seek peace—mostly likely peace with slavery—quickly seemed out of step with the pace of events: On September 2, Union soldiers marched into Atlanta. In Washington, Lincoln and his generals read with elation the telegram from General William Tecumseh Sherman that said it all: "Atlanta is ours and fairly won."[19] The war seemed suddenly winnable and Union spirits rose.

The shift in the tides of war lifted Lincoln's electoral fortunes, carrying him, on November 8, 1864, to another term in office, when he soundly defeated McClellan, with an Electoral College vote of 212 to

21.* Douglass had become a firm supporter of Lincoln, telling John Eaton, "I am satisfied now that he is doing all that circumstances will permit him to do." But he chose to stay relatively quiet during the campaign, aware that some voters might be less likely to vote for Lincoln if they saw the candidate too closely allied to a controversial Black man.

A HOMECOMING

For Frederick Douglass, as for Lincoln, November 1864 would be memorable. On the first day of the month, the state of Maryland formally abolished slavery within its borders, and Douglass decided to go to his childhood home for the first time in more than a quarter century.

Among the first to welcome him in Baltimore was his sister Eliza Bailey Mitchell. They had been unable to communicate since his departure, since slave correspondence was forbidden and, in any case, she was unlettered. Even so, Douglass found, she had made her way in the world. "Mammy Liza" and her husband, Peter Mitchell, had purchased her freedom—for one hundred dollars—and she birthed nine children. From afar she followed her brother's rise to fame and even named a daughter, Mary Douglass Mitchell, after him. Hearing of his well-publicized return to Maryland, she made the sixty-mile journey from Talbot County for a reunion with Fred, her elder by four years.

Arm in arm, brother and sister walked down the aisle on November 19 at what had once been Douglass's spiritual home, Baltimore's Bethel African Methodist Church, in his old Fells Point neighborhood. Other

* In the Fourth Ward in Tennessee, Blacks held an election; though their ballots would not be counted in the official results, their tally is worth noting. Lincoln's ticket garnered 3,193 votes to McClellan's 1. Quarles, *Lincoln and the Negro* (1962), p. 217.

Black Marylanders welcomed their famous son at the first of a series of lectures he gave during his two-week visit to Maryland. The church was packed with both African American and White citizens. In itself that was a revelation to Douglass, since the Baltimore he knew was a place where the divisions between the races had been rigid. Some of the faces looking up at him he recognized, now aged but familiar; others were strangers who came to hear the famous orator.

Speaking from the pulpit, surrounded by American flags, Douglass delivered an address unlike the many speeches for which he had become famous. He exhibited neither anger nor bitterness, but told his audience he came "not to condemn the past but to commend and rejoice over the present."[20] He offered not an indictment of slavery and slaveholders; instead, he spoke of himself in a personal way, something he

When Frederick Douglass returned to his childhood place in November 1864, he asked to visit the Lloyd family graveyard. "Everything about it [was] impressive," he wrote, "and suggestive of the transient character of human life and glory."

had rarely done in recent years. He had departed, he told them, "in the full fresh bloom of early manhood [when] . . . not one lock of all my hair was tinged by time or sorrow." Now he returned older and wiser, with "the early frost of winter . . . beginning to thicken visibly on my head."

He told them his life was bookended by this place. "My life has been distinguished by two important events, dated about twenty-six years apart. One was my running away from Maryland, and the other is returning to Maryland to-night." He left, he explained, "not because I loved Maryland less, but freedom more."

He talked for three hours, but his audience remained riveted to the son of Maryland who returned to walk among them. "Not a man or woman left the church," reported one Maryland newspaper, "[nor] evinced the faintest indications of weariness."[21]

He made his pitch to the White citizens present to permit their Black neighbors to vote. "If the negro knows enough to pay taxes, he knows enough to vote." But he finished his speech with the best advice he could muster for a room packed with people who had been enslaved.

> You are in one sense free. But you must not think that freedom means absence from work. Bear that in mind. I would impress it upon your minds, that if you would be prosperous, you must be industrious. . . . The black man is just as capable of being great as the white. All he needs is an effort—a persistent, untiring effort. You have now the opportunity, and I trust you will improve it.

The world as Douglass knew it in his boyhood was changing, though slowly, since another six years would pass before Maryland permitted African Americans to vote. But Douglass, along with his friend Lincoln, was helping to change it.

FIFTEEN

MY FRIEND DOUGLASS

One eighth of the whole population were colored slaves, not distributed generally over the Union, but localized in the Southern part of it. These slaves constituted a peculiar and powerful interest. All knew that this interest was, somehow, the cause of the war.

—Abraham Lincoln, Second Inaugural Address, March 4, 1865

Abraham Lincoln could neither win the war nor abolish slavery by himself. But after emerging as the clear victor in the November 1864 election, he rode the momentum of the moment, driving the country in the direction he wanted.

On the battlefront, General Sherman made his contribution. As the year approached its end, his army succeeded in dividing the Confederacy at its waistline, completing the "March to the Sea" across Georgia. From his new headquarters in a fine coastal mansion, he wrote to Lincoln, "I beg to present you as a Christmas gift, the city of Savannah."[1] In the new year, Sherman would continue the march, moving north to cinch the noose around the shrinking Confederacy. Rebel soldiers, seeing their cause was doomed, began deserting in droves.

Lincoln's policy of soft persuasion in convincing the border states to free those in bondage was slowly producing results. When Unionist counties in northwest Virginia became West Virginia, in 1863, the new state's constitution provided for emancipation. After Maryland freed its enslaved citizens in November 1864, Missouri followed suit, in January 1865; the recaptured Rebel states of Arkansas and Louisiana were also slave-free. But Kentucky and Delaware lagged and some three million men, women, and children in Rebel territory remained enslaved. A permanent national solution was necessary.

Lincoln threw his full support behind the Thirteenth Amendment, a measure that would prohibit slavery anywhere in the United States and its territories. Though passed by the Senate in the spring of 1864 and incorporated into the Republican platform for the November election, the amendment still awaited approval in the House of Representatives. The climactic vote took place on January 31, 1865.

The House chamber was packed. Dozens of senators came to watch, as did five Supreme Court justices. The press gallery was invaded by a mob of well-dressed women, some of whom had launched a petition drive for universal emancipation two years earlier. Black citizens came in force, too, and one spectator remarked, "It was quite a pepper and salt mixture."[2] Although Lincoln did not attend—he waited for news, nervous and attentive, at the White House—his intense lobbying and dealmaking looked to have broken the earlier impasse.*

Douglass, too, was elsewhere, delivering a lecture in upstate New

* The process of passing the Thirteenth Amendment is memorably dramatized in Steven Spielberg's movie *Lincoln* (2012).

York, but his son Charles, no longer a soldier but working at the Freedmen's Hospital in Washington, witnessed the calling of the roll. The result of the three o'clock vote was not a foregone conclusion since the measure twice before failed to reach the two-thirds vote required for passage. Republicans still worried about the wavering Democrats they needed to reach the two-thirds threshold, but the result—119 yeas to 56 nays—approved the amendment with two votes to spare.

When Speaker of the House Schuyler Colfax announced the official tally, "there was a pause of utter silence, as if the voices of the dense mass of spectators were choked by strong emotion."[3] Then followed an eruption of cheers, applause, and foot stomping. The deafening bedlam

The passage of the Thirteenth Amendment, as portrayed here in Harper's Weekly, *was cause for a raucous celebration on the floor of the House of Representatives.*

ended only when the shouts of an Illinois congressman for adjournment were seconded. The motion quickly carried, and a mass exodus took the celebration to the streets.

"I wish you could have been there," Douglass's youngest son wrote to his father. "Such rejoicing I never before witnessed, cannon firing, people hugging and shaking hands (white people I mean), flags flying all over the city."[4]

When a large procession of celebrants accompanied by a marching band arrived that evening at the executive mansion, Lincoln made an appearance in one of the tall windows over the entrance portico to give a brief speech. Looking down upon the faces lit by the flickering light of the torches they carried, he agreed that this was indeed a suitable moment to celebrate this "indispensable adjunct to the winding up of the great difficulty." Slavery caused the war; now they were about to abolish it. He congratulated everyone, himself and the crowd included, along with "the country and the whole world on this moral victory."[5]

"TEARFUL SKIES"[6]

One month later, on March 4, 1865, inauguration day dawned wet and windy. General Sherman's army was marching north through the Carolinas, and few doubted the war was rapidly coming to a close. But the work of freeing those who were enslaved wasn't complete; eighteen states had quickly ratified the Thirteenth Amendment in February; nine more were needed for it to become valid.* Still, there was much to celebrate with Lincoln about to take the oath of office for the second

* Lincoln would not live to see ratification. The Thirteenth Amendment reached the constitutionally required threshold of three-fourths of the states only after Georgia passed the measure nine months later, on December 6, 1865.

time. Thousands of spectators came to town. The hotels were full, with guests sleeping on cots in hotel hallways. A large crowd of some thirty thousand people began to assemble at the Capitol.

When Lincoln climbed aboard his carriage for the midmorning ride from the White House, he saw a bedraggled crowd lining Pennsylvania Avenue, their lace, velvet, and woolens dripping wet from a hard rain. The spectators watched as for the first time African American soldiers marched with the other troops, bands, and floats in the inaugural parade, with four companies of the Fifty-Fourth Regiment of the United States Colored Troops joining the presidential escort.

Roughly half of those in attendance at Lincoln's second inaugural were Black; perhaps foremost among them was Frederick Douglass, who had arrived in the city a few days before. When he caught sight of the president's carriage and its four horses, he worked his way as close to the president's barouche as he could. He would recall years later that on inauguration day he had felt a "vague presentiment" that there might be "murder in the air."[7] His memory may well have been enhanced by a knowledge of later events, but the fear was by no means his alone. In response to rumors of a conspiracy to take the president's life, the secretary of war increased security in the city, detailing extra guards to protect Lincoln and deploying detectives to observe the many Rebel deserters in Washington.

Inside the Capitol, out of sight of the milling throng outdoors, Lincoln watched as a red-faced Vice President Andrew Johnson took the oath of office in the Senate Chamber. Smelling strongly of whiskey, Johnson delivered a rambling speech. Meanwhile, Douglass found himself a prime spot outside from which to view Lincoln's swearing in.

When the president stepped onto the east portico, "a tremendous shout, prolonged and loud, arose from the surging ocean of humanity around the Capitol building."[8] As the noise of the crowd faded, Douglass was watching when Lincoln, who stood beside his tipsy vice president, pointed at him. Lincoln was clearly pleased to observe that his friend was in attendance, but Douglass, near enough to read Johnson's expression, could also clearly see the "contempt and aversion" that rippled across Johnson's features. When the new vice president registered that Douglass returned his gaze, he quickly erased his scowl to assume a more neutral expression.

"I got a peep into his soul," Douglass thought. "There are moments in the lives of most men when the doors of their souls are open, and unconsciously to themselves, their true characters may be read by the observant eye. It was at such an instant I caught a glimpse of the real nature of this man." As subsequent developments would prove, Douglass's insight at that moment was accurate. "I felt that, whatever else this man might be," Douglass said, "he was no friend to my people."[9]

Lincoln delivered his second inaugural speech from the east portico of the Capitol. In this image, he sits just left of center, his white shirt visible. Much more difficult to discern are John Wilkes Booth, in the balcony above, and Frederick Douglass, in the crowd below.

Another man watching the ceremony with keen interest was a well-known actor named John Wilkes Booth. Given a ticket to a balcony seat by a woman he was courting—she happened to be a senator's daughter—he would be no more than a witness, just as he was at the execution of John Brown five years before. His performance on the national stage came soon enough.

703 WORDS

Tall and gaunt, Lincoln stepped forward just as the sun penetrated the leaden gray clouds for the first time that day. In an extraordinary meteorological coincidence, the actor at center stage stood bathed in brilliant sunlight.

He held not a sheaf of papers but a single sheet, and from his close vantage, Douglass could see this would not be a long speech.

In the four paragraphs of what would be the shortest inaugural address ever delivered by an American president, Lincoln chose not to be triumphalist.[10] He could have celebrated the victory that now seemed assured, but instead Lincoln reminded the crowd this was a war that no one wanted. Rather than claiming victory in what he called "the great contest," he acknowledged only that "the progress of our arms" was "encouraging to all." He wished instead to talk about other matters.

He looked back four years in time to his first inauguration when the country was not yet at war with itself. The national division had been enormous, Lincoln noted, between those who "would *make* war rather than let the nation survive" and those who "would *accept* war rather than let it perish." With sadness he noted his negotiating attempts had failed and then "the war came."

For his listeners, this speech already defied the conventions of the day. People expected orators to present them with a long-form rhetorical argument, and in the past Lincoln typically delivered performances that persuaded, cajoled, and entertained. But as at Gettysburg sixteen months earlier, Lincoln took a different approach, one rich in religious reverence rather than politics.

Slavery, he said bluntly, was "the cause of the war." It might seem "strange that any men should dare to ask a just God's assistance in wringing their bread from the sweat of other men's faces," Lincoln observed, for that *was* what slavery amounted to. But he again shifted his tack: On this national day, he wished not to place blame or seek retribution; "let us judge not, that we be not judged." Instead, he raised the promise of peace and appealed to a higher power to bring about an end to the war. He looked not to reason but to conscience and faith, pointing out that both sides "read the same Bible and pray to the same God."

"Fervently do we pray," he said, "that this mighty scourge of war may speedily pass away." The bloodletting of the terrible war was "true and righteous," the high price paid for ending enslavement. Yet there was, Lincoln believed, a divine balance in the war's events.

Neither Lincoln nor Frederick Douglass, who stood listening intently, put great stock in church rituals. But each held a deep faith in the Almighty's capacity for goodness and justice. Douglass recognized that, rather than reading a state paper, Lincoln was, in effect, delivering a sermon. Hearing the president acknowledge slavery as the war's cause was, for Douglass, both a notion he had insisted upon for years and a particular balm coming from Lincoln as he spoke forthrightly to the nation and the world.

Lincoln wrapped up. With the dignity and economy of lines from Shakespeare, his final, intricate sentence rang out over the crowd. It was a plea for the war's end, one full of phrases destined to enter the American vernacular.

> With malice toward none; with charity for all; with firmness in the right, as God gives us to see the right, let us strive on to finish the work we are in; to bind up the nation's wounds; to care for him who shall have borne the battle, and for his widow, and his orphan—to do all which may achieve and cherish a just and a lasting peace, among ourselves, and with all nations.

He was done. After listening for a bit more than six minutes, the crowd, standing ankle-deep in mud, responded with a roar of applause. To Frederick Douglass, as perhaps never before, Lincoln spoke as prophet, delivering a radical speech that blended prayer and sermon, one that equated the blood spilled by the swords of war with the bleeding suffered by the enslaved at the hands of slaveholders. Having linked slavery and war as cause and effect, Lincoln asked for reconciliation.

He turned to the chief justice of the Supreme Court to take the oath of office. After he agreed to its terms with the words "So help me God," a cannon salute boomed. Abraham Lincoln's second term in office, abbreviated though it would be, officially began.

A FINAL MEETING

At the urging of Black friends, Frederick Douglass went to the White House that evening. Inspired by Lincoln's speech, he felt himself "a man among men" and ignored the taboo that forbade mingling with Whites at a presidential reception. He decided that "a colored man [might] offer his congratulations to the president."[11]

After waiting in a long line with thousands of other citizens, Douglass reached the entrance of the executive mansion. Then two guards suddenly interrupted the flow of people, taking Douglass firmly by the arms.

"No persons of color," they warned.

Unwilling to go quietly, Douglass objected. He insisted that President Lincoln issued no such order. The two burly men in blue uniforms relented—or they seemed to—and "with an air of politeness" permitted him to enter. Once inside, however, they marched him directly to a tall window adapted as an exit.

Douglass halted. "You have deceived me," he said. "I shall not go out of this building till I see President Lincoln." Before the standoff could escalate, Douglass recognized a passerby allowed to enter. "Be so kind as to say to Mr. Lincoln that Frederick Douglass is detained by officers at the door." The man hurried away, and the message, promptly delivered, meant Douglass soon got his wish.

"I walked into the spacious East Room, amid a scene of elegance such as in this country I had never witnessed before." The Marine Band played as women in silk gowns, generals in their uniforms, and members of the cabinet conversed.

Douglass spotted Lincoln in a receiving line, shaking hands and accepting congratulations, towering "like a mountain pine high above all others." Lincoln saw Douglass, too, and in a loud voice that carried over the noise in the room, he called out, "And here comes my friend, Frederick Douglass." Many eyes turned to see the broad-shouldered man, his hair streaked with silver, maneuvering his way through the elegantly dressed guests.

When he approached, Lincoln greeted him, taking Douglass's hand between his two. "Douglass, I saw you in the crowd to-day, listening to my inaugural address. How did you like it?" As one young Union officer noted in his diary that night, "The reception of Douglass was the most cordial of any I saw."[12]

With a long line of well-wishers waiting to greet the president, Douglass respectfully demurred. "Mr. Lincoln, I must not detain you with my poor opinion, when there are thousands wanting to shake hands with you." But in this very public moment, the president wanted more from his new friend.

"You must stop a little Douglass; there is no one in the country whose opinion I value more than yours."

Given no choice, Douglass obliged. Rarely a man of few words, he spoke from the heart as he gave Lincoln what he asked for.

"Mr. Lincoln, that was a sacred effort."

Some of those surrounding the two men smiled; others frowned. A few looked astonished that such an exchange could even happen, since few politicians would have dared welcome Douglass out of fear that mere association with his name would prove toxic in his next campaign. But Lincoln was delighted with Douglass's words: This listener, at least, grasped his essential point.

"I am glad you liked it," he replied. Douglass, deeply honored by Lincoln's attentions but caught by the tide of the crowd, moved on.

Not everyone admired or even understood the speech. The *New York World* thought it an "odious libel" to compare the carnage of the war with the beatings of Negroes and complained that Lincoln had substituted "religion for statesmanship." Other editors found it perplexing that Lincoln didn't celebrate the coming victory. Horace Greeley thought it lacked "generosity." *The Daily Ohio Statesman* described the address "as chilly and dreary as the day on which it was delivered."[13]

Lincoln appeared not to mind, acknowledging that he had anticipated that not everyone would like it, though he thought it would "wear as well as—perhaps better than—anything I have produced."[14] Over time, people would indeed come to agree with his offhanded assessment—"Lots of wisdom in that document, I suspect"—and with Douglass's admiration for it.[15]

Lincoln and Douglass's brief exchange on March 4, 1865, would be their last.

In this hand-tinted lithograph, titled Lincoln's Last Reception, *the Lincolns greet members of his cabinet, Union generals, and other visitors at a White House reception in the spring of 1865.*

SIXTEEN

APRIL IS THE CRUELEST MONTH

[Douglass is] one of the most meritorious men in America.

—Abraham Lincoln

With the war ending, Lincoln's spirits rose. In late March 1865, he decided to return to Grant's headquarters at City Point, Virginia, and for more than a week he occupied a cabin on the steamboat *River Queen.* He descended her gangplank daily to visit the wounded, mingle with the troops, and confer with Grant and even General Sherman, who made a turnaround visit from his campaign in North Carolina. Lincoln wanted to be near when Richmond finally fell.

Then it happened. On Sunday, April 2, Jefferson Davis and his government abandoned their capital on a midnight train, headed for Danville near the North Carolina border. Before dawn on Monday, explosions shook Richmond when Rebel soldiers lit powder magazines aboard ironclads in the harbor. They fired bridges and warehouses, too, and soon the flames spread from the waterfront, climbing the hillside into the city. By the time Union troops moved in a few hours later, the conflagration, sweeping from block to block, engulfed much of the Confederate capital. Union troops and military police raced to cordon off the fire, and by nightfall, had it under control.

Overall victory in the Civil War now seemed imminent, and Lincoln could contain himself no longer. "Thank God I have lived to see this!" he told Admiral David Porter. "It seems to me I have been dreaming a horrid dream for years, and now the nightmare is gone. I want to see Richmond."[1]

Against the advice of the secretary of war, on Tuesday morning, April 4, 1865, the *River Queen* slipped her moorings and steamed up the James River, accompanied by Porter's flagship, the USS *Malvern*; a gunboat named the *Bat*; a transport; and a tug. The flotilla, however, halted a few miles south of its destination, its path blocked by several rows of heavy pilings the Rebels had driven into the riverbed.

Lincoln would not be deterred. Together with son Tad, who was celebrating his twelfth birthday, three junior officers, and Admiral Porter, Lincoln clambered into a twelve-oar barge. The much smaller craft then safely maneuvered through a narrow channel between the piles. Lincoln's arrival in Richmond would not be a grand affair with flags flying and cannons firing in salute to the commander in chief. But Lincoln didn't mind. "It is well to be humble," he told Admiral Porter.

With twelve marines pulling against the current, the small, solitary craft reached Rockett's Landing, roughly a mile south of central Richmond, about 11:00 A.M.[2] Lincoln knew this visit to Richmond had its perils. He entered what, for four years, had been enemy territory, a city where people spoke his name as if it were a curse. The Union general now in charge at Richmond had no idea he was coming, so no escort awaited the barge to convoy Lincoln to U.S. Army headquarters, more than a mile away.

A group of a dozen Black laborers were working nearby when Lincoln stepped ashore. The day was warm and sunny, and the elderly leader of the crew shaded his eyes to peer at the new arrivals. He immediately recognized the gangly figure of Lincoln wearing his stovepipe hat. Glimpses they had seen of woodcuts and caricatures told them this was "the Great Emancipator."*

"Bless the Lord, here is the great messiah!" the man called, as he and his coworkers dropped their shovels. They ran to meet Lincoln, and the leader fell to his knees, crying "Glory, Hallelujah!"

"It was a touching sight," Porter reported in a memoir twenty years later, "that aged negro kneeling at the feet of the tall, gaunt-looking man who seemed in himself to be bearing the grief of the nation."[3]

Lincoln was plainly embarrassed.

"Don't kneel to me," Lincoln instructed the men. "That is not right. You must kneel to God only, and thank him for the liberty you will hereafter enjoy."

Surrounding the president, the formerly enslaved men broke into song, serenading Lincoln with a hymn of thanks. In the quiet city, their voices carried and nearby streets, deserted minutes before, began to come to life. Patrolled by Black troops, Richmond was under martial law, but as word of Lincoln's arrival spread, men came running from the waterfront. Families emerged from nearby buildings. "Some rushed forward to try and touch the man they had talked of and dreamed of for four long years, others stood off a little way and looked on in awe and wonder."

As the crowd closed in, Admiral Porter worried the crush of happy people posed a danger. He ordered the marines into formation, surrounding the president, who held Tad's hand. To shouts of "Bless the

---

* The origin of the title "Great Emancipator" is not clear. The Italian revolutionary Giuseppe Garibaldi used it in a letter to Lincoln in August 1863, although there is also conjecture that it is a coinage devised by Blacks after Lincoln issued the Emancipation Proclamation.

As Lincoln walked into the Confederate capital of Richmond after it fell to Union troops in April 1865, an artist traveling with the president's entourage captured him extending his hand to a Black Union soldier.

Lord, Father Abraham's Come," the outnumbered phalanx moved toward the Virginia capitol more than a mile away.

They made slow progress as the day grew warmer. They walked past Libby Prison, where many Union prisoners had been held. From time to time, Lincoln removed his hat to fan himself. Perspiration dampened his brow.

The freedmen trailing Lincoln saw him as Frederick Douglass could not. Douglass had been the exception: He was among the minority of enslaved people who, through some mix of bravery, resourcefulness, and opportunity, made their own way to freedom. In contrast, most of the men, women, and children who surrounded Lincoln had remained enslaved until just days or even hours before. They had been considered property, wholly owned chattels. Some would have managed to escape at great personal peril; others would over time have bought their freedom. If not for the war, the Thirteenth Amendment, and Abraham

Lincoln, most feared a fate of permanent servitude, with a good chance their children would be sold away from them. This man seemed the personification of their freedom.

As one point, having collected his thoughts, Lincoln spoke to the throng. "My poor friends," he said, "you are free—free as air. . . . Liberty is your birthright. God gave it to you as he gave it to others, and it is a sin that you have been deprived of it for so many years." The crowd shouted in joyful approval, but when Lincoln resumed speaking, they fell silent.

"You must try to deserve this priceless boon," Lincoln continued. "Let the world see that you merit it, and are able to maintain it by your good works." The advice might have been spoken by Douglass; it closely resembled his unrehearsed remarks at Baltimore on the emotional day the previous fall when he returned to Maryland after more than two decades in exile. "Don't let your joy carry you into excesses," Lincoln continued. "Learn the laws and obey them; obey God's commandments and thank him for giving you liberty."

With the speech over, the crowd opened to permit Lincoln to pass.

As Lincoln and company got closer to the Virginia capitol, the sidewalks were lined with people. Many Whites were framed in windows and doorways, but the procession became a largely African American parade, with Lincoln leading the dense crowd that trailed in his wake, the president's tall form recognizable to all. They passed the Tredegar Iron Works, where smoke still spiraled into the sky. Finally, after more than an hour, the entourage arrived at Jefferson Davis's residence. Lincoln, looking pale and haggard, disappeared into the tall home.

Once inside, hot and tired, he flopped into an easy chair in Davis's study. One of Davis's enslaved house servants brought him a welcome glass of cool water. After eating lunch, Lincoln toured the remains of the captured city, riding in an open carriage, accompanied by an entourage of Black cavalry.

. . .

FREDERICK DOUGLASS TOOK THE STAGE at Faneuil Hall after the news of Richmond's capture reached the Massachusetts capital. Boston reverberated with celebrations, and Douglass, one among numerous speakers at the famous hall, kept his remarks brief. With shared pride, he pointed out that the Fifth Massachusetts Cavalry, a Black regiment, had been the first to enter Richmond. And he reminded the crowd of the war's purpose.

"I rejoice, fellow-citizens—for now we are *citizens*. . . . What I want, now that the black men are citizens in war, is, that shall be made fully and entirely, all over this land, citizens in peace."[4]

A WEEK LATER, on April 11, General Robert E. Lee met General Ulysses S. Grant at Appomattox Court House, Virginia, to discuss terms and affix his signature to a document of surrender. Robert Todd Lincoln, the president's oldest son, stood by as a Grant adjutant. On returning to his men after the agreement was signed, Grant told them, "The war is over; the rebels are our countrymen again." The words were Grant's, the sentiment Lincoln's.

That evening, when the news reached him in Washington, the president ordered the firing of five hundred cannons at daylight in celebration. Lincoln, Douglass, and the nation regained peace with what, for practical purposes, was the end of the Civil War.*

---

* Three contingents of Rebel soldiers would fight on for a time, but on April 26, May 4, and May 12, Generals Joseph E. Johnston in North Carolina, Richard Taylor in Alabama, and Edmund Kirby Smith in Texas, respectively, surrendered their commands.

A contemporary woodcut depicting Lee's surrender. "The memorable event," reads the caption, "terminated the Great Rebellion." Ulysses S. Grant stands at left, with the terms of surrender in hand, and Robert E. Lee on the right. Officers and troops of both armies are arrayed in formation behind them.

DEATH OF A PRESIDENT

The Civil War took the lives of nearly three-quarters of a million soldiers, but the death of just one man, shot on Good Friday 1865, reverberated like no other. His killer used neither an army rifle nor one of the dreaded minié balls that proved so deadly in the war's killing fields at distances of two hundred yards and more. The murder weapon would be a single-shot pistol fired at point-blank range.

The man armed with the Philadelphia Derringer spent his adolescence in slaveholding Maryland, his family home across the Chesapeake

Bay from Frederick Douglass's birthplace. The shooter grew up believing slavery was "one of the greatest blessings that God ever bestowed upon a favored nation," and that its goodness extended not only to Whites but to Blacks, whom he regarded as "thick-skulled darkies."[5] The son and brother of world-renowned actors, he established himself as a performer by age twenty, favoring such Shakespearean roles as the murderous king in *Richard III* and Brutus, the trusted confidant who joins the conspiracy to kill Julius Caesar. By April 1865, a month short of his twenty-seventh birthday, he was regarded as the "handsomest man in Washington." His swashbuckling performances possessed "the fire, the dash, [and] the touch of *strangeness*."[6]

That strangeness extended to his private life. Years earlier, Booth temporarily abandoned his acting company in Richmond and patched together a military uniform so he could impersonate a militiaman and stand guard as the Commonwealth of Virginia hanged John Brown; he knew a theatrical moment when he saw one and did not want to be offstage. Known as a prodigious drinker—however many brandies he consumed, he never seemed intoxicated—he was a man whose sexual appetites ran the gamut from prostitutes to the women of Washington high society.

With the war on, he nurtured a deepening well of anger toward the Union in general and Abraham Lincoln in particular. He continued his acting career, finding the theater provided him with, in effect, a free pass to cross Union lines in order to play to both Rebel and Union audiences. Those travels also meant opportunity, and he found common cause with other Rebel sympathizers. Working with Confederate spies and a mix of other conspirators in the North and South, he worked up an elaborate plan to kidnap Lincoln and smuggle him across the Potomac in order to exchange him for Confederate prisoners of war. After Appomattox, however, the kidnapping no longer made sense because of Grant's and Lincoln's generous terms. Lee's soldiers had not been interned in P.O.W. camps but were permitted to keep their sidearms and horses and head for home.

Then, on April 11, 1865, Lincoln delivered a victory speech at the White House. Looking to the future and, in particular, to what would come to be known as Reconstruction, he recognized that changes needed to be made in the South. Having won the war, he wished to win the peace, and he spoke of forming new state governments. Lincoln made clear the right to vote would be granted to Black men who were literate or who'd fought for the Union.

The idea was anathema to John Wilkes Booth. Listening that evening to Lincoln's words amid a sea of umbrellas, he turned to a companion and said, "That is the last speech he will ever make."

The approaching peace did nothing to diminish his dangerous ambition. Booth remained intent upon doing "something that would bring his name forward in history."[7]

John Wilkes Booth, actor and man-about-town during the war years, posing with gloves and cane in hand.

GOOD FRIDAY

Lincoln was in better spirits than he had been in months. He rose early on Good Friday, April 14, 1865, and breakfasted with his son Captain Robert Todd Lincoln. Robert recounted the details of Lee's surrender, a scene he'd witnessed; his father relished the story. Over the course of the morning, the elder Lincoln tended to the nation's business, conversing with the Speaker of the House and convening a cabinet meeting. The good-humored Lincoln seemed restored, and his secretaries later described him as "singularly happy . . . after four years of trouble and tumult he looked forward to four years of comparative quiet and normal work."[8]

In the afternoon, Lincoln abandoned his desk for a carriage ride with Mary. She, too, was struck by his mood and asked about it. "I consider this day, the war, has come to a close," he responded. He told her he was ready to put the death of their son Willie behind them, too. "We must both," he told Mary, "be more cheerful in the future."[9]

The president and first lady expected to attend the theater that evening in the company of Julia and Ulysses Grant. Although the morning papers reported the plan, General Grant, after initially accepting the invitation, begged off; the Grants instead boarded a train for Burlington, New Jersey, to visit their children. Several of Lincoln's advisers counseled him to stay home, arguing that a widely publicized visit to Ford's Theatre might put him at risk in a city full of former Confederates. But after four years of threats that came to nothing, Lincoln wished to go anyway; the theater was a favorite amusement and he looked forward to a healthful dose of laughter. A younger couple of the Lincolns' acquaintance, Major Henry Rathbone and his fiancée, Clara Harris, the daughter of a New York senator, agreed to join the party.

The foursome arrived at about 8:30 P.M., well after the curtain went up on a forgettable farce called *My American Friend.* But the audience didn't mind the interruption. One of the actors, seeing the commotion in the presidential box, ad-libbed a line—"This reminds me of a story, as Mr. Lincoln would say"—and the crowd cheered.[10] The house orchestra played "Hail to the Chief." Lincoln stepped to the velvet balustrade at the front of the box, which looked down on the stage from a height of almost twelve feet. His bow and smile met with a fresh ovation from the nearly full house. Only then did he take a seat in the rocking chair considerately provided by the management for his comfort. The play resumed.

THE CONSPIRATORS

Word of the Lincolns' night out reached John Wilkes Booth early that afternoon when he went to check his mail at Ford's Theatre.

His larger band of kidnappers had disbanded but Booth, unwilling to accept the outcome of the war, maintained a cell of conspirators. They continued to meet, though when he shifted the terms of engagement—he now wanted to kill the president rather than kidnap him—the crew shrank further. Lincoln's secretaries John Hay and John Nicolay would later call his little circle "a small number of loose fish," but Booth could still call upon the violent intentions of Lewis Powell and George Atzerodt.[11] Powell was an Alabama native and former Confederate soldier wounded and captured at Gettysburg. He was battle-hardened and wanted passionately to the avenge the South. Atzerodt was a disgruntled Prussian immigrant.

On the evening of April 14, Booth summoned them to a meeting at eight o'clock. He issued his instructions: Powell was to proceed to Lafayette Square and the home of William Seward, where the secretary of state lay convalescing from a carriage accident the previous week. With doctors coming and going, Powell might gain access by posing as a deliveryman bringing medicines from an apothecary. The task of ending the life of Vice President Andrew Johnson went to Atzerodt; Booth's plan called for him to go to Kirkwood House, Johnson's hotel, to deliver an official-looking envelope. Booth kept the largest target for himself, and the men agreed that two hours later they would "decapitate the government" by simultaneously killing Seward, Johnson, and Lincoln at 10:14 P.M.

Riding a rented horse, Booth made his way to the familiar confines of Ford's Theatre. He tied up the animal in the alley behind the theater, anticipating a speedy escape, and entered through the stage door. At ten minutes after ten, he made his way from the parquet level of the theater to the dress circle, humming as he went. With most eyes on the performers, he moved quietly toward the State Box. Dressed in a dark suit, hat, and boots with spurs, he carried a small gun—the length from the walnut stock to the mouth of the rifled barrel was less than six inches—and a bowie knife with a long, well-honed blade.

The unreliable Washington policeman assigned to Lincoln was not at his post so only Lincoln's footman stood between killer and quarry. Booth produced a senator's calling card—the name on it was familiar to the guard, though to which senator it belonged is a detail lost to history—and the president's man ushered Booth into a little hallway.* Now only an unlocked door separated him from the double box where the president and his party watched the play. Looking through a peephole, Booth saw Lincoln clearly, with Mrs. Lincoln sitting to the president's right. He was in position.

---

* The best guess as to the senator's identity is John Hale of New Hampshire, since Booth and Hale's daughter Lucy had only recently ended a love affair; when Booth died, he had a photograph of Lucy Hale on his person. For a discussion of this notion and why the name Hale did not surface in the subsequent investigation, see Achorn, *Every Drop of Blood* (2020), pp. 280–82, 284–85, 288.

The executioner did not hesitate long. The audience's guffaws at a guaranteed laugh line was his cue, and Booth stepped forward to fire the derringer at the back of Lincoln's head. For many in the theater, the gunshot went unnoticed beneath the explosion of hilarity.

As Lincoln slumped forward in his chair, Booth dropped his gun and brushed past the president's rocker. Always the actor looking to make a dramatic exit, Booth rested a boot on the box's railing, ready to leap onto the stage below and run for the exit. Major Rathbone, though surprised by the half-heard gunshot and the veil of gun smoke, recovered quickly and went for Booth. But the actor managed to pull his knife from its sheath and swung it at Rathbone, tearing a great gash in the major's left arm.

Then Booth jumped, delivering his chosen line, "*Sic semper tyrannis*," the Virginia state motto, which translates as "Thus always to tyrants." When one of his spurs caught in the bunting that decorated the box, his

The moment the shot was fired at Ford's Theater, as represented in a Currier & Ives print released in the weeks afterward.

athletic leap became a tumble. Booth crash-landed on the stage, his right leg twisted beneath him, fracturing the smaller bone in his lower leg. Then he was up and, in a limping run, heading for center stage, where he delivered one last line.

Confused whether the unexpected arrival played a part in the play, the stunned audience watched as Booth raised both arms, brandishing the broad and bloodied blade as if he had just killed Caesar. "The south is avenged!" he cried before disappearing into the familiar back passages of the theater, bound for the stage door, his horse, and his escape.[12]

What had happened became clear only when a terrible shriek rang out. "They have shot the president!" screamed Mary Lincoln. "They have shot the president!"

Unconscious but breathing, Lincoln was soon carried to a house across Tenth Street; according to Victorian morality, his memory would have been sullied by dying in an unholy theater on Good Friday. His bearers laid the mortally wounded man diagonally across the bed, since he was taller than the bed was long.

A deathwatch assembled, including members of his cabinet. Senator Charles Sumner arrived with Lincoln's son Robert; Sumner remained at Lincoln's bedside all night, holding the limp hand of the dying president. Mary Lincoln, comforted by her son, wept in a nearby room when Lincoln stopped breathing, at 7:22 A.M. on Saturday, April 15, 1865.

Seward was not in attendance. Lewis Powell had attacked him, though the secretary of state would miraculously survive despite serious stab wounds. Andrew Johnson appeared briefly at Lincoln's deathbed scene; he had escaped the night's violence unscathed because George Atzerodt lost his nerve and made no attempt on the life of the vice president.

Even before Lincoln died, the Black citizens of Washington mourned him. In the dark before dawn, a member of Lincoln's cabinet stepped

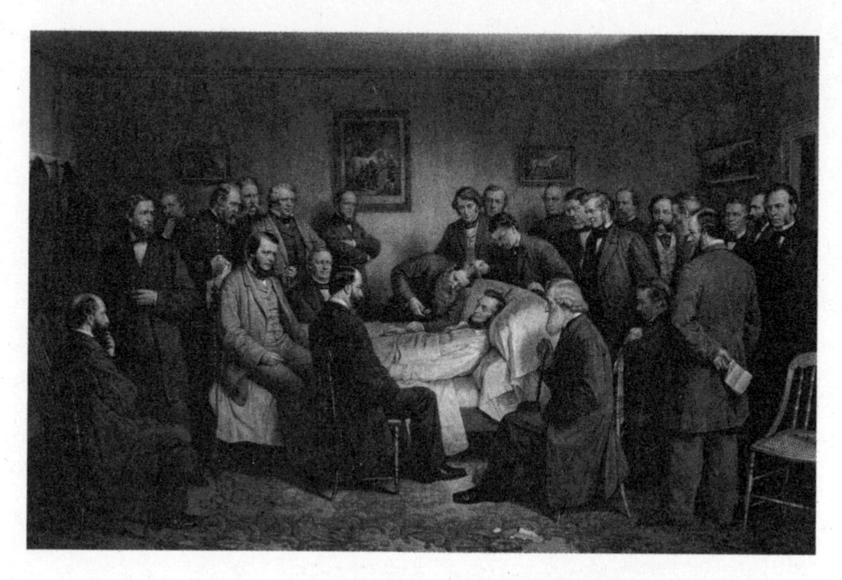

One of the many commemorative images of the Lincoln deathbed scene published a few years after the assassination. The dying man is surrounded by doctors, members of his cabinet, military men, and others. Eldest son Robert Todd Lincoln leans over his father's pillow, and to his right stands Charles Sumner.

outside and saw crowds had already begun to fill the streets. "The colored people," the secretary of the navy recorded, ". . . were painfully affected."[13]

When he stopped breathing, one of the several doctors on duty confirmed he was dead. Secretary of War Stanton broke the terrible silence that followed.

"Now he belongs to the ages."[14]

MOURNING IN ROCHESTER

The dot-dash of the telegraph brought the news to Rochester that the great and good Abraham Lincoln had joined the Union dead.

A mere six weeks earlier, at the White House, Lincoln had hailed "my friend, Frederick Douglass"; in return, Douglass thanked the now-deceased president, offering a heartfelt compliment for his inaugural speech as a "sacred effort."

Theirs had been a budding friendship. It was not a political courtship nor a convenient or collegial collaboration, but a true meeting of minds. Now, after the crack of a pistol shot, Lincoln once more would be found in the East Room, where the two men last spoke. This time Lincoln lay in state.

On the afternoon of Lincoln's death, Frederick Douglass and many other mourning Rochesterians "betook themselves to the City Hall." At 3:00 P.M. a public memorial service convened, and as Douglass later remembered, "though all hearts ached for utterance, few felt like speaking. We were stunned and overwhelmed by a crime and calamity hitherto unknown to our country and our government."[15]

Douglass found a seat in the back of the auditorium; the audience was largely White, and Douglass hadn't been invited to sit on the dais. He listened quietly as the city's mayor, a judge, and an Episcopal rector spoke. However, his presence noted, he soon heard his neighbors chanting, *Douglass! Douglass! Douglass!*

The mayor beckoned him to the stage. The crowd made way.

Hundreds of people had been turned away at the doors, and Douglass, looking out over the overflow house, admitted he was ill prepared to speak. "This is not an occasion for speech making, but for silence," he told the men and women of his city. "I have scarcely been able to say a word to any of those friends who have taken my hand and looked sadly in my eyes to-day."[16]

Grasping the importance of the moment, Douglass searched his mind for "some good that may be born of the tremendous evil."

In what would be one of the most direct and heartfelt speeches of his life, he suggested that Lincoln's death was a reminder to a nation that, until the assassination, had been "in danger of losing a just appreciation of the awful crimes of this rebellion." Perhaps out of relief, people had begun to put the war behind them; in that case, Lincoln's death might be seen as a necessary reminder "to bring us back to that equilibrium which we must maintain if the Republic was to be permanently redeemed."

The crowd applauded.

Speaking without notes but warming to his subject, Douglass recalled shaking Lincoln's "brave, honest hand, and look[ing] into his gentle eye." From memory, he quoted the president's inaugural at length, reminding his listeners that the "scourge of war" should not speedily be forgotten. "Let us not be in too much haste in the work of restoration," he advised. "Let us not be in a hurry to clasp to our bosom that spirit which gave birth to Booth." Most important of all, he urged, "Let us not forget that justice to the negro is safety to the Nation."

Speaking in Rochester that funereal afternoon, Douglass felt a sense of belonging, a stark contrast to the outsider status his nation had shown him as a child and young man. He invited the audience to embrace those who cared about the remade nation. "Know every man by his loyalty, and wherever there is a patriot in the North or South, white or black, helping on the good cause, hail him as a citizen, a kinsman, a clansman, a brother beloved." The crowd showered him with his largest ovation of the afternoon. He hit the note he wanted; it had struck a chord with his listeners.

Writing of the moment years later, Douglass would remember the afternoon of grief, shared by the White and Black alike; their sense of loss had made them "more than countrymen, it made us Kin."[17]

He might have said the same thing of his recently dead kinsman, Abraham Lincoln. "Though Abraham Lincoln dies," said Douglass, "the Republic lives." His profound belief and hope were that, with the war to abolish slavery finally at an end, Lincoln's promise of a "new birth of freedom" was indeed aborning, leading to equality and reconciliation between White and Black.

EPILOGUE

A BONE-HANDLED CANE

> [T]he best man, truest patriot, and wisest statesman of his time and country . . . [Mr. Lincoln's] name should never be spoken but with reverence, gratitude, and affection.
>
> —FREDERICK DOUGLASS ON ABRAHAM LINCOLN

Abraham Lincoln went home to Illinois in a box. After lying in state in the East Room and the Capitol Rotunda, his remains retraced the route that the living Lincoln took four years before as president-elect.

The somber, twenty-mile-per-hour train journey was interrupted by stops in major cities like Baltimore, New York, and Chicago, where processions carried Lincoln's casket along crowd-lined streets to important public buildings, including Philadelphia's Independence Hall, birthplace of the Declaration of Independence. The exhumed remains of son Willie, dead three years earlier, accompanied his father on the train, though surviving sons Robert and Tad remained in Washington. They consoled their bereft mother, who suffered "the wails of a broken heart, the unearthly shrieks, the terrible convulsions, the wild, tempestuous outbursts of grief from the soul."[1]

Millions of Americans paid their respects along the 1,654-mile route traveled by the "Lincoln Special." One stop on the twelve-day journey was Rochester, New York, where the funeral train halted for fifteen minutes on April 27 at 3:20 A.M. The coffin never left the presidential car, but despite the early hour, the tracks were lined with people carrying torches. An artillery salute was fired as "an immense crowd" watched at the depot.[2]

Later that same morning, the War Department released the news that John Wilkes Booth, after eluding capture for twelve days, was dead. The manhunt ended in Port Royal, Virginia, where Booth took refuge in a tobacco barn. He refused to give himself up, and his pursuers, looking to flush him out, set fire to the building. Despite orders to take him alive, a soldier, sighting Booth through a gap in the rough siding, shot the fugitive in the neck. The slug severed Booth's spinal cord. Alive but paralyzed, Lincoln's killer was dragged to safety and died of asphyxia three hours later.

When the funeral train arrived in Springfield, Lincoln's open coffin again lay in state, this time in the Illinois capitol, the great limestone building shrouded "roof to basement in black velvet and silver fringe."[3] Two days later, on May 4, 1865, he was buried at a simple ceremony. Accompanying the prayers and hymns, the words of his Second Inaugural Address were recited over the grave.

AN OBLONG BOX

A few weeks later, a package arrived at the Douglass home on Rochester's South Avenue. Mary Todd Lincoln had sent the long, oddly shaped box, and inside Douglass found the dead president's favorite walking stick. Noting in an accompanying letter that her late husband considered him a special friend, Mary Lincoln wrote, "I know of no one that would appreciate this more than Fred. Douglass."[4]

Douglass replied, sending the widow Lincoln a note of thanks. "I assure you that this inestimable memento of his excellency will be retained in my possession while I live." He called the maple stick with the well-worn bone handle "an object of sacred interest." Perhaps more important still, he interpreted the gift as an "indication of [Lincoln's] human interest in the welfare of my whole race."[5]

President Andrew Johnson exhibited no such concern. After taking office, he instituted a range of policies that allowed the passage of so-called "Black Codes." These permitted former Confederate states to avoid granting freed men new civil and political rights. The laws also limited freedom of movement, leaving few options other than to return to work for former slaveholders. A brutal battle over such issues broke out between the racist president and congressional Republicans. Douglass, attempting to have his voice heard, bulldozed his way into the White House on February 7, 1866, along with a delegation of other influential Black men.

The unwanted meeting with Johnson did not go well. After listening to a forty-five-minute tirade in which Johnson described himself as "the friend of the colored man," Douglass argued with the president. Their bitter exchange—Douglass refused to kowtow to Johnson—left the president livid. "Those d____d sons of b____s thought they had me in a trap," sputtered Johnson after his visitors departed. "I know that d____d Douglass; he's just like any n____r, & would sooner cut a white man's throat than not."[6]

The remainder of Johnson's term would be consumed by a legislative war with Congress, which, despite the president's opposition, managed to pass the Fourteenth Amendment to the Constitution, in June 1866, guaranteeing "equal protection" of "life, liberty, or property" for all citizens. Congress would also attempt to impeach Johnson, in 1868; though the House of Representatives approved articles of impeachment, the trial in the Senate fell one vote short of the two-thirds majority required to remove him from office. Following Ulysses S. Grant's election as president that November, the Fifteenth Amendment was enacted and ratified, guaranteeing all citizens the right to vote, regardless of "their race, color, or previous condition of servitude."

Newly empowered Black voters asserted themselves in the late 1860s in the South, voting their brethren into local, state, and even national offices. But Frederick Douglass's days as an agitator and firebrand at the ramparts faded; looking back a few years later, he wrote, "The anti-slavery platform had performed its work, and my voice was no longer needed."[7] He moved to Washington, D.C., in 1870, and returned for a time to the newspaper business as owner of the *New National Era.* But he lost interest in the short-lived paper, handing over the reins to Lewis and Frederick Jr., before the publication folded in a cascade of debts a few years later.

In 1872, his family home in Rochester burned, probably as the result of arson. His wife, Anna, who still resided there, escaped safely, and Douglass hurried back to the city that had welcomed him twenty-five years before. On arrival, however, he was turned away from two hotels, treatment that the angry Douglass attributed to "Northern colorphobia." He complained bitterly that even in a liberal Northern city, "that Ku Klux spirit . . . makes anything owned by a colored man a little less respected and secure."[8] Slavery had ended; racial discrimination did not, and the years that followed would see a gradual erosion of the rights only recently granted to African Americans.

Many of the abolitionists he'd collaborated with began to fade into the past. Douglass's first mentor, William Lloyd Garrison, ceased publication of *The Liberator* in 1865; though he involved himself with the women's suffrage movement for some years, he would die in 1879 after a long illness. On the morning of Charles Sumner's death, on March 11, 1874, Frederick Douglass was summoned to his bedside. Still preoccupied with the rights of freedmen, the senator urged Douglass to advance his Civil Rights Act, then pending in Congress. "Don't let that bill fail," he pleaded.[9] A watered-down version did pass, but the Supreme Court struck it down a few years later. Another eight decades would go by before the Civil Rights Act of 1964 was enacted during Lyndon Johnson's presidency. But in 1876 Douglass would have perhaps his proudest moment as a public man.

THE FREEDMAN'S MEMORIAL

Flags flew at half-mast. Congress had declared April 14, 1876, a public holiday, and dozens of senators and congressmen, Supreme Court justices, and members of the cabinet made the one-mile trip from the Capitol to Lincoln Square. Even President Ulysses S. Grant arrived to take a seat on the temporary platform. Next to the stage, hidden by a drapery of red, white, and blue bunting and American flags, a large but shapeless form rose some twenty feet in the air.

Although the men of the government came to remember a dead president, the scene that unfolded was no routine commemoration of a great man. The honoree would be Abraham Lincoln, but the inspiration for the day's events belonged not to a politician but to a formerly enslaved woman, now elderly, who stood nearby. Upon hearing of Abraham Lincoln's death exactly eleven years before, Charlotte Scott had announced she wished to dedicate her life savings, the modest but hard-earned sum of five dollars, to build a monument to "the best friend of the colored people," the man who authored the Emancipation Proclamation and engineered passage of the Thirteenth Amendment, thereby ending American slavery forever. Scott's spontaneous gesture triggered a campaign among grateful African Americans, many of them former soldiers and freedmen. They contributed their own dollars to fulfill her wish.

The realization of Charlotte Scott's idea was about to be unveiled as the Freedman's Memorial.[10]

The sound of music wafted in as a parade high-stepped into the square. The marchers included National Guardsmen, the Knights of Saint Augustine, and the Sons of Purity—all told, some twenty Negro organizations—accompanied by banners and flags, cornet bands and drum corps. As their colorful uniforms blended into a crowd of some twenty-five thousand, roughly half of whom were Black, the scheduled speakers at the day's event walked from their carriages in the procession to join President Grant and the others on the platform. The most recognizable of the new arrivals was Frederick Douglass. Neither his generous mane of swept-back white hair nor his full beard could obscure the intense flashing eyes that, as in every photograph he ever took, not only engaged those looking at him but seemed to challenge them, too. Douglass had been the inevitable choice to deliver the keynote address.

At two o'clock, the program in Lincoln Square opened with a mini-

Mathew Brady took Douglass's picture, too. This image, made at roughly the time Douglass dedicated the Freedman's Memorial, captures his still-commanding presence, though he has evidently aged, with his mane gone to white.

ster's invocation, followed by a recitation of the Emancipation Proclamation. The crowd also heard the story of the statue, how Scott's five dollars grew to the seventeen thousand dollars paid to American sculptor Thomas Ball. Then the moment came to unveil Ball's bronze.

General Grant rose from his chair. The silent crowd watched, still and expectant. The president stepped forward, paused, and then without uttering a word, pulled a velvet cord. It released the massive curtain of fabric, which slumped to the ground. In the brief silence that followed, all eyes lifted to gaze upon the towering statue.

A nine-foot-tall Lincoln dominated the composition, his right hand resting on a podium, the fingers grasping a copy of the Emancipation Proclamation. The somber-faced figure looked down, his other arm extended as if in a gesture of benediction over a kneeling man. Naked from the waist up, the well-muscled former bondsman, shackles behind him, appeared to be rising with one knee off the ground. Cast into the base in twelve-inch letters was the word EMANCIPATION.

The Freedman's Memorial, also known as the Emancipation Memorial and the Emancipation Group. At its dedication in 1876, Douglass chose his words carefully, seeking to honor Lincoln despite disliking the portrayal of the figure at the feet of Lincoln's likeness.

The quiet suddenly gave way to a deafening explosion of sound, with spontaneous cries of admiration, a brassy burst of music from the Marine Band, and a booming cannon salute.

Now came Frederick Douglass's turn.

Of all the thousands of speeches he'd delivered over the years, this was the only time he—or, for that matter, any Black man of his century—would command the full and immediate attention of the three branches of the American government.* Douglass promptly recognized the congressional leadership, the chief justice, and "the honored and truest President of the United States," but he just as quickly made clear that today they were guests. The hosts of this gathering, he said, were "we, the colored people, newly emancipated and rejoicing in our blood-bought freedom." Black folk had underwritten this monument of remembrance for the great services rendered by Lincoln to "ourselves, our race, to our country and to the whole world."[11]

Some in the vast audience had arrived thinking that Douglass, now in his fifty-ninth year, was aging into irrelevance. He himself would later admit that, with the end of the war a decade before and abolition won, "I had reached the end of the noblest and best part of my life."[12] But before he faded away, Douglass wanted to get a few things straight.

To the surprise of those who came expecting a celebration of national pride and patriotism, Douglass pointed out that Lincoln, "the Great Emancipator," had not always been his people's best friend. "Truth compels me to admit," Douglass told the crowd, "even here in the presence of the monument we have erected to his memory, Abraham Lincoln was not, in the fullest sense of the word, either our man or our model.

---

* As Douglass's biographer David Blight has pointed out, another 133 years would elapse before a second Black man could lay claim to so distinguished an audience; that was in 2009 when Barack Obama delivered his first inaugural speech.

In his interests, in his association, in his habits of thought, and in his prejudices, he was a white man."

In case he hadn't been clear enough, Douglass spoke separately to the two halves of the audience. "You are the children of Abraham," he said to the powerful men on the stage and the other White listeners. Addressing his Black fellow citizens, Douglass said, "We are at best only his step-children."

Yet he had made his peace with Lincoln's slow rise to the cause of freeing the Black man. "Had he put the abolition of slavery before the salvation of the Union," Douglass acknowledged, Lincoln would have alienated Whites and "rendered resistance to rebellion impossible. Viewed from the genuine abolition ground, Mr. Lincoln seemed tardy, cold, dull, and indifferent; but measuring him by the sentiment of his country, a sentiment he was bound as a statesman to consult, he was swift, zealous, radical, and determined."

Douglass spent an hour delivering his history lesson, praising Lincoln, speaking respectfully of his memory. But he refused to do the sixteenth president the disservice of reducing him, as history had already begun to do, to a saintly and uncomplicated figure who, out of the sheer goodness in his heart, ended slavery. Douglass knew Lincoln as a man of complex and shifting intellect. At first, "his great mission was . . . to save the country from dismemberment and ruin." Only later, Douglass pointed out, had Lincoln devoted himself to the great work of "free[ing] his country from the great crime of slavery."

Standing in the shadow of the tall monument on that sunny April afternoon, Douglass called Ball's sculpture a "highly interesting object," a description that, to the perceptive listener, suggested the speaker's disappointment; as Douglass told a Washington newspaperman, he would have preferred the freedman "erect on his feet like a man" rather than "on his knees like a four-footed animal."[13] It was a criticism that, in our time, has led to calls for the statue's removal.

Douglass cherished his personal relations with Lincoln, but he could not offer a humble hymn of praise. Perhaps more than anyone present, Frederick Douglass—a freedman himself, a writer of great gifts, a fighter for freedom—recognized that monuments are about memory. What he wished to convey above all else to his impressive audience was that Lincoln must be remembered fully.

That day would not be the last of his public life. The next year President Rutherford B. Hayes rewarded Douglass for his continuing service to the party, appointing him U.S. Marshal of the District of Columbia. It was a first: No Black man had ever been nominated for a position that required Senate approval. Though largely ceremonial, the job cemented Douglass's national standing as Black America's most visible and admired man.

He purchased a handsome brick mansion called Cedar Hill. Located in Southeast Washington, D.C., on a fifteen-acre parcel, it overlooked the Anacostia River and the Capitol beyond. Douglass saw with a mix of satisfaction and irony that the original deed specified the property was "for the use, benefit and enjoyment of white persons only." Lincoln and the Civil War had changed that.[14]

Douglass gave fewer lectures as he became a man of means, with holdings and directorships in banks, insurance, and manufacturing.[15] After Anna died, in 1882, he married Helen Pitts, a White woman who'd worked as his secretary when he served as Recorder of Deeds in the Garfield administration, another political appointment. In contrast to Anna, who never learned to read or write, Helen was a graduate of Mount Holyoke Female Seminary. The couple shared many interests. They welcomed students from Howard University for Sunday tea and traveled widely to Europe and the Middle East. Disapproval of his marriage was widespread among Blacks, Whites, and even his children, but Douglass refused to take the objections seriously. He observed that his first wife "was the color of my mother, and the second, the color of my father."[16]

Douglass, together with his second wife, Helen Pitts Douglass (seated), and sister-in-law Eva Pitts, ca. 1885.

Douglass worked at a third and expanded version of his autobiography. Published in 1881, *Life and Times of Frederick Douglass* reflected how far its author had come. If it lacked some of the shock value of his earlier slave narratives, it put Douglass's remarkable rise into its historic context as he recounted his relations with John Brown, Mrs. Stowe, Gerrit Smith, and other notables. Above all, as he wrote of his life at Cedar Hill, he mused upon his relationship with Abraham Lincoln. In drafting those pages, he cast an occasional glance at a posthumous likeness of Lincoln. "His picture, now before me in my study," Douglass noted in *Life and Times*, "corresponds well with the impression I have of him."[17]

Douglass reengaged in the fight to get the vote for women; back in 1848, he had been one of the few men present at the landmark Seneca Falls convention. On February 20, 1895, he attended a women's rights rally, where he entered the hall arm in arm with his fellow warrior Susan B. Anthony. They were greeted with tumultuous applause.

"Men called him homely," Frederick Douglass wrote after Lincoln's death, "and homely he was; but it was manifestly a human homeliness." He liked this picture of the dead president, which was painted and engraved by William E. Marshall in 1866, and hung a copy in his library.

He returned to Cedar Hill shortly after dark. While dining with Helen, he recounted the events of the exhilarating day, then abruptly rose from his chair. A moment later, he fell to his knees and collapsed to the floor. Although he outlived his friend Lincoln by almost thirty years, Frederick Douglass, age seventy-seven, was dead.

LAYING CLAIM TO LINCOLN

Abraham Lincoln got no chance to write an autobiography. The job of chronicling his life was left to friends, colleagues, and newspapermen. An old law partner, his long-time bodyguard, and dozens of others produced an avalanche of reminiscences and biographies that culminated with the exhaustively researched and authorized *Abraham Lincoln: A*

History. Written by John Hay and John Nicolay, Lincoln's former secretaries, it appeared in ten volumes in 1890.

The fate of Frederick Douglass was the opposite. During the years that born-in-a-log-cabin "Honest Abe" gained almost mythic status, Douglass began a steep fade into obscurity.

He lived long enough to worry about new inequalities, boldly telling a Washington, D.C., audience late in his life, "I admit that the negro, and especially the plantation negro, . . . is worse off, in many respects, than when he was a slave."[18] But Douglass would die a year before the Supreme Court made this inequality official. In 1896, the landmark case *Plessy v. Ferguson* established the doctrine of "separate but equal," upholding recent "Jim Crow" laws that, in many southern states, disenfranchised Black voters and permitted racial segregation in public places. The result was more than half of a century of separate-but-far-from-equal, since it wasn't until 1954 that the court ruled the doctrine unconstitutional in the momentous *Brown v. Board of Education* decision.

That the memory of Frederick Douglass was suppressed alongside the rights of African American citizens during those years is no surprise.

During those decades, Lincoln's lionization continued, particularly in what became his most memorable biography. Written by a Chicago journalist and poet named Carl Sandburg, *Abraham Lincoln* was published in two parts, with *The Prairie Years* appearing in 1926, followed thirteen years later by *The War Years.* A better writer than historian, Sandburg won a Pulitzer Prize and wide readership for his poetic prose. Lincoln, whom so many loved to hate in his own time, emerged as a man posterity worshipped. Sandburg made him a hero at the center of the American story.

Meanwhile, Douglass was mentioned rarely or not at all in history texts. Then, after World War II, a young scholar named Philip Foner began looking for a commercial publisher or a university press to pub-

The elderly Frederick Douglass at his home, Cedar Hill, where he would die in 1895. He purchased the desk at which he sits from the estate of his old abolitionist ally Charles Sumner.

lish Douglass's papers. But most of the editors he approached had never heard of Douglass.[19] Foner eventually persuaded the Marxist press International Publishers to release *The Life and Writings of Frederick Douglass*; its five volumes came off the presses between 1950 and 1955. The books helped launched a rediscovery of Douglass, as did an early biography by pioneering African American studies professor Benjamin Quarles.[20] The return of Douglass's voice to the national conversation fueled the civil rights movement and the activism of the 1950s and '60s, which led to the Civil Rights Acts of 1964 and 1968.

During those years, Lincoln's saintly status lost some of its luster. *Ebony* magazine put the question in stark terms in an article titled "Was Lincoln a White Supremacist?"[21] The discussion that followed resulted in a more measured view of the man.

By the end of the twentieth century, Lincoln came off his pedestal and Douglass rose from obscurity. Today these two good men once again occupy a shared space in the ongoing story of the American experiment.

THE FOG OF MEMORY

To see Lincoln clearly, the name "Great Emancipator" needed to be retired. To know Douglass, we cannot use the rosy lens that is the "Lost Cause."

To explain: Shortly after the Civil War ended, a Virginia newspaper editor promoted the idea of the Lost Cause. According to this "moonlight and magnolias" view, later perpetuated by the book and movie *Gone With the Wind*, the antebellum South was a genteel culture that, faced with Northern aggression, had no choice but to defend itself. Embedded in this sentimental distortion of history is the acceptance of slavery as a paternalistic good that was of mutual benefit to Whites and the Blacks they enslaved.

More than perhaps anyone else, Douglass understood the utter dishonesty of the Lost Cause perspective on the Civil War. Douglass spent the war years telling people that the Civil War was about slavery. The politicians in the states that withdrew from the Union understood that then; slavery, not states' rights, had been at the center of debate at secession conventions.[22] Ironically, Abraham Lincoln was among the many in the North who needed to be persuaded that slavery was the bleeding heart of the conflict.

Recollecting his younger days floating down the Mississippi, Lincoln once observed, "The pilots on our Western rivers steer from *point to point* as they call it—setting the course of the boat no farther than they can see; and that is all I propose to myself in this great problem."[23] The largely self-taught Lincoln approached many challenges—most of all slavery—in the same way. He was a pragmatic problem-solver confident that solutions, like destinations, would offer themselves with the passage of time.

During Lincoln's boyhood, the inferiority of the Black man was received wisdom, a belief he shared with almost all Whites in the North as well as the South. Then, as a young man, Lincoln's coordinates changed as his values evolved and he accepted that slavery was a clear moral wrong. In manhood he began to imagine a route that, in time, could lead to abolition; and he gradually accepted that Blacks were the equal of Whites and must be made full citizens. Only in the last several years of his life did Lincoln manage to steer the country, once and for all, away from slavery.

Frederick Douglass's personal history was very different. In his youth, he smelled and tasted the blood of slavery; sometimes that blood was his own. He knew from a very young age what he wanted: *freedom*, plain and simple. After he gained freedom for himself, on his own, his goal became to free all of those enslaved in his country. Unlike Lincoln, from youth he possessed a North Star—it was freedom for his people—a constant he could rely upon. His bearings never changed as he made himself famous, on his own terms, and played an immense role in achieving freedom for his people.

In this book we have looked at how Lincoln—lawyer, legislator, careful reader of the nation's founding documents—came face-to-face in the 1850s with the Kansas–Nebraska Act, then *Dred Scott v. Sandford.* We have seen how the politics of slavery rapidly changed, an evolution that accelerated after John Brown's raid, the secession of Southern states following Lincoln's election, and the advent of war. Lincoln did not direct any of these occurrences: As he admitted to a Kentucky newspaperman, "I claim not to have controlled events, but confess plainly that events have controlled me."[24] Still, his actions did shape the course of that war.

He started out thinking the fight was for Union. Although he never abandoned that belief, he slowly moved to the view that the only way to save the Union was to banish slavery, and, later, that enlisting Black men was essential. By the end of the war, the flood of African American volunteers numbered roughly 10 percent of the total soldiers and sailors who fought for the Union, providing a key, even decisive, advantage in the war. The agency of the Black Union soldier would long be overlooked by most historians, just as Douglass was for so many decades. But in his time Lincoln recognized both. He adopted the belief that the Black man should vote and that, in short, Black lives meant as much as White ones did and that the laws of the Union needed to be brought in line with the founding promises of equality.

At the beginning of his public life, Frederick Douglass opened the eyes of audiences to the injustices of slavery. Before the terms were coined, he was a "poster boy" for abolitionism, and later, when he broke from Garrison, he became a "public intellectual." Initially his had been a voice that most White Americans did not want to hear, but in his middle years, many of those same people—most prominently among them, Abraham Lincoln—listened. Douglass possessed a raw and original intellectual power unlike any contemporary, Black or White.

The fault lines of the Civil War era threw Douglass and Lincoln together. When needed the most, these two unlikely people appeared on the scene, as Andrew Jackson had done during the War of 1812, each of them born with nothing. They were underdogs who somehow rose to become national figures at a time of crisis. Like other American saviors, Lincoln and Douglass would transcend their era, standing taller than their contemporaries.

Lincoln did what no other politician of his era could do: free the enslaved. Since Lincoln left us little commentary on their friendship, how large an influence Douglass was on his thinking cannot be proved. Yet it is clear that Douglass was an alter ego to Lincoln, a conscience to the president and their nation, one that nagged and prodded and never bent. Douglass played to Lincoln's better angels, helping guide the president through a dark time toward a brighter, fairer, better one.

Both Douglass and Lincoln, as they worked to change the future of Black Americans, influenced their nation's direction toward a more perfect union. Their immense and shared accomplishment was ending slavery; their next goal, however, to achieve true equality, was left unfinished. In their honor, the work must go on.

ACKNOWLEDGMENTS

So many people made valuable contributions to the idea, execution, and promotion of this book that it's hard to fit them all into just one section, but I will try. First off, a salute to Fox Nation's great executives Jason Klarman, John Finley, and Jennifer Hegseth for letting me crisscross the country telling great stories about America's history for the series *What Made America Great.* Doing so allows me to talk to the best and brightest who are keeping our past alive, which is how the idea for *The President and the Freedom Fighter* first took root. I wanted to tell a Civil War story that didn't further divide the nation while confronting the horrific era of slavery, and I hope you, the reader, feel we did just that. The fact that I will be able to talk about this story on Fox News and in a Fox Nation TV special is invaluable, and all in thanks to Executive Chairman and CEO of Fox Corporation Lachlan Murdoch, CEO of Fox News Suzanne Scott, and President and Executive Editor Jay Wallace. As far as the actual shooting, editing, and writing of the special goes, Fox Senior Producer Carrie Flatley produced every detail, neither too big nor too small, in every shoot. In fact, it was she who helped me come up with the concept of interweaving Douglass's and Lincoln's stories into a single narrative. Dan Cohen helped tell the story of two people in just one hour. Bud Knapp, Monica Mari, and Mary Drabich are also indispensable.

In helping to hone the concept of the project into a fast-moving, reader-friendly book, credit goes to the best publishing president in the country, Adrian Zackheim. He not only oversaw the entire book process, but he became so intrigued by this story that he visited the physical sites where it all took place over 150 years ago. Adrian's greatest move was to keep Bria Sandford on as Supreme Allied Commander and Editor of All Things Brian. Without her clear, steady hand, intelligence, and ability to spackle together my sometimes fractured thoughts, this book would not have happened, especially with the challenges brought on by this pandemic.

The team at Sentinel that promotes the book, makes sure it arrives at and is actually *seen* in bookstores, and lines up a twelve-week tour around my TV and radio schedule is simply first class. They are led by Tara Gilbride, Regina Andreoni in marketing, and lead publicist Marisol Salaman, who never leaves a loose end loose; who, to my knowledge, has never made a mistake; and who sets every book up for success. With new addition DeQuan Foster, they are even better! The book production team has always been fantastic and had to pull out all the stops to get this book printed in time coming out of a pandemic. Paul and Amanda Guest, who head up social media promotion, do a remarkable job of getting word out through very creative and resourceful digital campaigns.

Providing the fuel that ignites, supports, and sustains this book in the public's mind is the *Fox and Friends* team. They make every book launch feel like graduation, prom, and wedding combined! Led by Vice President Gavin Hadden, along with the steady hand of Lauren Petterson, the team allows me to tell these American stories on our show and let the most patriotic audience in TV know that there is yet another reason to believe that our nation is indeed great. Executive Producer Tami Radabaugh and senior producers Kelly Nish and Sara Sonnack always find a way to skillfully fold in book mentions and appearances in order to allow these passion projects to experience such success, and their support goes way beyond the broadcast. A. J. "Mr. Everything" Hall is always finding new ways to condense and edit the book into smaller chunks that let our audience know what they can expect when they dive into the chapters. Also infusing brilliance into those packages is Megan Macdonald. Jayleen Murray has been a big plus for all three *Fox and Friends* anchors, and whether it's through uploading videos or coordinating events, she astounds us with her energy, workload, and productivity. Of course, in front of the camera I am truly moved by the stalwart support of my co-anchors Ainsley Earhardt and Steve Doocy, along with Jillian Mele and Janice Dean. I will also never forget the weekend team: Pete Hegseth, Will Cain, Rachel Campos Duffy, and Lawrence Jones. The support of the other stellar Fox shows is so vital and appreciated. They always find room in their rundown for an appearance, which gives us a coveted opportunity to reach other Fox fans days, nights, and weekends.

No one does more with and for me on a daily basis than Coordinating Producer Alyson Mansfield. While her main job is executive producer of *The Brian Kilmeade Show* on Fox Radio, she also books all my TV and radio responsibilities with book tours, fill-in hosting requests, Fox Nation shoots, and live stage shows. Behind the scenes, but shining through on a daily basis, is Eric Albeen. His creativity astounds me and his adjustments for my many road shows, and for my pre-tapes so that I can go on the road early—an extra burden he never complains about (at least to me)—is truly appreciated. Pete Caterina's boundless energy and willingness to book and adjust for the tours have been invaluable to the success of the show and this project. Radio executives Reynard Erney, John Sylvester, and Maria Donovan have also been huge assets to the books, the show, and me. They always go out of their way to promote and support, along with Willie Sanchez, Tamara Karcev, and Dave Manning. I will always be in awe of the separate support of my radio affiliates who spread the word in their markets out of goodwill and respect. I don't take it for granted. One of the most special things about our book tours is getting to meet all the listeners and station staff in the separate markets and letting them know how much they matter to all of us—so thank you!

Finally, I'd like to thank UTA super-agents Adam Leibner, Jerry Silbowitz, and Byrd Leavell for making all of the moving parts work in sync on a daily basis, and Jay Sures for forming the best agency in the country and putting me on the roster.

NOTES

PREAMBLE

1. Declaration of Independence, original of rough draft, in *The Papers of Thomas Jefferson*, vol. 1, Julian P. Boyd, ed. (Princeton, NJ: Princeton University Press, 1950), pp. 243–47.
2. *Notes of Debates in the Federal Convention of 1787.* Madison did not publish that journal until decades later, however, by which time he had made many emendations and corrections. See especially Mary Sarah Bilder, *Madison's Hand: Revising the Constitutional Convention* (Cambridge, MA: Harvard University Press, 2015). https://slavery.princeton.edu/stories/james-madison.
3. "Address to the Republic," November 9, 1789.

CHAPTER I: FROM THE BOTTOM UP

1. Henry Onstot, quoted in Wilson, *Honor's Voice* (1998), p. 53.
2. Caleb Carman to William H. Herndon, November 30, 1866.
3. Lamon, *The Life of Lincoln* (1872). The flatboat story also appears in *Herndon's Lincoln* (1890) and other early biographies.
4. Much of this section is based upon the notes Lincoln prepared for a campaign biography at the request of the *Chicago Press and Tribune.* The several thousand words he put on paper would be the closest he ever came to writing an autobiography. "Autobiography Written for John L. Scripps," June 1860. *Collected Works of Abraham Lincoln*, vol. 4 (1953), pp. 61–67.

5. Donald, *Lincoln* (1995), p. 28.
6. Lincoln, "Autobiography Written for John L. Scripps" (1860), p. 62.
7. Sarah Bush Lincoln interview with William H. Herndon, September 8, 1865.
8. John Hanks interview with William H. Herndon-1865–1866.
9. Wilson, *Honor's Voice* (1998), p. 57.
10. Lincoln, "Autobiography Written for John L. Scripps" (1860), p. 63.
11. Douglass, *Narrative of the Life of Frederick Douglass, an American Slave* (1845), p. 6.
12. Douglass, *My Bondage and My Freedom* (1855), p. 77.
13. Douglass, *My Bondage* (1855), p. 135.
14. Douglass, *My Bondage* (1855), p. 143.
15. Douglass, *Narrative* (1845), p. 32.
16. Douglass, *My Bondage* (1855), p. 147.
17. Douglass, *Narrative* (1845), p. 31.
18. Douglass, *My Bondage* (1855), p. 146.
19. Douglass, *My Bondage* (1855), p. 146.
20. Douglass, *Narrative* (1845), p. 38.
21. *The Liberator*, July 4, 1854.
22. Adams, *Memoirs* (1876), December 24, 1832.

CHAPTER 2: A FIGHTING CHANCE

1. Herndon and Weik, *Herndon's Lincoln* (1890), p. 1.
2. Mentor Graham interview with William H. Herndon, October 10, 1866.
3. Henry McHenry to William H. Herndon, October 10, 1866.
4. Robert B. Rutledge to William H. Herndon, ca. November 1, 1866.
5. James Short to William H. Herndon, July 7, 1865.
6. Robert B. Rutledge to William H. Herndon, ca. November 1, 1866.
7. James Short to William H. Herndon, July 7, 1865.
8. Robert B. Rutledge to William H. Herndon, ca. November 1, 1866.
9. Henry McHenry, quoted in Wilson, *Honor's Voice* (1998), p. 31.
10. Nicolay and Hay, *Abraham Lincoln* (1890), vol. 1, p. 81.
11. Jason Duncan to William H. Herndon, late 1866, early 1867.
12. Mentor Graham interview with William H. Herndon, May 29, 1865.
13. AL, Communication to the People of Sangamon County, March 9, 1832.
14. AL, Communication to the People of Sangamon County, March 9, 1832.
15. Elizabeth Edwards interview with William H. Herndon, January 10, 1866.
16. Robert L. Wilson to William H. Herndon, February 10, 1866.
17. W. G. Greene to William H. Herndon, May 29, 1865.
18. James McGrady Rutledge, quoted in Wilson, *Honor's Voice* (1998), p. 103.
19. Herndon and Weik, *Herndon's Lincoln* (1890), p. 126.
20. Herndon and Weik, *Herndon's Lincoln* (1890), p. 162.
21. Lincoln, "Autobiography Written for John L. Scripps" (1860), p. 61.

22. Protest in the Illinois Legislature on Slavery, March 3, 1837.
23. Lincoln, "Autobiography Written for John L. Scripps" (1860), p. 65.
24. Herndon and Weik, *Herndon's Lincoln* (1890), p. 375.
25. Douglass, *My Bondage* (1855), p. 203.
26. Douglass, *Narrative* (1845), p. 66.
27. Douglass, *Narrative* (1845), p. 63.
28. Douglass, *Narrative* (1845), p. 65.
29. Douglass, *My Bondage* (1855), p. 228.
30. Douglass, *My Bondage* (1855), p. 246.
31. Douglass, *My Bondage* (1855), p. 246.
32. Douglass, *My Bondage* (1855), p. 275.
33. "Dialogue Between a Master and Slave," in Bingham, *The Columbian Orator* (1797).
34. Douglass, *My Bondage* (1855), p. 159.
35. Douglass, *My Bondage* (1855), p. 158.
36. Douglass, *My Bondage* (1855), p. 303.
37. Preston, *Young Frederick Douglass* (2018), p. 177.
38. Douglass, *My Bondage* (1855), p. 314.
39. Douglass, *Life and Times* (1881), p. 198.
40. Douglass, *Life and Times* (1881), p. 200.
41. Douglass, *Life and Times* (1881), p. 200.
42. Douglass, *My Bondage* (1855), p. 340.

CHAPTER 3: SELF-MADE MEN

1. Douglass, *Narrative* (1845), p. 116.
2. Douglass, *Life and Times* (1881), p. 210.
3. Douglass, *My Bondage* (1855), p. 356.
4. Douglass, *My Bondage* (1855), p. 357.
5. *The Liberator*, January 1, 1831.
6. *The Liberator*, August 20, 1841.
7. Douglass, *Narrative* (1845), p. 117.
8. Accounts of Douglass's speech vary considerably. See Garrison, preface to Douglass, *Narrative* (1835), pp. iv–vi; McFeely, *Frederick Douglass* (1990), pp. 86–90; Blight, *Frederick Douglass* (2018), pp. 98–100; Lampe, *Frederick Douglass* (1998), pp. 59–63; Douglass, *Narrative* (1845), p. 117.
9. Douglass, *My Bondage* (1855), p. 358.
10. May, *Some Reflections of Our Antislavery Conflict* (1869), p. 122.
11. Garrison, preface to Douglass, *Narrative* (1835), p. iv.
12. Garrison, preface to Douglass, *Narrative* (1835), p. v.
13. Garrison, preface to Douglass, *Narrative* (1845), p. vi.
14. AL, "The Perpetuation of Our Legal Institutions," January 27, 1838.

15. Helm, *The True Story of Mary, Wife of Lincoln* (Los Angeles, CA: McCall Publishing Co., 1928), p. 41.
16. Speed, *Reminiscences* (1896), p. 34.

CHAPTER 4: ON THE ROAD

1. Douglass, *My Bondage* (1855), p. 359.
2. *The Liberator,* October 15, 1841.
3. "Like two very brothers," Douglass would later write to White, "[we] were ready to dare—do, and even die for each other." FD to William White, July 30, 1846.
4. Several versions of the events at Pendleton appeared in contemporary slavery publications, including one in *The Liberator*, October 13, 1843. See also Douglass, *Life and Times* (1882), p. 234.
5. FD to Richard Josiah Hinton, quoted in Lampe, *Frederick Douglass* (1998), p. 189.
6. *The Liberator*, August 30, 1844.
7. Foner, *The Life and Writings of Frederick Douglass*, vol. 1 (1950), p. 60.
8. *Pennsylvania Freeman*, March 6, 1846.
9. Douglass, *Life and Times* (1881), p. 262.
10. McFeely, *Frederick Douglass* (1990), p. 144.
11. FD to Ellen Richardson, April 29, 1847.
12. *Sheffield Mercury*, September 11, 1846.
13. *The North Star*, December 22, 1848.
14. *The Liberator*, July 23, 1847.
15. Douglass, *Life and Times* (1881), p. 270.
16. *The North Star*, December 3, 1847.
17. Findley, *A. Lincoln*, (1979), p. 130.
18. Abraham Lincoln, speech in House of Representatives, January 12, 1848.
19. AL to Joshua F. Speed, August 24, 1855.
20. Abraham Lincoln, "Peoria Speech," Illinois, October 16, 1854.
21. AL to Joshua F. Speed, August 24, 1855.
22. Beveridge, *Abraham Lincoln: 1809–1858*, vol. 1 (1928), p. 482.
23. Lincoln, "Autobiography Written for John L. Scripps" (1860), p. 67.
24. Gerrit Smith to FD, December 8, 1847.
25. Gerrit Smith to Elizabeth Livingston Smith, July 16, 1815.
26. Gerrit Smith obituary, *New York Times*, December 29, 1874.
27. Foner, *The Life and Writings of Frederick Douglass*, vol. 2 (1950), p. 54.

CHAPTER 5: WHERE THERE IS SMOKE

1. Harriet Beecher Stowe to FD, July 9, 1851.
2. Douglass, "Resistance to Blood-Houndism," January 8, 1851.
3. Hochman, "'Uncle Tom's Cabin' in the 'National Era'" (2004), p. 144.

4. Hedrick, *Harriet Beecher Stowe: A Life* (1994), p. 235.
5. Harriet Beecher Stowe to William Lloyd Garrison, December 19, 1853.
6. Douglass, *Life and Times* (1881), p. 289.
7. In 1860, he recalled (in the third person) that in the early 1850s "his profession had almost superseded the thought of politics in his mind." Lincoln, "Autobiography Written for John L. Scripps" (1860), p. 67.
8. Lincoln, "Notes for a Law Lecture" (ca. 1850).
9. Unless otherwise noted, Lincoln's words in this passage are drawn from his "Peoria Speech," October 16, 1854.
10. See Burlingame, *The Inner World of Abraham Lincoln* (1994), p. 27. Chapter 2, "'I Used to Be a Slave'" offers a cogent look at Lincoln's thinking regarding slavery.
11. Browne, *Abraham Lincoln and the Men of His Time*, vol. 1 (1907), p. 285.
12. Beveridge, *Abraham Lincoln: 1809–1858*, vol. 3 (1928), p. 238.
13. William H. Pierce, quoted in Lehrman, *Lincoln at Peoria* (2008), p. 57.
14. Horace White, "Address before Illinois State Historical Society," January 1908.
15. Here and after, Charles Sumner, "The Crimes Against Slavery," May 19–20, 1856.
16. Here and after, see "Alleged Assault Upon Senator Sumner." Senate Report no. 182, 34th Congress, Session 1, June 2, 1856; and Senate Report no. 191, 34th Congress, Session 1, May 28, 1856. See especially the testimony of Sumner and Brooks. See also Donald, *Charles Sumner and the Coming of the Civil War* (1960), pp. 278–311.
17. Donald, *Charles Sumner and the Coming of the Civil War* (1960), p. 301.
18. Donald, *Charles Sumner and the Coming of the Civil War* (1960), p. 305.
19. Salmon Brown, quoted in Reynolds, *John Brown, Abolitionist* (2005), p. 158.

CHAPTER 6: A SUBTERRANEAN PASSWAY

1. Douglass, *Life and Times* (1881), p. 278.
2. Douglass, *Life and Times* (1881), p. 280.
3. *The North Star*, February 18, 1848.
4. John Brown to Henry L. Stearns, July 15, 1857.
5. Lyman Epps Jr., quoted in Reynolds, *John Brown, Abolitionist* (2005), p. 127.
6. *Dred Scott v. John F. A. Sandford*, March 6, 1857.
7. Douglass, *Life and Times* (1881), pp. 323–24.
8. Douglass, *Life and Times* (1881), p. 325.
9. A. J. Philips to W. P. Smith, October 17, 1859.
10. *New-York Times*, October 18, 1859.
11. Reynolds, *John Brown, Abolitionist* (2005), p. 320.
12. Horwitz, *Midnight Rising* (2011), p. 176.
13. Villard, *John Brown, 1800–1859* (1909), p. 453.
14. Alexander R. Boteler, "Recollections of the John Brown Raid," *Century Magazine*, July 1883, p. 409.

15. Israel Green, quoted in Horwitz, *Midnight Rising* (2011), p. 180.
16. Brown interview with Mason, Vallandingham, and others, conducted on October 18, 1859, published in the *New York Herald*, October 21, 1859.
17. Douglass, *Life and Times* (1881), p. 311.
18. *Frederick Douglass' Paper*, November 11, 1859.
19. *Frederick Douglass' Paper*, December 16, 1859.
20. Henry David Thoreau, "A Plea for Captain John Brown," October 30, 1859.
21. In his poem "The Portent," Herman Melville wrote, "[T]he streaming bear is shown / (Weird John Brown), / The Meteor of the War." *Battle-Pieces and Aspects of the Civil War*, 1866.

CHAPTER 7: THE DIVIDED HOUSE

1. James A. Briggs to AL, October 12, 1859. The telegram arrived while Lincoln was out of town; he read it on his return on October 15.
2. AL, speech to Republican State Convention, Springfield, IL, June 16, 1856.
3. Donald, *Lincoln* (1995), p. 209.
4. Lincoln-Douglas debate, Springfield, Illinois, July 17, 1858.
5. Lincoln-Douglas debate, Alton, Illinois, October 15, 1858.
6. William Henry Seward, March 11, 1850.
7. *American Journal of Photography*, cited in Holzer, *Lincoln at Cooper Union* (2004), p. 88.
8. Richard C. McCormick, "Lincoln's Visit to New York in 1860," *New-York Post*, May 3, 1865. Reprinted in Wilson, ed. *Intimate Memories* (1945), pp. 250–55.
9. Meredith, *Mr. Lincoln's Camera Man: Mathew B. Brady* (1974), p. 57.
10. Meredith, *Mr. Lincoln's Camera Man: Mathew B. Brady* (1974), p. 59.
11. Ward Lamon's description, from his *Life of Lincoln* (1872, 2012), p. 384.
12. Meredith, *Mr. Lincoln's Camera Man: Mathew B. Brady* (1974), p. 59.
13. Holzer, *Lincoln at Cooper Union* (2004), p. 107.
14. McCormick, "Lincoln's Visit to New York in 1860," *New-York Post*, May 3, 1865.
15. AL, Cooper Union speech, February 27, 1860.
16. AL, Cooper Union speech, February 27, 1860.
17. McCormick, "Lincoln's Visit to New York in 1860," *New-York Post*, May 3, 1865.
18. AL to Lyman Trumbull, April 29, 1860.
19. FD, "The Trials and Triumphs of Self-Made Men," Halifax, England, January 4, 1860.
20. Kendrick and Kendrick, *Douglas and Lincoln* (2008), p. 46; FD, "To My British Anti-Slavery Friends," *Douglass' Monthly*, June 1860.
21. Annie Douglass to FD, December 7, 1859.
22. Douglass, *Life and Times* (1881), p. 328.
23. FD, "To My British Anti-Slavery Friends," *Douglass' Monthly*, June 1860.
24. Kendrick and Kendrick, *Douglas and Lincoln* (2008), p. 46.

25. Temple, "Lincoln's Fence Rails" (1954), p. 22.
26. Temple, "Lincoln's Fence Rails" (1954), p. 26.
27. John Farnsworth to Elihu B. Washburne, May 18, 1860.
28. Donald, *Lincoln* (1995), p. 251.

CHAPTER 8: THE ELECTION OF 1860

1. AL to Harvey G. Eastman, April 7, 1860.
2. *Illinois State Journal*, May 7, 1860.
3. Donald, *Lincoln* (1995), p. 254.
4. Quote in Donald, *Lincoln* (1995), p. 256.
5. AL to Samuel Haycraft, November 10, 1860.
6. AL to Lyman Trumbull, December 10, 1860.
7. *Douglass' Monthly*, June 1860.
8. FD to Gerrit Smith, July 2, 1860.
9. *Douglass' Monthly*, September 1860.
10. *Douglass' Monthly*, June 1860.
11. Headnote to "Speech on John Brown" in Foner and Taylor, eds. *Frederick Douglass, Selected Speeches and Writings* (1999), p. 417.
12. Tremont Temple Resolution, *Harper's Weekly*, December 15, 1860.
13. Tremont Temple Resolution, *Harper's Weekly*, December 15, 1860.
14. Kendrick and Kendrick, *Douglas and Lincoln* (2008), pp. 66–67.

CHAPTER 9: MR. LINCOLN'S WAR

1. Lamon, *Life of Lincoln* (1872, 2012), p. 306. See also Searcher, *Lincoln's Journey to Greatness* (1960).
2. AL address at Springfield, February 11, 1861.
3. AL to Alexander Stephens, December 22, 1861.
4. *Douglass' Monthly*, March 1861.
5. Rutherford B. Hayes to Laura Platt, February 13, 1861.
6. AL, speech at Cincinnati, February 12, 1861.
7. AL, speech at Cleveland, February 15, 1861.
8. Grace Bedell to AL, October 15, 1860; AL to Grace Bedell, October 19, 1861.
9. Searcher, *Lincoln's Journey to Greatness* (1960), p. 116.
10. Here and after, unless otherwise specified, the account of the Baltimore conspiracy is Lincoln's own version, as recounted by Benson Lossing in his book *The Pictorial History of the Civil War*, vol. 1 (1868), pp. 279–82. See also Waller, *Lincoln's Spies* (2019), pp. 4–7.
11. AL, speech at Philadelphia, February 22, 1861.
12. *Harper's Weekly*, March 9, 1861.
13. *Douglass' Monthly*, April 1861.

14. AL, First Inaugural Address, March 4, 1861.
15. *Douglass' Monthly,* April 1861.
16. *Douglass' Monthly,* May 1861.
17. For a more detailed and highly readable account of the fall of Fort Sumter, see McPherson, *Battle Cry of Freedom* (1988), pp. 264–75.
18. Detzer, *Allegiance* (2001), p. 264.
19. Quoted in McPherson, *Battle Cry of Freedom* (1989), p. 274.
20. John Ellis, quoted in Nicolay and Hay, *Abraham Lincoln* (1890), vol. 4, p. 90.
21. *Douglass' Monthly*, May 1861.
22. *Douglass' Monthly*, May 1861.
23. FD, Rochester, New York, address, May 5, 1861.
24. *Douglass' Monthly*, May 1861.
25. *Douglass' Monthly*, May 1861.
26. Bland, *Life of Benjamin F. Butler* (1879), pp. 51–52.
27. *Douglass' Monthly*, September 1861.
28. Quoted in Robertson, James Jr., *Stonewall Jackson* (New York: Simon and Schuster Macmillan, 1997), p. 263.
29. Perret, *Lincoln's War* (2004), p. 67.

CHAPTER 10: WAR IN THE WEST

1. Denton, *Passion and Principle* (2007), p. 293.
2. Frémont, *Memoirs of My Life* (2001), pp. 221–22.
3. Rebecca Harding Davis, quoted in Denton, *Passion and Principle* (2007), p. 309.
4. Frémont's Declaration, August 30, 1861.
5. *New York Evening Post,* September 1861.
6. Stowe, *The Independent*, September 21, 1861.
7. *Douglass' Monthly*, October 1861.
8. AL to John Frémont, September 2, 1861.
9. John Frémont to AL, September 8, 1861.
10. Hay, *Inside Lincoln's White House* (1997), p. 123.
11. *Douglass' Monthly*, October 1861.
12. *Douglass' Monthly*, September 1861.
13. Handbill reprinted in *Douglass' Monthly*, December 1861.

CHAPTER 11: TO PROCLAIM OR NOT TO PROCLAIM

1. Donald, *Lincoln* (1995), p. 336.
2. Foner, *The Life and Writings of Frederick Douglass*, vol. 3 (1952), p. 20.
3. Nevins, *Diary of the Civil War: George Templeton Strong* (1952), p. 302.
4. Sumner, "The Barbarism of Slavery," June 4, 1860.
5. FD to Charles Sumner, April 8, 1862.

6. Douglass, "The War and How to End It," March 25, 1862.
7. Donald, *Charles Sumner and the Coming of the Civil War* (1960), p. 383.
8. Poore, "Lincoln and the Newspaper Correspondents," in Rice, *Reminiscences of Abraham Lincoln by Distinguished Men of His Time* (1888), p. 333.
9. Mary Lincoln to Mrs. Orne, November 28, 1869.
10. Randall, *Mary Lincoln* (1953), p. 356.
11. Donald, *Charles Sumner and the Coming of the Civil War* (1960), p. 388.
12. Mary Lincoln to Mrs. Orne, November 28, 1869.
13. Charles Sumner, letter, *The Independent,* June 19, 1862.
14. Nevins, *The War for the Union* (1960), p. 5.
15. Douglass, "What to the Slave Is the Fourth of July," Rochester, New York, July 4, 1852.
16. *The Liberator,* July 7, 1854.
17. AL to Orville Hickman Browning, September 22, 1861.
18. Starr, *Bohemian Brigade* (1954), p. 125; Donald, *Charles Sumner and the Rights of Man* (1970), p. 60. Italics added.
19. Here and after, the story comes from painter Francis B. Carpenter; see Fehrenbacher and Fehrenbacher, *Recollected Words of Abraham Lincoln* (1996), pp. 79–80.
20. Lyman Trumbull quoted in Foner, *Fiery Trial* (2010), p. 222.
21. AL, "Address on Colonization to a Deputation of Negroes," August 14, 1862.
22. *Douglass' Monthly,* September 1862.
23. FD to Montgomery Blair, September 16, 1862.
24. AL to Horace Greeley, August 22, 1862.
25. See Kendrick and Kendrick, *Douglas and Lincoln* (2008), p. 111; FD to Gerrit Smith, September 8, 1862.
26. George S. Boutwell to J. G. Holland, June 10, 1865.
27. AL to George McClellan, September 15, 1862.
28. Edwards Pierrepont, quoted in Fehrenbacher and Fehrenbacher, *Recollected Words of Abraham Lincoln* (1996), p. 360.
29. *Douglass' Monthly,* October 1862.
30. Douglass, *Life and Times* (1881), p. 356.
31. Stowe and Stowe, *Harriet Beecher Stowe: The Story of Her Life* (1911), p. 203. See also Quarles, *Lincoln and the Negro* (1962), p. 134.
32. Guelzo, *Lincoln's Emancipation Proclamation* (2004), pp. 182–83.
33. *The Liberator,* January 16, 1863; and Douglass, *Life and Times* (1881), pp. 358–60.

CHAPTER 12: TURNING POINT AT GETTYSBURG

1. Emancipation Proclamation, January 1, 1863.
2. Emilio, *History of the Fifty-Fourth Regiment* (1894), p. 12. See also Charles Heller, *Portrait of an Abolitionist: A Biography of George Luther Stearns* (Westport, CT: Greenwood Press, 1996).

3. *Douglass' Monthly*, May 1861.
4. Charles Sumner to Carl Schurz, July 5, 1862.
5. Edwin Stanton, quoted in Foner, *Fiery Trial* (2010), p. 230.
6. *Douglass' Monthly*, January 1863.
7. *Douglass' Monthly*, March 1863.
8. Douglass, "Men of Color, to Arms," March 21, 1863.
9. James Shepherd Pike to William H. Seward, December 31, 1862.
10. *Die Presse*, October 12, 1862.
11. AL to Andrew Johnson, March 26, 1863.
12. William Henry Wadsworth to S. L. M. Barlow, December 18, 1863.
13. *Chicago Tribune*, June 1, 1863.
14. Brooks, *Washington in the Time of Lincoln* (1896), pp. 47–58.
15. Nathaniel Banks, quoted in Cornish, *The Sable Arm* (1956, 1987), p. 143.
16. Charles Dana quoted in *War of the Rebellion: The Official Records of the Union and Confederate Armies*. Series 1, vol. 24, pt. 1, War Department, p. 106.
17. George Meade to Henry Halleck, July 2, 1863.
18. Rusling, *Men and Things I Saw in Civil War Days* (1894), p. 15.
19. AL to Ulysses S. Grant, August 9, 1863.
20. *Douglass' Monthly*, August 1863.
21. Douglass, *Life and Times* (1881), p. 362.
22. Lewis Douglass to Helen Amelia Loguen, July 20, 1863.
23. FD to George Stearns, August 1, 1863.
24. FD to George Stearns, August 12, 1863.

CHAPTER 13: A BLACK VISITOR TO THE WHITE HOUSE

1. Ottilie Assing, quoted in Blight, *Frederick Douglass* (2018), p. 292.
2. FD recounted this first visit to Lincoln almost immediately in a letter to George Stearns, dated August 12, 1863. He did so repeatedly over the years, including in a Philadelphia speech, delivered on December 4, 1863; in a Brooklyn, New York, speech delivered near the end of his life, on February 13, 1893; and in *Life and Times* (1881). The quotations here are drawn from those sources unless otherwise noted.
3. David Wills to Abraham Lincoln, November 2, 1863.
4. David Wills to Andrew Curtin, July 24, 1863.
5. *New-York Tribune*, July 8, 1863.
6. Brooks, *Washington in the Time of Lincoln* (1896), pp. 252–53.
7. AL, The Gettysburg Address, November 19, 1863.

CHAPTER 14: THE MISSION OF THE WAR

1. Douglass, *Life and Times* (1881), pp. 354–55.
2. FD to anonymous, February 17, 1864.

3. Douglass, *Life and Times* (1881), p. 358.
4. Here and after: FD, "The Mission of War," Cooper Union, January 13, 1864. Italics added.
5. Blassingame and McKivigan, eds. *The Frederick Douglass Papers*, vol. 4 (1991), p. 3.
6. *New York Times*, January 14, 1864.
7. The details of Lincoln's visit to City Point here and after are based on the account in Lieutenant Colonel Horace Porter's *Campaigning with Grant* (1907), pp. 216ff.
8. Horace Porter in *Campaigning with Grant* (1907), p. 281.
9. Welles, *Diary of Gideon Welles* (1911), p. 58.
10. AL to Albert G. Hodges, April 4, 1864.
11. AL to Ulysses S. Grant, August 9, 1863.
12. Ulysses S. Grant to AL, August 23, 1863; Bunting, *Ulysses S. Grant* (2004), p. 49.
13. Ulysses Grant to H. W. Halleck, August 1, 1864.
14. Eaton, *Grant, Lincoln, and the Freedman* (1907), pp. 167–75.
15. Diary of Joseph T. Mills, in *Collected Works of Abraham Lincoln*, vol. 7 (1953), p. 508.
16. Here and after, the several key sources for the details of Douglass and Lincoln's August 19 meeting are Douglass's own writings, including his letter to Theodore Tilton, October 15, 1874; *Life and Times* (1881), pp. 363–365; and "Abraham Lincoln, the Great Man of Our Century," an address given in Brooklyn, New York, on February 13, 1893.
17. Douglass, *Life and Times* (1881), p. 364.
18. FD to AL, August 29, 1864.
19. William T. Sherman to Will Halleck, September 3, 1864.
20. Here and after, quotations are from FD, "A Friendly Word to Maryland," November 17, 1864.
21. *Cambridge Intelligencer*, undated clipping, cited in *Frederick Douglass Papers*, vol. 4 (1991), p. 38.

CHAPTER 15: MY FRIEND DOUGLASS

1. William T. Sherman to AL, December 22, 1864. Because the message had to be carried by ship back to Union territory, Lincoln received it on Christmas Day.
2. Hugh Highland Grant, quoted in Quarles, *Lincoln and the Negro* (1962), p. 223.
3. Brooks, *Washington in the Time of Lincoln* (1896), p. 207.
4. Charles R. Douglass to FD, February 9, 1865.
5. AL, February 1, 1865.
6. The poetic words are those of attendee Noah Brooks, Lincoln's friend and sometime confidant.
7. Douglass, "Lincoln and the Colored Troops" (1888), p. 320.
8. Brooks, *Washington in the Time of Lincoln* (1896), p. 238.

9. Douglass recounted two versions of this moment: In one it is Lincoln who indicates Douglass; in another it is Chief Justice Salmon Chase. Douglass, "Lincoln and the Colored Troops" (1888), p. 321; Douglass, *Life and Times* (1881), p. 370.
10. Here and after, Lincoln's words are drawn from his Second Inaugural Speech, March 4, 1865.
11. This vignette was recounted by Douglass in his *Life and Times* (1881), pp. 371ff, and in his address, delivered in Brooklyn, "Abraham Lincoln, the Great Man of Our Century," on February 13, 1893. The diaries of Henry Clay Warmoth, a young Union officer who overheard their conversation, corroborate the Douglass-Lincoln encounter.
12. Warmoth, *The Diary of Henry Clay Warmoth* (1960), p. 126.
13. *New York World,* March 6, 1865; *New-York Tribune*, March 5, 1865; and the *Daily Ohio Statesman*, March 6, 1865.
14. AL to Thurlow Weed, quoted in Donald, *Lincoln* (1995), p. 568.
15. Carpenter, *Six Months at the White House* (1866), p. 234.

CHAPTER 16: APRIL IS THE CRUELEST MONTH

1. Porter, *Incidents and Anecdotes of the Civil War* (1885), p. 294.
2. Sources for Lincoln's April 4, 1865, visit to Richmond include Bruce, *The Capture and Occupation of Richmond* (1927), pp. 31–33; Patrick, *The Fall of Richmond* (1960), pp. 127–33; and Porter, *Incidents and Anecdotes of the Civil War* (1885), pp. 294–302. Porter's recollections seem, at times, a bit too richly remembered to be entirely accurate.
3. Porter, *Incidents and Anecdotes of the Civil War* (1885), p. 295.
4. FD, "The Fall of Richmond," April 4, 1865.
5. Booth, *Right or Wrong* (2001), p. 169.
6. Achorn, *Every Drop of Blood* (2020), pp. 113, 117.
7. James F. Moulton to his uncle, April 17, 1865.
8. Nicolay and Hay, *Abraham Lincoln*, vol. 10 (1890), pp. 285–86.
9. Donald, *Lincoln* (1995), p. 593.
10. Oates, *With Malice Toward None* (1977), p. 468.
11. Nicolay and Hay, "The Fourteenth of April," *Century Magazine*, vol. 40 (January 1890), p. 432.
12. Of the many, many versions of the assassination story, one of the most complete is Terry Alford's *Fortune's Fool* (2015).
13. Gideon Welles, quoted in Foner, *Fiery Trial* (2010), p. 332.
14. Nicolay and Hay, *Abraham Lincoln*, vol. 10 (1890), p. 302.
15. Douglass, *Life and Times* (1881), pp. 378–79.
16. Here and after, FD, "Our Martyred President," April 15, 1865.
17. Douglass, *Life and Times* (1881), p. 379.

EPILOGUE: A BONE-HANDLED CANE

1. Keckley, *Behind the Scenes* (1868), p. 191.
2. *Rochester Democrat and Chronicle*, April 28, 1865.
3. Nicolay and Hay, *Abraham Lincoln*, vol. 10 (1890), p. 323.
4. FD, "Lincoln and the Colored Troops" (1909), p. 325.
5. FD to Mrs. Abraham Lincoln, August 17, 1865.
6. As reported in the *New York World*; see Blight, *Frederick Douglass* (2018), pp. 473–75.
7. Douglass, *Life and Times* (1881), p. 380.
8. *New National Era*, June 6 and 17, 1872.
9. Donald, *Charles Sumner and the Rights of Man* (1970), pp. 586–87.
10. The stories of Charlotte Scott, the Freedman's Memorial, and the events of April 14, 1876, have been recounted often. Minor details differ from one version to another, but the most reliable sources include Quarles, *Lincoln and the Negro* (1962), pp. 3–14; Stauffer, *Giants* (2008), pp. 302–6; and Blight, *Frederick Douglass* (2018), pp. 1–9.
11. Here and after, Douglass, "Oration in Memory of Abraham Lincoln," April 14, 1876.
12. Douglass, *Life and Times* (1881), p. 380.
13. FD to editor of the *National Republican*, April 19, 1876.
14. Office of Record of Deeds, District of Columbia, folio 77.
15. Quarles, *Frederick Douglass* (1948), pp. 336, 339.
16. FD to anonymous, n.d. *The Life and Writings of Frederick Douglass*, vol. 4 (1955), p. 115.
17. Douglass, *Life and Times* (1881), p. 378.
18. FD, "In Law Free: In fact, A Slave," April 16, 1888.
19. Foner and Taylor, eds. *Frederick Douglass, Selected Speeches and Writings* (1999), p. xiii.
20. Quarles's *Frederick Douglass* (Washington, D.C.: Associated Publishers) appeared in 1948.
21. Lerone Bennett Jr., "Was Lincoln a White Supremacist?," *Ebony*, February 1968.
22. For a detailed dissection of secessionist conventions, see Charles B. Dew, *Apostles of Disunion* (Charlottesville: University Press of Virginia, 2001).
23. James Gillespie Blaine, *Twenty Years of Congress: From Lincoln to Garfield*, vol. 2. (Norwich, CT: Henry Bill, 1884), p. 49.
24. AL to A. G. Hodges, April 4, 1864.

FOR FURTHER READING

Achorn, Edward. *Every Drop of Blood.* New York: Grove-Atlantic, 2020.

Adams, John Quincy. *Memoirs of John Quincy Adams: His Diary from 1795–1848.* Charles Francis Adams, ed. Philadelphia, PA: J. B. Lippincott and Co., 1876.

Alford, Terry. *Fortune's Fool: The Life of John Wilkes Booth.* New York: Oxford University Press, 2015.

Barnes, L. Diane. *Frederick Douglass: A Life in Documents.* Charlottesville: University of Virginia Press, 2013.

Basler, Roy P. "Did President Lincoln Give the Smallpox to William Johnson?" *Huntington Library Quarterly* 35, no. 3 (May 1972), pp. 279–84.

Berlin, Ira. *The Long Emancipation: The Demise of Slavery in the United States.* Cambridge, MA: Harvard University Press, 2015.

Beveridge, Albert J. *Abraham Lincoln: 1809–1858.* Boston: Houghton Mifflin Co., 1928.

Bingham, Caleb, ed. *The Columbian Orator.* Boston: Manning and Loring, 1797.

Bland, T. A. *Life of Benjamin F. Butler.* Boston: Lee and Shepard, 1879.

Blassingame, John W., and John McKivigan, eds. *The Frederick Douglass Papers.* 5 vols. New Haven: Yale University Press, 1979, 1982, 1985, 1991, 1992.

Blight, David. *Frederick Douglass' Civil War.* Baton Rouge: Louisiana State University Press, 1989.

———. *Frederick Douglass: Prophet of Freedom.* New York: Simon and Schuster, 2018.

Booth, John Wilkes. *Right or Wrong, God Judge Me: The Writing of John Wilkes Booth.* Urbana: University of Illinois Press, 2001.

Boteler, Alexander R. "Recollections of the John Brown Raid," *Century Magazine,* July 1883, pp. 399–411.

Breiseth, Christopher N. "Lincoln and Douglass: Another Debate." *Journal of the Illinois Historical Society* 68, no. 1 (February 1975), pp. 9–26.

Brooks, Noah. *Washington in the Time of Lincoln.* New York: The Century Company, 1896.

Browne, Robert H. *Lincoln and the Men of His Time.* 2 vols. Chicago: Blakely-Oswald Printing Co., 1907.

Bruce, George A. *The Capture and Occupation of Richmond.* Richmond, VA: 1927.

Bunting, Josiah, III. *Ulysses S. Grant.* New York: Times Books, 2004.

Burlingame, Michael. *The Inner World of Abraham Lincoln.* Urbana: University of Illinois Press, 1994.

Carpenter, Francis. *Six Months at the White House: The Story of a Picture.* New York: Hurd and Houghton, 1866.

Cornish, Dudley Taylor. *The Sable Arm: Black Troops in the Union Army, 1861–1865.* Lawrence: University of Kansas Press, 1987.

Davis, Cullom, et al., eds. *The Public and the Private Lincoln: Contemporary Perspectives.* Carbondale: Southern Illinois University Press, 1979.

Denton, Sally. *Passion and Principle: John and Jessie Frémont, the Couple Whose Power, Politics, and Love Shaped Nineteenth-Century America.* New York: Bloomsbury, 2007.

Detzer, David. *Allegiance: Fort Sumter, Charleston, and the Beginning of the Civil War.* New York: Harcourt, 2001.

Dew, Charles B. *Apostles of Disunion: Southern Secession Commissioners and the Causes of the Civil War.* Charlottesville: University of Virginia Press, 2001.

Donald, David [Herbert]. *Charles Sumner and the Coming of the Civil War.* New York: Alfred A. Knopf, 1960.

———. *Charles Sumner and the Rights of Man.* New York: Alfred A. Knopf, 1970.

Donald, David Herbert. *Lincoln.* New York: Simon and Schuster, 1995.

Douglass, Frederick. *Douglass' Monthly.* Available online at the Smithsonian Institution. See https://transcription.si.edu/project/13034.

———. *Narrative of the Life of Frederick Douglass, an American Slave.* Boston: Anti-Slavery Office, 1845.

———. *My Bondage and My Freedom.* New York: Miller, Orton and Mulligan, 1855.

———. *Life and Times of Frederick Douglass, Written by Himself, His Early Life as a Slave, His Escape from Bondage, and His Complete History to the Present Time.* Hartford, CT: Park Publishing Co., 1881.

———. "Lincoln and the Colored Troops" in *Reminiscences of Abraham Lincoln by Distinguished Men of His Time,* Allen Thorndike Rice, ed. New York: North American Review, 1888, pp. 315–25.

———. *The Frederick Douglass Papers.* Philip W. Foner, ed. 3 vols. New York: International Publishers, 1950, 1952.

———. *The Frederick Douglass Papers.* Blassingame, John W., and John McKivigan, eds. 5 vols. New Haven: Yale University Press, 1979, 1982, 1985, 1991, 1992.

———. *Autobiographies.* Henry Louis Gates Jr., ed. New York: Library of America, 1994.

Eaton, John. *Grant, Lincoln, and the Freedman: Reminiscences of the Civil War.* New York: Longmans, Green, and Company, 1907.

Emilio, Luis Fenolossa. *History of the Fifty-Fourth Regiment of Massachusetts Volunteer Infantry, 1862–1865.* Boston: Boston Book Company, 1894.

Fehrenbacher, Don E., and Virginia Fehrenbacher, eds. *Recollected Words of Abraham Lincoln.* Stanford, CA: Stanford University Press, 1996.

Findley, Paul. *A. Lincoln: The Crucible of Congress.* New York: Crown Publishers, 1979.

Foner, Eric. *The Fiery Trial: Abraham Lincoln and American Slavery.* New York: W. W. Norton and Co., 2010.

Foner, Philip, ed. *The Life and Writings of Frederick Douglass.* 3 vols. New York: International Publishers, 1950, 1952.

Foner, Philip S[heldon]. *Frederick Douglass: A Biography.* New York: Citadel Press, 1969.

Foner, Philip, and Yuval Taylor, eds. *Frederick Douglass, Selected Speeches and Writings.* Chicago: Lawrence Hill Books, 1999.

Freeman, Andrew A. *Abraham Lincoln Comes to New York.* New York: Coward, McCann, 1960.

Frémont, Jessie Benton. *Memoirs of My Life.* New York: Cooper Square Press, 2001.

Guelzo, Allen C. *Lincoln's Emancipation Proclamation: The End of Slavery in America.* Simon and Schuster, 2004.

Hay, John. *Inside Lincoln's White House: The Complete Civil War Diary of John Hay.* Michael Burlingame and John R. Turner Ettlinger, eds. Carbondale: South Illinois University Press, 1997.

Herndon, William H., and Jesse William Weik. *Herndon's Lincoln: The True Story of a Great Life.* Chicago: Bedford-Clarke, 1890.

Hochman, Barbara. "'Uncle Tom's Cabin' in the 'National Era.'" *Book History,* 2004, pp. 143–69.

Holzer, Harold. *Lincoln at Cooper Union: The Speech That Made Abraham Lincoln President.* New York: Simon and Schuster, 2004.

Horwitz, Tony. *Midnight Rising: John Brown and the Raid That Sparked the Civil War.* New York: Henry Holt and Company, 2011.

Huggins, Nathan Irvin. *Slave and Citizen: The Life of Frederick Douglass.* Boston: Little, Brown and Co., 1980.

Keckley, Elizabeth. *Behind the Scenes, or, Thirty Years a Slave, and Four Years at the White House.* New York: G. W. Carleton, 1868.

Kendrick, Paul, and Stephen Kendrick. *Douglas and Lincoln: How a Revolutionary Black Leader and a Reluctant Liberator Struggled to End Slavery and Save the Union.* New York: Walker, 2008.

Lamon, Ward H. *The Life of Lincoln, from His Birth to Inauguration as President.* Boston: James R. Osgood and Co., 1872.

Lampe, Gregory P. *Frederick Douglass: Freedom's Voice: 1818–1845.* East Lansing: Michigan State University Press, 1998.

Lehrman, Lewis F. *Lincoln at Peoria: The Turning Point.* Mechanicsburg, PA: Stackpole Books, 2008.

Lincoln, Abraham. *Collected Works of Abraham Lincoln.* Roy P. Basler, et al., eds. 9 vols. New Brunswick, NJ: Rutgers University Press, 1952–1955.

Lossing, Benson J. *The Pictorial History of the Civil War.* Vol. 1. Philadelphia, PA: G. W. Childs, 1868.

Mackey, Thomas C. "'That All Mankind Should Be Free': Lincoln and African Americans." *OAH Magazine of History* 21, no. 4 (October 2007), pp. 24–29.

May, Samuel J. *Some Reflections of Our Antislavery Conflict.* Boston: Fields, Osgood and Co., 1869.

McFeely, William S. *Frederick Douglass.* New York: Norton, 1990.

McPherson, James. *Battle Cry of Freedom.* New York: Oxford University Press, 1988.

———. *Lincoln and the Second American Revolution.* New York: Oxford University Press, 1991.

———. *This Mighty Scourge: Perspectives on the Civil War.* New York: Oxford University Press, 2007.

———. *Tried by War: Abraham Lincoln as Commander in Chief.* New York: Penguin, 2008.

———. *The Struggle for Equality: Abolitionists and the Negro in the Civil War and Reconstruction.* Princeton, NJ: Princeton University Press, 2014.

Meredith, Roy. *Mr. Lincoln's Camera Man: Mathew B. Brady.* 2nd ed. New York: Dover Publications, 1974.

Nevins, Allan, ed. *Diary of the Civil War: George Templeton Strong.* New York: Macmillan, 1952.

———. *The War for the Union: War Becomes Revolution.* New York: Charles Scribner's Sons, 1960.

Nicolay, John G., and John Hay. *Abraham Lincoln: A History.* 10 vols. New York: The Century Company, 1890.

Oakes, James. *The Radical and the Republican.* New York: W. W. Norton, 2007.

Oates, Stephen B. *To Purge This Land with Blood: A Biography of John Brown.* New York: Harper and Row, 1970.

———. *With Malice Toward None: The Life of Abraham Lincoln.* New York: Mentor, 1978.

Patrick, Rembert W. *The Fall of Richmond.* Baton Rouge: Louisiana State University Press, 1960.

Perret, Geoffrey. *Lincoln's War: The Untold Story of America's Greatest President as Commander in Chief.* New York: Random House, 2004.

Pierson, Michael D. "'All Southern Society Is Assailed by the Foulest Charges': Charles Sumner's 'The Crime against Kansas' and the Escalation of Republication Anti-Slavery Rhetoric." *New England Quarterly* 68, no. 4 (December 1995), pp. 531–57.

Poore, Benjamin Pearly. "Lincoln and the Newspaper Correspondents," in *Reminiscences of Abraham Lincoln by Distinguished Men of His Time*, Allen Thorndike Rice, ed. New York: North American Review, 1888, pp. 327–42.

Porter, David D. *Incidents and Anecdotes of the Civil War.* New York: D. Appleton and Co., 1885.

Porter, Horace. *Campaigning with Grant.* New York: The Century Company, 1907.

Preston, Dickson J. *Young Frederick Douglass.* Baltimore, MD: Johns Hopkins University Press, 2018.

Quarles, Benjamin. *Frederick Douglass.* Washington, D.C.: Associated Publishers, 1948.

———. *Lincoln and the Negro.* Oxford University Press, 1962.

———, ed. *Frederick Douglass.* Englewood, NJ: Prentice-Hall, Inc., 1968.

Randall, Ruth Painter. *Mary Lincoln: A Biography of Marriage.* Boston: Little, Brown, 1953.

Reynolds, David S. *John Brown, Abolitionist: The Man Who Killed Slavery, Sparked the Civil War, and Seeded Civil Rights.* New York: Alfred A. Knopf, 2005.

Ruchames, Louis, ed. *John Brown: The Making of a Revolutionary.* New York: Grosset and Dunlap, 1969.

Rusling, James F. *Men and Things I Saw in Civil War Days.* New York: Eaton and Mains, 1899.

Savage, Kirk. *Standing Soldiers, Kneeling Slaves.* Princeton, NJ: Princeton University Press, 2018.

Searcher, Victor. *Lincoln's Journey to Greatness: A Factual Account of the Twelve-Day Inaugural Trip.* Philadelphia, PA: John C. Winston Co., 1960.

Simpson, Brooks D., ed. *The Civil War: The Third Year Told by Those Who Lived It.* New York: Library of America, 2013.

Speed, Joshua. *Reminiscences of Abraham Lincoln.* Louisville, KY: Bradley and Gilbert Co., 1896.

Starr, Louis M. *Bohemian Brigade: Civil War Newsmen in Action.* New York: Alfred A. Knopf, 1954.

Stauffer, John. *The Black Hearts of Men.* Cambridge, MA: Harvard University Press, 2002.

———. *Giants: The Parallel Lives of Frederick Douglass and Abraham Lincoln.* New York: Twelve, 2008.

Stauffer, John, Zoe Trodd, and Celeste-Marie Bernier. *Picturing Frederick Douglass: An Illustrated Biography of the Nineteenth Century's Most Photographed American.* New York: Liveright, 2015.

Stowe, Charles Edward, and Lyman Beecher Stowe. *Harriet Beecher Stowe: The Story of Her Life.* Boston: Houghton Mifflin Co., 1911.

Sundquist, Eric J., ed. *Frederick Douglass: New Literary and Historical Essays.* Cambridge, UK: Cambridge University Press, 1990.

Temple, Wayne C. "Lincoln's Fence Rails." *Journal of the Illinois State Historical Society* 47, no. 1 (spring 1954), pp. 20–34.

Thomas, John L. *The Liberator, William Lloyd Garrison: A Biography.* Boston: Little, Brown and Co., 1963.

Villard, Oswald Garrison. *John Brown, 1800–1859: A Biography Fifty Years After.* Boston: Houghton Mifflin Co., 1909.

Vorenberg, Michael. *Final Freedom: The Civil War, the Abolition of Slavery, and the Thirteenth Amendment.* Cambridge, UK: Cambridge University Press, 2001.

Waller, Douglas. *Lincoln's Spies: Their Secret War to Save a Nation.* New York: Simon and Schuster, 2019.

Warmoth, Henry Clay. *The Diary of Henry Clay Warmoth, 1861–1867.* Paul H. Hass. MS thesis. University of Wisconsin, 1961.

Welles, Gideon. *Diary of Gideon Welles, Secretary of the Navy under Lincoln and Johnson.* Boston: Houghton Mifflin Co., 1911.

Wickenden, Dorothy. "Lincoln and Douglas: Dismantling the Peculiar Institution." *The Wilson Quarterly* 14, no. 4 (Autumn 1990), pp. 102–12.

Wills, Garry. *Lincoln at Gettysburg: The Words That Remade America.* New York: Simon and Schuster, 1992.

Wilson, Douglas L. *Honor's Voice: The Transformation of Abraham Lincoln.* New York: Alfred A. Knopf, 1998.

Wilson, Douglas L., and Rodney O. Davis, eds. *Herndon's Informant: Letters, Interviews, and Statements About Abraham Lincoln.* Urbana: University of Illinois Press, 1997.

Wilson, Rufus Rockwell. *Intimate Memories of Lincoln.* Elmira, NY: Primavera Press, 1945.

INDEX

Note: Page numbers in *italics* refer to photographs or illustrations.

IMAGE CREDITS

p. 15 A young Lincoln wrestles Jack Armstrong: Historical Images Archive / Alamy Stock Photo.
p. 27 Anna Murray Douglass: Frederick Douglass National Historic Site / NPS
p. 32 William Lloyd Garrison: Library of Congress, Prints & Photographs Division [LC-USZ62-10320]
p. 35 First photo of Frederick Douglass: Collection of Greg French
p. 38 (Left) Abraham Lincoln, Congressman-elect from Illinois: Library of Congress, Prints & Photographs Division, photograph by Nicholas H. Shepherd [LC-DIG-ppmsca-53842]
p. 38 (Right) Mary Todd Lincoln, wife of Abraham Lincoln: Library of Congress, Prints & Photographs Division, photograph by Nicholas H. Shepherd [LC-USZ6-300]
p. 43 *Fighting the Mob in Indiana:* Frederick Douglass, *Life and Times of Frederick Douglass, Written by Himself, His Early Life as a Slave, His Escape from Bondage, and His Complete History to the Present Time*, by Frederick Douglass (Hartford, CT: Park Publishing Co., 1881), p. 235
p. 45 *The Fugitive's Song*: Library of Congress, Prints & Photographs Division [LC-DIG-ppmsca-07616]
p. 54 Portrait of Gerrit Smith: Boston Public Library
p. 59 Title page from volume 1, U.S. first edition of *Uncle Tom's Cabin: or, Life Among the Lowly* by Harriet Beecher Stowe (1811–1896) published in 1852: AF Fotografie / Alamy Stock Photo
p. 60 Author and abolitionist Harriet Beecher Stowe: Library of Congress, Prints & Photographs Division, Liljenquist Family collection [LC-DIG-ppmsca-49807]
p. 67 Charles Sumner: National Archives
p. 68 *Arguments of the Chivalry* print: Library of Congress, Prints & Photographs Division [LC-USZ62-38851]
p. 70 John Brown: Library of Congress, Prints & Photographs Division [LC-USZ62-106337]
p. 74 Dred Scott: Library of Congress, Prints & Photographs Division by Century Company [LC-USZ62-5092]

p. 75 John Brown, three-quarter length portrait, facing left, holding *New York Tribune*: Library of Congress, Prints & Photographs Division [LC-USZ62-89569]

p. 81 *Harper's Ferry insurrection—Interior of the Engine-House, just before the gate is broken down by the storming party—Col. Washington and his associates as captives, held by Brown as hostages*: Library of Congress, Prints & Photographs Division [LC-USZ62-132541]

p. 88 Lincoln at a debate with Stephen Douglas: Photograph of Robert Marshall Root's *Abraham Lincoln and Stephen A. Douglas debating at Charleston, Illinois on 18th September 1858, 1918* (oil on canvas) by Cool10191/Wikimedia Commons

p. 89 Stephen A. Douglas: Library of Congress, Prints & Photographs Division [LC-USZ62-135560]

p. 95 Abraham Lincoln, candidate for U.S. president, delivering his Cooper Union address in New York City: Library of Congress, Prints & Photographs Division [LC-USZ62-5803]

p. 99 Frederick Douglass and his daughter Annie: John B. Cade Library, Shade Collection, South University and A&M College, Baton Rouge, LA

p. 105 Abraham Lincoln engraving after Mathew Brady photograph: *Harper's Weekly*, May 26, 1860 with image of Abraham Lincoln. Engraving after Brady photograph. National Portrait Gallery, Smithsonian Institution, Washington, D.C.

p. 107 *Lincoln the Rail Splitter*: Library of Congress, Prints & Photographs Division [LC-USZC4-2472]

p. 111 *Expulsion of Negroes and abolitionists from Tremont Temple, Boston, Mass., on Dec. 3, 1860*: Boston Athenæum

p. 119 Inauguration of Abraham Lincoln at the U.S. Capitol, 1861: Library of Congress, Prints & Photographs Division [LC-DIG-ppmsca-35445]

p. 120 *The Flight of Abraham* cartoon: Library of Congress, Serial and Government Publications Division

p. 125 Jefferson Davis, three-quarter length portrait, facing right: Library of Congress, Prints & Photographs Division, by Mathew B. Brady [LC-DIG-ppmsca-23852]

p. 128 Civil war envelope showing fugitive slaves working as sappers and miners with General Benjamin Butler: Library of Congress, Prints & Photographs Division, Liljenquist Family Collection [LOT 14043-6, no. 123]

p. 133 William Henry Seward: Library of Congress, Prints & Photographs Division [LC-USZ62-21907]

p. 136 John C. Frémont and wife, Jessie Benton Frémont: San Francisco History Center, San Francisco Public Library

p. 143 *Mrs. Abraham Lincoln*: Library of Congress, Prints & Photographs Division [LC-DIG-cwpbh-01028]

p. 151 *Abe Lincoln's Last Card; or, Rouge-et-Noir* cartoon: Album / Alamy Stock Photo

p. 158 *Dead in Front of Dunker Church, Antietam, MD*: Library of Congress, Prints & Photographs Division, Civil War Photographs [LC-DIG-cwpbh-03384]

p. 162 *The First Reading of the Emancipation Proclamation before the Cabinet*: Library of Congress, Prints & Photographs Division, by A. H. Ritchie and F. B. Carpenter [LC-DIG-pga-03452]

p. 166 George Luther Stearns: Collection of the Massachusetts Historical Society

p. 169 Lewis Douglass: Courtesy of the Moorland-Spingarn Research Center, Howard University Archives, Howard University, Washington DC

p. 170 Robert E. Lee at Chancellorsville: Library of Congress, Prints & Photographs Division [LC-USZ62-51832]
p. 174 *The Battle of Gettysburg:* Library of Congress, Prints & Photographs Division [LC-DIG-pga-03266]
p. 176 *Storming Fort Wagner:* Library of Congress, Prints & Photographs Division by Kurz & Allison [LC-DIG-pga-01949]
p, 180 Secretary of War Edwin Stanton: Library of Congress, Prints & Photographs Division, Liljenquist Family Collection [LC-DIG-ppmsca-52235]
p. 184 Frederick Douglass, 1863: Edwin Burke Ives (1832–1906) and Reuben L. Andrews, January 21 1863, Howell Street, Hillsdale, MI, Carte-de-viste (2½ x 4 in), Hillsdale College. Photograph courtesy of Hillsdale College
p. 191 Lincoln portrait by Alexander Gardner: Library of Congress, Prints & Photographs Division by Alexander Gardner [LC-DIG-npcc-01932]
p. 194 *Long Abraham Lincoln a Little Longer* cartoon: *Harper's Weekly*, November 26, 1864
p. 196 *President Lincoln, General Grant, and Tad Lincoln at a Railway Station* sketch: Purchase from the J. H. Wade Fund, the Cleveland Museum of Art
p. 199 President Lincoln at Fort Stevens: Courtesy DC Public Library, The People's Archive
p. 208 Frederick Douglass at Lloyd family graveyard: From the New York Public Library
p. 212 Passage of the Thirteenth Amendment: *Harper's Weekly*, 18 February 1865
p. 215 President Abraham Lincoln delivering second inaugural address in front of the U.S. Capitol, March 4, 1865: Library of Congress, Prints & Photographs Division by Alexander Gardner [LC-USZ62-8122]
p. 221 *Abraham Lincoln's Last Reception*: Library of Congress, Prints & Photographs Division [LC-DIG-pga-01590]
p. 225 Lincoln shaking hands with a Black Union soldier (*City Point/65--"Good god, you goin to shake with me, Uncle Abe"*): Library of Congress, Miscellaneous Items in High Demand by Charles W. Reed [LC-USZ62-75182]
p. 228 *The Surrender of General Lee. And His Entire Army to Lieut General Grant, April 9th, 1865*: Library of Congress, Prints & Photographs Division by John Smith [LC-DIG-pga-08989]
p. 230 John Wilkes Booth: Library of Congress, Prints & Photographs Division [LC-USZ62-25166]
p. 234 *Assassination of Lincoln*: Library of Congress National Photo Company Collection [LC-F81-2117]
p. 236 Lincoln on his deathbed surrounded by mourners: Library of Congress, Prints & Photographs Division [LC-DIG-ppmsca-46775]
p. 244 Fred K. Douglas (i.e., Frederick Douglass): Library of Congress, Prints & Photographs Division by Mathew B. Brady [LC-DIG-ppmsca-70869]
p. 245 Emancipation statue, Washington, D.C.: Library of Congress, Prints & Photographs Division, Detroit Publishing Company [LC-DIG-det-4a05594]
p. 249 Frederick Douglass, Helen Pitts Douglass and Eva: National Park Service (FRDO 3912)
p. 250 Abraham Lincoln, portrait by William E. Marshall: Library of Congress, Prints & Photographs Division [LC-DIG-ppmsca-46791]
p. 252 Frederick Douglass in His Study at Cedar Hill: National Park Service (FRDO3886)

PENNSYLVANIA
GETTYSBURG
Wilmington
N.J.
ANTIETAM/SHARPSBURG
Harper's Ferry
Baltimore
Lloyd Plantation
WEST VIRGINIA (1863)
Washington, DC
DEL.
BULL RUN/ MANASSAS
Appalachian Mountains
CHANCELLORSVILLE
MARYLAND
FREDERICKSBURG
VIRGINIA
James River
Richmond
Petersburg
APPOMATTOX
CITY POINT
FORT MONROE
0 Miles 100
0 Kilometers 100
MINNESOTA
Lake Super
WISCONSIN
Minneapolis
Mississippi River
Madison
IOWA
Des Moines
Chicago
ILLINOIS
Sangamon River
New Salem
Springfield
COLORADO TERRITORY
KANSAS
Topeka
LAWRENCE
MISSOURI
St. Louis
NEW MEXICO TERRITORY
FORT DONELSO
INDIAN TERRITORY
FORT HENRY
FORT PILLOW
ARKANSAS
SHIL
Little Rock
Memphis
Mississippi River
MISSISSIPPI
MILLIKEN'S BEND
TEXAS
VICKSBURG
LOUISIANA
Mobile
PORT HUDSON
Austin
New Orleans
MEXICO
Gulf of Mexico
© 2021 Jeffrey L. Ward